A LEVEL
PHYSICS

Jim Breithaupt
Head of Science and Maths, Wigan and Leigh College, Wigan

Ken Dunn
Head of Physics and Electronics, Taunton's College, Southampton

EDUCATIONAL

Every effort has been made to trace copyright holders and to obtain their permission for the use of copyright material. The authors and publishers will gladly receive information enabling them to rectify any error or omission in subsequent editions.

First published 1983
Revised 1985, 1988, 1991,1993, 1995
Reprinted 1983, 1985, 1986, 1992, 1994, 1996

Letts Educational
Aldine House
Aldine Place
London W12 8AW
Tel: 0181-740 2266

British Library Cataloguing in Publication Data
A CIP record for this book is available from the British Library.

ISBN 1 85758 339 6

Printed in Great Britain by Ashford Colour Press, Gosport, Hants

Letts Educational is the trading name of BPP (Letts Educational) Ltd

PREFACE

Most Examination Boards have revised their syllabuses at least twice since the first publication of this book in 1983. Previous revisions of this book have catered for these changes and this sixth edition also takes account of the latest syllabus updates. The emphasis has always been on helping students to understand, explain and apply the central concepts of A-level Physics. Since these concepts remain major features of the latest A-level syllabuses, little change in the text has been necessary. We hope that the new layout of this sixth edition makes the book even better than before. All A-level and AS-level Physics syllabuses include a common core of key principles and facts. This book is based around the common core, as well as the more popular optional topics.

Questions styles *have* changed since 1983. The introduction of examination data sheets by most Boards means more demands are made in terms of understanding and explaining concepts and theories. New questions have therefore been introduced to replace old questions on topics no longer of central importance at A level and all are relevant in style and content to the current range of syllabuses.

Several boards now offer modular syllabuses, enabling marks to be accumulated during the course. Module tests are at A-level standard and therefore the questions in this book may be used as appropriate to revise for module tests as well as for final examinations.

We would like to thank the staff of Letts Educational for their excellent advice, unfailing courtesy and determination to 'get things right', regardless of the time and effort involved. We are most grateful to the following Examination Boards for permission to use questions from their examinations: the Associated Examining Board, the University of Cambridge Local Examinations Syndicate, the University of London Examinations and Assessment Council, the Northern Examinations and Assessment Board, the Northern Ireland Council for the Curriculum Examinations and Assessment, the Oxford and Cambridge Schools Examination Board (including Nuffield questions), the University of Oxford Delegacy of Local Examinations, the Scottish Examination Board and the Welsh Joint Education Committee.

Jim Breithaupt and
Ken Dunn, 1995

CONTENTS

SECTION 1: STARTING POINTS

SECTION 2: A-LEVEL PHYSICS

STARTING POINTS

In this section

How to use this book
   The structure of this book
   Using your syllabus checklist

Syllabus checklists and paper analysis
   Examination boards and addresses

Studying and revising Physics
   The difference between GCSE and A/AS Level
   Modular courses
   Study strategy and techniques
   Coursework
   Revision techniques

The examination
   Question styles
   Examination techniques
   Taking module tests
   Final preparation

HOW TO USE THIS BOOK

The structure of this book

The key aim of this book is to guide you in the way you tackle A-level Physics. It should serve as a study guide, work book and revision aid throughout any A-level/AS-level Physics course, no matter what syllabus you are following. It is not intended to be a complete guide to the subject and should be used as a companion to your textbooks, which it is designed to complement rather than duplicate.

We have divided the book into three sections. **Section One, Starting Points**, contains study tips and syllabus information – all the material you need to get started on your A-level study – plus advice on planning your revision and tips on how to tackle the exam itself. Use the **Syllabus Checklists** to find out exactly where you can find the study units which are relevant to your particular syllabus.

Section Two, the main body of the text, contains the core of A-level Physics. It has been devised to make study as easy – and enjoyable – as possible, and has been divided into chapters which cover the themes you will encounter on your syllabus. The chapters are split into units, each covering a topic of study.

A **list of objectives** at the beginning of each chapter directs you towards the key points of the chapter. The **Chapter Roundup** at the end brings the topics of the chapter into focus and links them to other themes of study. To reinforce what you have just read, there is a **Question Bank** at the end of each chapter. Recent examinations from all the boards (including Scottish Higher) provide the question practice. The information in brackets after the question indicates whether the question is relevant to your examining board; an asterisk indicates the question is relevant if you are taking the appropriate option. The tutorial notes give you practical guidance on how to answer A-level questions, highlighting important points and possible pitfalls, and provide additional information relevant to that particular topic.

In **Section Three, Test Run**, we turn our attention to the examination you will face at the end of your course. First, you can assess your progress using the **Test Your Knowledge Quiz** and analysis chart. Then, as a final test, you should attempt the **Mock Exam Questions**, under timed conditions. This will give you invaluable examination practice in a variety of question types and, together with the answers and hints specially written by the author, will help you to judge how close you are to achieving your A-level pass.

The table of constants and numerical values provides the data needed for the questions, together with all the major physical constants.

Using your syllabus checklist

Whether you are using this book to work step-by-step through the syllabus or to structure your revision campaign, you will find it useful to keep a checklist of what you have covered. Keep the checklist at hand when you are doing your revision; it will remind you of the chapters you have revised, and those still to be done.

The checklist for each examination – A, AS or Higher Grade – is in two parts. First there is a list of topics covered by this book which are part of the syllabus. Although the checklists are detailed, it is not possible to print entire syllabuses. **You are therefore strongly recommended to obtain an official copy of the syllabus for your examination and consult it when the need arises.** The examination board addresses are given after the syllabus checklists.

When you have revised a topic make a tick in the column provided and, if there are questions elsewhere in the book, try to answer them.

The second part of the checklist gives you information about the examination, providing details about the time allocated for each written paper and the weighting of the questions of each paper. The different types of questions which may be set are explained in detail later in this section under the heading The Examination.

SYLLABUS CHECKLISTS AND PAPER ANALYSIS

ASSOCIATED EXAMINING BOARD
1996 A-level Syllabus 0635
1997 A-level Syllabus 0635 (modular or terminal)
AS-level Syllabus 0997 (modular or terminal)

Syllabus topic		Covered in Unit No	✔
1 Structures and forces at rest			
Scalar and vector quantities	(P1)	1.5	
Moments and torques	(P1)	1.5	
Equilibrium	(P1)	1.5	
Structures	(P1)	5.1–5.3	
2 Forces at work			
Linear motion	(P1)	1.1	
Linear dynamics	(P1)	1.2, 1.3	
Work, energy and power	(P1)	1.4	
Rotation: circular motion	(P1)	1.7	
Rotational dynamics	(P3)	1.8	
3 Vibrations and waves			
Oscillating systems	(P2)	3.1	
Simple harmonic motion	(P2)	1.6	
Waves	(P2)	3.4, 4.1, 4.2	
Wave properties	(P2)	3.5, 3.6	
Reflection and refraction	(P2)	4.3	
The electromagnetic spectrum	(P2)	3.7	
4 Fields and forces			
Gravitational potential	(P3)	2.1–2.3, 2.5–2.7	
Electric fields	(P3)	2.8, 2.9	
Magnetic fields	(P4)	2.10, 2.11	
5 dc electricity			
dc circuits with resistance	(P1)	7.1–7.5	
Capacitors	(P1)	7.6	

Syllabus topic		Covered in Unit No	✔
6 Electric forces and electromagnetism			
Force on moving charges	(P3)	2.10, 8.1	
Electromagnetic induction	(P3)	2.12–2.14	
Alternating current	(P3)	7.7, 7.8	
Use of oscilloscope	(P3)	7.7	
7 Transfer phenomena			
Transfer mechanisms	(P4)		
Mass transfer	(P4)	5.4–5.6	
Charge transfer	(P4)	7.1, 7.3	
Heat transfer	(P4)	6.4–6.5	
Transfer equations	(P4)	5.4, 6.4, 7.3	
8 Systems and processes			
Systems, boundaries and energy change		6.7	
The laws of thermodynamics	(P4)	6.7	
Ideal gases	(P4)	6.6	
Empirical temperature scales	(P4)	6.1	
9 Microscopic properties of materials			
Atoms, molecules and kinetic theory	(P4)	6.6, 6.7	
Radioactivity	(P2)	8.6	
Structure of the atom	(P2)	8.5, 8.7	
Quantum phenomena	(P2)	1.10, 8.2, 8.3, 8.4	

From 1997, the syllabus detailed above is examined either on a modular assessment route or a terminal assessment route, as described below. The modular A level route is based on 4 compulsory modules, an objective test paper and either practical coursework or a practical examination. The content of each module P1, P2, P3, P4 is denoted in the syllabus analysis above by the relevant module code.

1996 Paper analysis   All topics are compulsory

Paper 1	*1½ hours*	40 multiple-choice questions	30% of total mark
Paper 2	*3 hours*		
Section A		6 to 8 short compulsory questions	25% of total mark
Section B, C, D		2 questions in each section, candidates must answer 1 question from each section and 1 further question from any section.	25% of total mark
Either			
Paper 3	*2 hours*	Practical examination	20% of total mark
or			
Paper 4		Centre-based assessment of practical work	20% of total mark

1997 Paper analysis

Paper 1	1½ hours	30%	3–5 compulsory questions on P1 topics
			3–5 compulsory questions on P2 topics
Paper 2	1½ hours	30%	3–5 compulsory questions on P3 topics
			3–5 compulsory questions on P4 topics
Paper 3	1½ hours	20%	40 compulsory objective questions on topics P1–P4 inclusive
Either practical examination	2 hours	20%	
or centre-based assessment of practical work		20%	

A-level Modular assessment
Module tests on modules P1–P4

	¾ hour	15%	3–5 compulsory questions on each test
Paper 3	1½ hours	20%	40 compulsory objective questions on modules P1–P4 inclusive
Either practical examination	2 hours	20%	
or centre-based assessment of practical work		20%	

AS-level assessment

Module tests on P1 and P2	¾ hour	30%	3–5 compulsory questions on each test
Practical examination	2 hours	40%	

Note: module tests P1 and P3 only are available in the winter examination.

ASSOCIATED EXAMINING BOARD
1996 AS-level Syllabus 0997

Syllabus topic	Covered in Unit No	✔	Syllabus topic	Covered in Unit No	✔
1 Structures and forces at rest			5 dc electricity		
Scalar and vector quantities	1.5		dc circuits with resistance	7.1–7.5	
Moments and torques	1.5		Capacitors	7.6	
Equilibrium	1.5		6 Electric forces and electromagnetism		
Structures	5.1–5.3		Magnetic field patterns	2.10	
2 Forces at work			Force on current carrying conductors	2.10, 8.1	
Linear motion	1.1		Electromagnetic induction	2.12–2.14	
Linear dynamics	1.2, 1.3		Alternating current	7.7, 7.8	
Work, energy and power	1.4		Use of oscilloscope	7.7	
Circular motion	1.7		7 Transfer phenomena (Not applicable to AS Physics)		
3 Vibrations and waves			8 Systems and processes (Not applicable to AS physics)		
Oscillating systems	3.1				
Simple harmonic motion	1.6		9 Microscopic properties of materials		
Waves	3.4, 4.1, 4.2		Atoms and molecules (Kinetic theory of gases not in AS Physics)	6.6	
Wave properties	3.5, 3.6				
Reflection and refraction	4.3		Radioactivity	8.6	
The electromagnetic spectrum	3.7		Structure of the atom	8.5, 8.7	
4 Fields and forces (Not applicable to AS Physics)			Quantum phenomena	1.10, 8.2, 8.3, 8.4	

All topics are compulsory.

Paper analysis

Paper 1   *2¼ hours*
    Section A   5 to 8 compulsory short questions   49% of total mark
    Section B   2 out of 3 questions   21% of total mark
Paper 2 (Practical) *2 hours*
    Section A   2 questions of ½ hour each to test planning, measurement
                        and analysis skills.   15% of total mark
    Section B   1 compulsory experiment   15% of total mark

UNIVERSITY OF CAMBRIDGE LOCAL EXAMINATIONS SYNDICATE
A-level Syllabus 9244

Syllabus topic	Covered in Unit No	✔	Syllabus topic	Covered in Unit No	✔
I General physics			15 dc circuits	7.4	
1 Physical quantities and units	9.1		16 Electric fields	2.1–2.3, 2.5, 2.6	
2 Measurement techniques	9.2, 9.3		17 Capacitance	7.6	
II Newtonian mechanics			18 Magnetic fields	2.10	
3 Kinematics	1.1, 1.2		19 Electromagnetism	2.10, 2.11	
4 Dynamics	1.3		20 Electromagnetic induction	2.12	
5 Forces	1.5		21 Alternating current	7.7	
6 Work, energy and power	1.4		V Matter		
7 Gravitational fields	2.1–2.3, 2.5, 2.6		22 Phases of matter	5.3, 5.4	
8 Motion in a circle	1.7, 2.7		23 Deformation of solids	5.1, 5.2	
III Oscillations and waves			24 Temperature	6.1	
9 Oscillations	1.6, 3.2, 3.3		25 Thermal properties	6.2, 6.7	
10 Waves	3.4, 4.3		26 Ideal gases	6.6	
11 Superposition	3.5, 3.6, 4.1, 4.2		27 Transfer of thermal energy	6.4, 6.5	
12 Electromagnetic waves	3.7		28 Charged particles	8.1	
IV Electricity and magnetism			29 Quantum physics	1.10, 8.2–8.4	
13 Electrostatics	7.1		30 Atomic structure	8.5	
14 Current electricity	7.1–7.3		31 Radioactivity	8.6	

There are seven optional topics; candidates take two options only from: A – Astrophysics and cosmology; C – The physics of materials; E – Electronics; F – The physics of fluids; M – Medical physics; P – Environmental physics; T – Telecommunications.

Paper analysis

Paper 1   *1 hour*   30 multiple-choice questions   19% of total mark

Paper 2   *1¾ hours*   Compulsory structured questions and
                             1 data analysis question   29% of total mark

Paper 3   *2½ hours*   36% of total mark
    Section A   4 out of 6 longer questions   26% of total mark
    Section B   Options: there are 7 questions, 1 on each option.
                             Candidates are required to answer 2 questions.
                                                           10% of total mark

Practical assessment   Either coursework assessed internally; or
                                       3-hour practical examination   16% of total mark

UNIVERSITY OF CAMBRIDGE LOCAL EXAMINATIONS SYNDICATE
AS-level Syllabus 8494

Syllabus topic	Covered in Unit No	✔
I General physics		
1 Physical quantities and units	9.1	
II Newtonian mechanics		
2 Kinematics	1.1, 1.2	
3 Dynamics	1.3	
4 Forces	1.5	
5 Energy, work and power	1.4	
6 Motion in a circle	1.7, 2.7	
III Oscillations and waves		
7 Oscillations	1.6, 3.2, 3.3	
8 Waves	3.4, 4.3	
9 Superposition	3.5, 3.6, 4.1, 4.2	
10 Electromagnetic waves	3.7	

Syllabus topic	Covered in Unit No	✔
IV Electricity and magnetism		
11 Current electricity	7.1–7.3	
12 dc circuits	7.4, 7.5	
13 Capacitance	7.6	
14 Magnetic fields	2.10	
15 Electromagnetism	2.10, 2.11	
V Matter		
16 Deformation of solids	5.1, 5.2	
17 Thermal properties of materials	6.2, 6.7	
18 Quantum physics	1.10, 8.2–8.4	
19 Atomic structure	8.5	
20 Radioactivity	8.6	

There are seven optional topics; candidates are required to study one option from:
A – Astrophysics and cosmology; C – The physics of materials; E – Electronics; F – The physics of fluids; M – Medical physics; P – Environmental physics; T – Telecommunications.

Paper analysis

Paper 1	*1 hour*	30 multiple-choice questions 24% of total mark
Paper 2	*1¼ hours*	Compulsory short questions, including 1 data analysis question 29% of total mark
Paper 3	*1½ hours*	31% of total mark
Section A		3 out of 5 longer questions 25% of total mark
Section B		1 question on each option. Candidates must answer 1 question. 6% of total mark
Practical skills		Either internal assessment of practical skills; or 1½-hour practical examination 16% of total mark

UNIVERSITY OF CAMBRIDGE LOCAL EXAMINATIONS SYNDICATE
A-level Syllabus 9536 (modular)

All candidates must study five content-based modules including Physics foundation, Basic 1, Basic 2 and two further optional modules. In addition, all candidates for A-level Physics must study the Experimental skills module.

Syllabus topic	Covered in Unit No	✔
Module 4830 Physics foundation		
1 Physical quantities and units	9.1	
2 Kinematics	1.1, 1.2	
3 Dynamics	1.3	
4 Forces	1.5	
5 Work, energy and power	1.4	
6 Waves	3.4, 4.3	
7 Current electricity	7.1–7.3	
8 dc circuits	7.4, 7.5	
9 Atomic structure	8.5	
Module 4831 Basic 1		
1 Motion in a circle	1.7	
2 Oscillations	1.6, 3.2, 3.3	
3 Superposition	3.5, 3.6, 4.1, 4.2	
4 Electromotive force and pd	7.2, 7.4	
5 Capacitance	7.6	
6 Magnetic fields	2.10	
7 Electromagnetism	2.10, 2.11	
8 Electromagnetic induction	2.12	
9 Phases of matter	5.3, 5.4	
10 Solids and their deformation	5.1, 5.2	
11 Quantum physics	1.10, 8.2–8.4	
12 Nuclear physics	8.5, 8.7	
13 Radioactivity	8.6	

Syllabus topic	Covered in Unit No	✔
Module 4832 Basic 2		
1 Electric field	2.1–2.3, 2.5, 2.8	
2 Gravitational field	2.1–2.3, 2.5, 2.6	
3 Electromagnetism	2.10, 2.11	
4 Electromagnetic induction	2.12	
5 Thermodynamics	6.7	
6 Ideal gases	6.6	
7 Transfer of thermal energy		
Option module 4833 Further physics		
1 Motion in a circle	1.7, 2.7	
2 Oscillations	1.6, 3.2, 3.3	
3 Electromagnetic waves	3.7	
4 Superposition	3.5, 3.6, 4.1, 4.2	
5 Electrostatics	7.1	
6 Capacitance	7.6	
7 dc circuits	7.4	
8 Alternating currents	7.7, 7.8	
Option module 4834 Nuclear physics		
1 Charged particles	8.1	
2 Radioactivity	8.6	
3 X-rays	8.4	
4 Temperature	6.1	
5 Nuclear fission and fusion	8.7	

There are four further optional modules, which are: 4835 – Health physics; 4836 – Physics of transport; 4837 – Cosmology; 4838 – Telecommunications. Candidates may also choose not more than one of the following additional optional modules: 4841 – Instrumentation electronics; 4842 – Scientific communication.

Paper analysis

Content-based modules
Each content-based module is assessed by means of a 1½-hour written paper consisting of:

Section A	*1 hour* *(advised)*	Compulsory short-answer questions and/or comprehension questions
Section B	*½ hour* *(advised)*	1 from 2 free-response questions

$5 \times 16\frac{2}{3}\%$ of total mark

Experimental skills module
Either   Teacher assessment (Module 4853); or
                Practical examination (Module 4859)
                        $16\frac{2}{3}\%$ of total mark

UNIVERSITY OF CAMBRIDGE LOCAL EXAMINATIONS SYNDICATE
AS-level Syllabus 8536 (modular)

All candidates must study the Physics foundation module and the Basic 1 module. In addition, all candidates must study module 4847 Fields and gases, which is unique to AS Physics. This module is a half-module of subject content and a half-module of experimental skills assessment. The three modules are equally weighted.

Syllabus topic	Covered in Unit No	✓	Syllabus topic	Covered in Unit No	✓
Module 4830 Physics foundation			5 Capacitance	7.6	
1 Physical quantities and units	9.1		6 Magnetic fields	2.10	
2 Kinematics	1.1, 1.2		7 Electromagnetism	2.10, 2.11	
3 Dynamics	1.3		8 Electromagnetic induction	2.12	
4 Forces	1.5		9 Phases of matter	5.3, 5.4	
5 Work, energy and power	1.4		10 Solids and their deformation	5.1, 5.2	
6 Waves	3.4, 4.3		11 Quantum physics	1.10, 8.2–8.4	
7 Current electricity	7.1–7.3		12 Nuclear physics	8.5, 8.7	
8 dc circuits	7.4, 7.5		13 Radioactivity	8.6	
9 Atomic structure	8.5		Module 4847 Fields and gases		
Module 4831 Basic 1			1 Electric field	2.1–2.3, 2.5, 2.8	
1 Motion in a circle	1.7		2 Gravitational field	2.1–2.3, 2.5, 2.6	
2 Oscillations	1.6, 3.2, 3.3				
3 Superposition	3.5, 3.6, 4.1, 4.2		3 Ideal gases	6.6	
4 Electromotive force and pd	7.2, 7.4		4 Thermodynamics	6.7	

Paper analysis

Modules 4830 and 4831 are assessed by means of 1½-hour written papers consisting of:

| Section A | *1 hour (advised)* | Compulsory short-answer questions and/or comprehension questions |
| Section B | *½ hour (advised)* | 1 from 2 free-response questions |

The subject content of module 4847 is assessed by means of a ¾-hour written paper consisting of short compulsory questions and/or comprehension questions.

| Practical skills | Either teacher assessment; or 1½-hour practical examination |

UNIVERSITY OF LONDON EXAMINATIONS AND ASSESSMENT COUNCIL
A-level Syllabus 9541 (modular or terminal examination)

Syllabus topic	Covered in Unit No	✔	Syllabus topic	Covered in Unit No	✔
Module 6541 PH1 Mechanics and electricity			3.2 Energy conservation	6.7	
1.1 Physical quantities	9.1		3.3 Heating matter	6.2	
1.2 SI units	9.1		3.4 Thermal transfer of energy	6.4, 6.5	
1.3 Derived (SI) units	9.1		3.5 Kinetic theory of gases	6.6, 6.7	
1.4 Scalar and vector quantities	1.5		3.6 Liquids and gases	5.4	
1.5 Force	1.5		3.7 Energy sources		
1.6 Kinematics	1.1, 1.2		3.8 The thermal power station	8.7	
1.7 Dynamics	1.3		Optional topics		
1.8 Mechanical energy	1.4		3A Electronics	7.10, 7.11	
1.9 Circular motion	1.7		3B Medical physics		
1.10 Statics	1.5		3C Astrophysics		
1.11 Electrical transfer of energy	7.1, 7.2		Module 6544 PH4 Fields		
1.12 Electrical resistance	7.3		4.1 Electrostatics	8.1	
1.13 Capacitors	7.6		4.2 Electric fields	2.1–2.3, 2.5, 2.8	
Module 6542 PH2 Matter and waves			4.3 Gravitational fields	2.1–2.3, 2.5, 2.6, 2.7	
2.1 The nuclear atom	8.5				
2.2 Radioactive decay	8.6		4.4 Capacitance	2.8	
2.3 Properties of solids	5.1, 5.2, 5.3		4.5 Magnetic fields	2.10, 2.11	
2.4 Oscillations	1.6, 3.1, 3.2		4.6 Electromagnetic induction	2.12, 2.13, 7.7	
2.5 Waves	3.4, 3.5				
2.6 Superposition of waves	3.5, 3.6, 4.1, 4.2		4.7 Sinusoidal variation	1.6	
2.7 Geometrical optics	4.3		Optional topics		
2.8 Wave–particle duality	1.10, 8.2, 8.3		4A Solid materials	2.4, 2.13, 5.1, 5.2, 5.3	
Module 6543 PH3 Thermal physics			4B Earth and atmosphere		
3.1 Temperature	6.1		4C Particle physics	2.10, 8.6, 8.8	

There are four compulsory modules, PH1–4. All the topics in modules PH1 and PH2 are compulsory. Modules PH3 and PH4 comprise compulsory and optional topics.

Paper analysis

1 Module tests PH1 and PH2   *1 hour 20 minutes each*
       Each test will consist of about 7 compulsory questions
       PH1 and PH2 each provide 15% of total mark

2 Module tests PH3 and PH4   *1 hour 20 minutes each*
       Each test will consist of:
       (a) 5 compulsory questions on the compulsory content of the module:
               10% *50 minutes*
       (b) 3 structured questions, one on each optional topic. Candidates will
         be expected to answer one question only:
               5% *30 minutes*
       PH3 and PH4 each provide 15% of total mark

3 Synoptic paper PH6   *2 hours*
This paper is intended to test accumulated understanding of the A-level syllabus as a

whole. It consists of three parts:

Passage analysis:	8%	*50 minutes*
Data handling question:	4%	*20 minutes*
Long structured questions:	8%	*50 minutes* (3 questions are set, based

on knowledge and understanding from the compulsory content of modules PH1, 2, 3 and 4. Each candidate is expected to answer 2 questions).

20% of total mark

4 Practical test PH8   *2½ hours*

Shorter questions: *1 hour 15 minutes*
10% of total mark

Long question: *1 hour 15 minutes*
10% of total mark

UNIVERSITY OF LONDON EXAMINATIONS AND ASSESSMENT COUNCIL
AS-level Syllabus 8541 (modular or terminal examination)

Syllabus topic	Covered in Unit No	✔	Syllabus topic	Covered in Unit No	✔
Module 6541 PH1 Mechanics and electricity			1.12 Electrical resistance	7.3	
1.1 Physical quantities	9.1		1.13 Capacitors	7.6	
1.2 SI units	9.1		Module 6542 PH2 Matter and waves		
1.3 Derived (SI) units	9.1		2.1 The nuclear atom	8.5	
1.4 Scalar and vector quantities	1.5		2.2 Radioactive decay	8.6	
1.5 Force	1.5		2.3 Properties of solids	5.1, 5.2, 5.3	
1.6 Kinematics	1.1, 1.2		2.4 Oscillations	1.6, 3.1, 3.2	
1.7 Dynamics	1.3		2.5 Waves	3.4, 3.5	
1.8 Mechanical energy	1.4		2.6 Superposition of waves	3.5, 3.6, 4.1, 4.2	
1.9 Circular motion	1.7				
1.10 Statics	1.5		2.7 Geometrical optics	4.3	
1.11 Electrical transfer of energy	7.1, 7.2		2.8 Wave–particle duality	1.10, 8.2, 8.3	

The syllabus consists of two compulsory modules, PH1 and PH2. All the topics in modules PH1 and PH2 are compulsory.

Paper analysis

1 Module tests PH1 and PH2   *1 hour 20 minutes each*

Each test will consist of about 7 compulsory questions
PH1 and PH2 each provide 30% of total mark

2 Synoptic paper PH5   *1¼ hours*

This paper is intended to test accumulated understanding of the AS-level syllabus as a whole. It consists of two parts:

Experimental skills and
data handling: 12% *50 minutes*
Long structured questions: 8% *25 minutes* (2 questions are set,

based on knowledge and understanding from the compulsory content of modules PH1 and 2. Each candidate is expected to answer 1 question)

20% of total mark

3 Practical test PH7   *1¼ hours*

Shorter questions identical to the shorter questions in PH8
20% of total mark

NORTHERN EXAMINATIONS AND ASSESSMENT BOARD
A-level Syllabus 4181 (terminal examination)

Syllabus topic	Covered in Unit No	✓
1 Mechanics		
1.1 Statics	1.5	
1.2 Dynamics	1.1–1.4	
1.3 Centripetal force and gravitation	1.7, 2.1–2.3, 2.5–2.7	
2 Oscillations and waves		
2.1 Oscillations	1.6, 3.1, 3.2	
2.2 Waves	3.4–3.7	
3 Optics		
3.1 Reflection and refraction	4.3	
3.2 Lenses	4.4	
3.3 Telescopes	4.4	
4 Electricity, fields and electromagnetism		
4.1 Current electricity	7.1–7.4	
4.2 Magnetic effects of current	2.10–2.14	
4.3 Electric fields and capacitance	2.1–2.3, 2.5, 2.8, 2.9, 7.6	
4.4 Alternating current	7.7, 7.8	
5 Matter		
5.1 Solids	5.1–5.3	
5.2 Thermal properties	6.2, 6.4–6.7	
6 Atomic and nuclear physics		
6.1 Structure of atoms	8.5	
6.2 Radioactivity	8.6	
6.3 Particles and accelerators	8.8	
6.4 Nuclear properties	8.7	
6.5 Nuclear power sources	8.7	
6.6 Wave–particle duality	1.10, 8.2, 8.3	

Syllabus topic	Covered in Unit No	✓
Options		
Ph7 Medical physics		
7.1 Physics of the human body		
7.2 Physics of biological measurement		
7.3 Physics of non-ionising radiation		
7.4 Ionising radiation		
Ph8 Applied physics		
8.1 Alternating currents	7.8	
8.2 Thermodynamics and engines	6.7	
8.3 Flow of fluids	5.4	
8.4 Rotational dynamics	1.8	
Ph9 Turning points in physics		
9.1 The discovery of the electron	8.1	
9.2 Wave–particle duality	1.10, 8.2	
9.3 Special relativity	1.10	
9.4 Towards absolute zero	6.9	
E11 Basic electronics		
E11.1 Basic electrical principles	2.14, 7.1–7.4, 7.6, 7.7	
E11.2 Devices	7.9	
E11.3 Analogue electronics	7.10	
E11.4 Digital electronics	7.11	

Paper analysis

Paper 1   *3 hours*
   Section 1   *1¼ hours*   40 multiple-choice questions
       27% of total mark
   Section 2   *1¾ hours*   4 out of 6 long questions
       20% of total mark

Paper 2   *3 hours*
   Section 1   *1¾ hours*   6 to 9 short-answer questions
       20% of total mark
   Section 2   *1¼ hours*   4 to 7 questions on each of 4 options. Candidates answer all questions from 1 option.
       15% of total mark

Paper 3   Either practical examination *3 hours*; or Internal assessment of practical skills
       18% of total mark

NORTHERN EXAMINATIONS AND ASSESSMENT BOARD
AS-level Syllabus 3181 (terminal examination)

Syllabus topic		Covered in Unit No	✓	Syllabus topic		Covered in Unit No	✓
1	Mechanics			4	Electricity, fields and electromagnetism		
1.1	Statics	1.5		4.1	Current electricity	7.1–7.4	
1.2	Dynamics	1.1–1.4		4.2	Magnetic effects of current	2.10–2.14	
1.3	Centripetal force and gravitation	1.7, 2.1–2.3, 2.5–2.7		4.3	Capacitance	7.6	
				5	Matter		
2	Oscillations and waves			5.1	Solids	5.1–5.3	
2.1	Oscillations	1.6, 3.1, 3.2		6	Atomic and nuclear physics		
2.2	Waves	3.4–3.7		6.1	Structure of atoms	8.5	
3	Optics			6.2	Radioactivity	8.2, 8.3, 8.6	
3.1	Reflection and refraction	4.3		6.3	Wave–particle duality	1.10, 8.2, 8.3	
3.2	Lenses	4.4					

All topics are compulsory.

Paper analysis

Paper 1	$1\frac{1}{2}$ hours	
Section 1	1 hour	28 multiple-choice questions 29% of total mark
Section 2	$\frac{1}{2}$ hour	1 out of 3 long questions 13% of total mark

Paper 2	$1\frac{3}{4}$ hours	
Section 1	$\frac{1}{2}$ hour	1 data processing question 11% of total mark
Section 2	$1\frac{1}{4}$ hours	5 to 7 short graded questions 29% of total mark

Paper 3   Either practical examination $1\frac{1}{2}$ hours; or
Internal assessment of practical skills
18% of total mark

NORTHERN EXAMINATIONS AND ASSESSMENT BOARD
A-level Syllabus 4183 (modular)/AS-level Syllabus 3183 (modular)

A-level syllabus 4183
All candidates must study three compulsory modules Ph1, Ph2 and Ph3 and either a further three modules from modules Ph4, 5, 6, 7, 8, 9;
or two modules from Ph4, 5, 6, 7, 8, 9 and one of El1, El2, El3 or El4.
Each module contributes $\frac{1}{6}$th (i.e. $16\frac{2}{3}$%) of the total mark.

AS-level syllabus 3183
All candidates must study two compulsory modules Ph1 and Ph2 and one further module from Ph3, 4, 5, 6, 7, 8 or 9.
Each module contributes $\frac{1}{3}$rd (i.e. $33\frac{1}{3}$%) of the total mark.

Syllabus topic	Covered in Unit No	✓
Module Ph1 Mechanics and electricity		
1.1 Mechanics		
Statics	1.5	
Dynamics	1.1–1.4	
Uniform motion in a circle	1.7	
1.2 Matter and materials		
Structure of solids	5.1	
Elasticity	5.2	
Plastic behaviour	5.3	
1.3 Current electricity		
dc circuits	7.1, 7.2	
Resistivity	7.3	
Potential divider	7.4	
CRO	7.7	
Capacitance	7.6	
Module Ph2 Particles and waves		
2.1 Nuclear physics		
The nucleus	8.5	
Radioactivity	8.6	
Mass and energy	8.7	
2.2 Waves and oscillations		
Simple harmonic motion	1.6	
Free and forced vibrations	3.1, 3.2	
Waves	3.4, 3.5	
Interference and diffraction	4.1, 4.2	
Stationary waves	3.6	
Reflection and refraction	4.3	
2.3 Quantum phenomena		
Electromagnetic spectrum	3.7	
Photoelectricity	8.2	
Energy levels	8.3	
Wave–particle duality	1.10	
Module Ph3 Further physics		
3.1 Magnetic effects of electric currents		
Magnetic flux density $F = BIl$	2.10, 2.11	
Force on moving charges	2.10	
Magnetic flux	2.13	
Electromagnetic induction	2.12	
Self-inductance	2.14	
Alternating current	7.7	
The transformer	2.14	

Syllabus topic	Covered in Unit No	✓
3.2 Gravitational and electric fields		
Newton's law of gravitation	2.1–2.3	
Gravitational field strength and potential	2.5, 2.6, 2.8	
Satellite motion	2.7	
Coulomb's law	2.5	
Electric field strength and potential	2.5	
Analogy between electric and gravitational fields	2.5	
Capacitor combinations	7.6	
3.3 Thermal physics		
Internal energy	6.2	
First law of thermodynamics	6.7	
Thermal conductivity	6.4	
Convection and radiation	6.5	
Ideal gas law	6.6	
Kinetic theory of gases	6.6	
Ph4 Astronomy and optics		
Lenses, optical fibres and detectors	4.3	
Telescopes	4.4	
Radiation	6.5	
Classification of stars		
Distances to the stars		
Doppler effect	3.6	
Ph5 Physics of materials		
Structure of materials	5.1, 5.3	
Microstructure		
Mechanical properties	5.2	
Ferromagnetic properties		
Electrical properties		
Optical properties		
Ph6 Particle physics		
Gross nuclear properties		
Particles and accelerators	8.8	
Nuclear stability	8.7	
Detectors	8.6	
Decay mechanisms		
Nuclear power sources	8.7	
Ph7 Medical physics – see p.11		
Ph8 Applied physics – see p.11		
Ph9 Turning points in physics – see p.11		
El1 Basic electronics – see p.11		
El2 Advanced electronics		
El3 Communication systems		
El4 Microprocessor and instrumentation systems		

Paper analysis

A-level (modular)
Six modules must be completed, including Ph1, 2 and 3. Each module is weighted at $16\frac{2}{3}\%$ and is assessed by means of a $1\frac{1}{2}$-hour written test ($13\frac{2}{3}\%$) and teacher-assessed practical skills (3%).

AS-level (modular)

Three modules must be completed, including Ph1 and 2. Each module is weighted at $33\frac{1}{3}\%$ and is assessed by means of a $1\frac{1}{2}$-hour written test ($27\frac{1}{3}\%$) and teacher-assessed practical skills (6%).

NORTHERN IRELAND COUNCIL FOR THE CURRICULUM EXAMINATIONS AND ASSESSMENT
A-level

Syllabus topic	Covered in Unit No	✔
Physical quantities and units	9.1	
1 Mechanics		
Static forces	1.5	
Dynamics	1.1–1.3	
Energy, work and power	1.4	
Uniform motion in a circle	1.7	
2 Oscillations and waves		
Simple harmonic motion	1.6	
Resonance and damping	3.2	
Wave properties	3.1, 3.4	
Polarisation	3.5	
Stationary waves	3.6	
Sound	3.6	
Interference and diffraction	4.1, 4.2	
Electromagnetic spectrum	3.7	
Refraction of light	4.3	
Lenses	4.3	
3 Matter		
Types of solids	5.1	
Elasticity	5.2, 5.3	

Syllabus topic	Covered in Unit No	✔
Ideal gas law	6.6	
Kinetic theory of gases	6.7	
Temperature	6.1	
Thermal conduction	6.4	
4 Fields		
Field strength and potential	2.1–2.3	
Gravitational fields	2.5–2.7	
Electric fields	2.5, 2.8, 2.9	
Capacitors	7.6	
5 Current electricity		
dc electricity	7.1–7.4	
Magnetic effects	2.10, 2.11	
Electromagnetic induction	2.12, 2.14	
Alternating currents	7.7, 7.8	
Electronics	7.10, 7.11	
6 Particles and photons		
Electron physics	8.1	
Atomic and nuclear physics	8.5–8.7	
Photons and energy levels	8.2, 8.4	

All topics are compulsory.

Paper analysis

Paper 1	*$1\frac{1}{2}$ hours*	40 multiple-choice questions 20% of total mark
Paper 2	*2 hours*	6 short structured questions 30% of total mark
Paper 3	*$2\frac{1}{2}$ hours*	
	$1\frac{3}{4}$ hours	4 out of 6 longer questions 7% each
	$\frac{3}{4}$ hour	1 comprehension question 9% of total mark
Practical test	*2 hours*	4 short compulsory tests 13% of total mark

NUFFIELD
A-level Syllabus 9661 (terminal examination)

Syllabus topic	Covered in Unit No	✓
Unit A Materials and mechanics		
The behaviour of materials	5.1, 5.2	
The structure of solid materials	5.3	
Structures and composite materials	1.5	
Momentum and the simple kinetic theory of gases	6.6	
Unit B Currents, circuits and charge		
Things which conduct	7.9	
Currents in circuits	7.1–7.5	
Electric charge	7.6, 9.1	
Unit C Digital electronic systems		
Combinational logic	7.11	
Sequential logic		
Designing digital systems		
Unit D Oscillations and waves		
Introduction to oscillations	3.1, 3.2	
Mechanical waves and superposition	3.4	
Uniformly accelerated motion	1.1, 1.2	
Mechanical oscillations	1.6, 3.1, 3.2	
Forced vibrations and resonance	3.2	
Unit E Field and potential		
Uniform electric field	2.1–2.3, 2.8	
Gravitational field and potential	2.1–2.3, 2.5–2.7	
The electrical inverse square law	2.5	
Unit F Radioactivity and the nuclear atom		
The Rutherford model of the atom	8.5	
Exponential decay	8.6	

Syllabus topic	Covered in Unit No	✓
The nucleus	8.7	
Unit G Energy sources		
Energy supplies		
Nuclear power	8.7	
Energy options		
Unit H Magnetic fields and ac		
Magnetic fields	2.10, 2.11	
Electromagnetic induction	2.12	
Alternating current	2.12, 2.14, 7.7	
Unit I Linear electronics, feedback and control		
Operational amplifiers	7.10	
Feedback and control		
Using electronics		
Unit J Electromagnetic waves		
Waves through an aperture	4.1, 4.2	
Waves through gratings	4.2	
Electromagnetic waves	2.15, 3.7	
Unit K Energy and entropy	6.10	
Unit L Waves, particles and atoms		
Photons	8.2	
Electrons	1.9	
Electron waves in atoms	8.3	
Unit M Thermal physics		
Internal energy	6.2	
First law of thermodynamics	6.7	
Energy transfer	6.4	

All units except C, G, I and K are compulsory. In addition, students must study either Units C and I or Units G and K.

Paper analysis

Paper 1   *1¼ hours*   40 multiple-choice questions on the compulsory units
  20% of total mark

Paper 2   *1½ hours*   About 7 short-answer questions on the compulsory units
  20% of total mark

Paper 3   *2½ hours*   Knowledge of the compulsory units only is required for this paper

  Section A   Comprehension paper
    10% of total mark

  Section B   Data handling paper
    8% of total mark

  Section C   3 out of 6 questions on physics applications or situations
    6% of total mark

Paper 4   *1½ hours*   8 compulsory practical questions
                        16% of total mark

Paper 5 (coursework)   Internally assessed practical investigation of about 2 weeks'
                        duration as part of the course
                        10% of total mark

Paper 6 (coursework)   Internally assessed research and analysis report on a topic
                        from either Units C and I or Units G and K
                        10% of total mark

NUFFIELD
AS-level Syllabus 8491 (terminal examination)

Syllabus topic	Covered in Unit No	✓	Syllabus topic	Covered in Unit No	✓
1 Materials and mechanics			Forced vibrations and resonance	3.2	
The behaviour of materials	5.1, 5.2		Waves through an aperture	4.1, 4.2	
The structure of solid materials	5.3		Waves through gratings	4.2	
Structures and composite materials	1.5		Electromagnetic waves	2.15, 3.7	
Momentum and the simple kinetic theory of gases	6.6		Photons	8.2	
			Electrons	1.9	
Uniformly accelerated motion	1.1, 1.2		4 Field, force and potential		
2 Currents, circuits and charge			What is a field?	2.1–2.3	
Things which conduct	7.9		Uniform fields	2.1–2.3, 2.8	
Currents in circuits	7.1–7.5		Inverse square law	2.5	
Electric charge	7.6, 9.1		5 Topics G and P		
The Rutherford model of the atom	8.5		Topic G Energy sources		
Exponential decay	8.6		Energy supplies		
The nucleus	8.7		Nuclear power	8.7	
3 Oscillations and waves			Energy options		
Introduction to oscillations	3.1, 3.2		Topic P Electronics		
Mechanical waves and superposition	3.4		Combinational and sequential logic	7.11	
Mechanical oscillations	1.6, 3.1, 3.2		Operational amplifiers	7.10	

All units except Unit 5 are compulsory. In addition, students must study either Topic
G or Topic P from Unit 5.

Paper analysis

Paper 1   *1 hour*   30 multiple-choice questions on the compulsory units
                       29% of total mark

Paper 2   *1 hour*   About 5 short-answer questions on the compulsory units
                       25.5% of total mark

Paper 3   *1½ hours*   Knowledge of the compulsory units only is required for this paper

  Section A   Data analysis/comprehension
                      19% of total mark

  Section B   2 out of 4 questions on physics applications or situations
                      7.5% of total mark

Paper 4   Internally assessed practical investigation of about 2 weeks' duration as part of the course
                      9.5% of total mark

Paper 5   Internally assessed research and analysis report on a topic from Unit 5
                      9.5% of total mark

NUFFIELD
A-level Syllabus 9664 (modular)
AS-level Syllabus 8364 (modular)

The modular A-level syllabus is based on the same content as the linear syllabus (9661) and consists of 4 compulsory modules and coursework consisting of one option module and practical work. The modular AS-level syllabus is a subset of the modular A-level syllabus.

Module M1: Units A, B, M
Module M2: Units D, F
Module M3: Units E, H
Module M4: Units J, L
Option module: one topic from either Units C and I or G and K

Paper analysis

A-level
Module tests on modules M1–M4   *1½ hours*
                  4–5 short structured compulsory questions and 1 longer question
                      16% for each module test

Coursework   As papers 4, 5 and 6 of the linear syllabus (9661)

AS-level
Module tests on modules M1 and M2 *1½ hours*
                  4–5 short structured compulsory questions and 1 longer question
                      32% for each module test

Coursework   Practical investigation   20%
                Practical problems paper   *1 hour*   16%

UNIVERSITY OF OXFORD DELEGACY OF LOCAL EXAMINATIONS
A-level (modular or terminal examination)

Syllabus topic	Covered in Unit No	✔
Module 1 Mechanics and basic electricity		
Scalars and vectors	1.5	
Linear motion	1.1, 1.2	
Newton's laws of motion	1.3	
Conservation of momentum	1.3	
Work, energy and power	1.4	
Pressure	5.4	
Friction and drag	1.5	
Uniform circular motion	1.7	
Statics	1.5	
Current, pd and resistance	7.1, 7.2	
dc circuits	7.3, 7.4	
Electrical energy and power	7.2	
Capacitors	7.6	
Module 2 Materials and waves		
Structure of solids	5.1, 5.3	
Elasticity	5.2	
Atomic structure	8.5	
Quantum theory	8.2, 8.3	
Radioactivity	8.6	
Fission and fusion	8.7	
Mass and energy	8.7	
Simple harmonic motion	1.6	
Free and forced vibrations	3.1, 3.2	
Waves	3.4, 3.5	
Diffraction	3.5	
Superposition	4.1	
Stationary waves	3.6	
Wave–particle duality	1.10	
Refraction	4.3	
Module 3 Thermal effects and fields		
Thermal energy flow	6.4	
Thermal conductivity	6.4, 6.5	
Thermal calculations	6.2	
Ideal gases	6.6	
Kinetic theory of gases	6.6	
Molecular energy	6.6	
Thermodynamics	6.7	
Gravitational fields	2.1–2.3, 2.5, 2.6	
Satellites	2.7	
Electrical forces and fields	2.1–2.3, 2.5, 2.8	
Potentials and potential gradient	2.3	

Syllabus topic	Covered in Unit No	✔
Potential difference	8.1	
Comparisons between gravitational and electric fields	2.5	
Module 4 Further electricity		
Magnetic fields	2.10	
Flux and flux densities	2.10, 2.11	
Electromagnetic forces	2.11	
Electromagnetic torque	2.11	
Electromagnetic induction	2.12	
Self-induction	2.14	
Mutual induction	2.14	
Capacitors	2.9, 7.6	
Capacitor discharge	7.6	
Alternating currents	7.7, 7.8	
Cathode ray oscilloscope	7.7	
Power generation		
Power distribution	2.14	
Module 5		
Section A Waves and optics		
Electromagnetic spectrum	3.7	
Spectra	8.4	
Reflection and refraction	4.3	
Single-slit diffraction	4.2	
Interference	4.1	
Diffraction grating	4.2	
Doppler effect	3.6	
Plane polarization	3.5	
Inverse square law	3.4	
Section B Nuclear physics		
The nucleus	8.5	
Mass, energy and stability	8.5	
Nuclear reaction equations	8.5	
Exponential decay	8.6	
Radioactive series	8.6	
Fission	8.7	
Fission reactors and safety	8.7	
Fusion	8.7	
Particle physics	8.8	
Section C Medical physics 1		
Section D Medical physics 2		
Section E Transport		
Section F Communications		
Module 6 Experimental skills		

The syllabus is organised in six modules. Modules 1–4 are compulsory, module 5 is a set of optional subjects and module 6 assesses experimental skills.

Paper analysis

Written papers
(modules 1–5) *1¼ hours each*   The paper for module 5 is set in 6 sections A to F, as listed above. Candidates must attempt all questions in any two sections.
16% of total mark each module

Experimental skills
(module 6)

Either coursework investigation of practical problems; or An examination paper consisting of a written exercise on planning an investigation, and practical exercises
20% of total mark

UNIVERSITY OF OXFORD DELEGACY OF LOCAL EXAMINATIONS
AS-level (modular or terminal examination)

Syllabus topic	Covered in Unit No	✔	Syllabus topic	Covered in Unit No	✔
Module 1 Mechanics and basic electricity			Wave–particle duality	1.10	
Scalars and vectors	1.5		Refraction	4.3	
Linear motion	1.1, 1.2		Module 5		
Newton's laws of motion	1.3		Section A Waves and optics		
Conservation of momentum	1.3		Electromagnetic spectrum	3.7	
Work, energy and power	1.4		Spectra	8.4	
Pressure	5.4		Reflection and refraction	4.3	
Friction and drag	1.5		Single-slit diffraction	4.2	
Uniform circular motion	1.7		Interference	4.1	
Statics	1.5		Diffraction grating	4.2	
Current, pd and resistance	7.1, 7.2		Doppler effect	3.6	
dc circuits	7.3, 7.4		Plane polarisation	3.5	
Electrical energy and power	7.2		Inverse square law	3.4	
Capacitors	7.6		Section B Nuclear physics		
Module 2 Materials and waves			The nucleus	8.5	
Structure of solids	5.1, 5.3		Mass, energy and stability	8.5	
Elasticity	5.2		Nuclear reaction equations	8.5	
Atomic structure	8.5		Exponential decay	8.6	
Quantum theory	8.2, 8.3		Radioactive series	8.6	
Radioactivity	8.6		Fission	8.7	
Fission and fusion	8.7		Fission reactors and safety	8.7	
Mass and energy	8.7		Fusion	8.7	
Simple harmonic motion	1.6		Particle physics	8.8	
Free and forced vibrations	3.1, 3.2		Section C Medical physics 1		
Waves	3.4, 3.5		Section D Medical physics 2		
Diffraction	3.5		Section E Transport		
Superposition	4.1		Section F Communications		
Stationary waves	3.6		Experimental skills		

The syllabus is organised in three modules. Modules 1 and 2 are compulsory, and the third module is an optional topic from module 5 and an assessment of practical skills by coursework.

Paper analysis

Modules 1 and 2   *1¼ hours each* Written paper
                            32% of total mark each module

Module 3
Written paper   *40 minutes*   6 sections A to F as listed on p. 18. Candidates
                            must attempt all questions in any one section
                            16% of total mark

Experimental skills   Coursework investigation of practical problems
                            20% of total mark

OXFORD AND CAMBRIDGE SCHOOLS EXAMINATION BOARD
A-level Syllabus 9685/AS-level Syllabus 8385 (modular or terminal)

Syllabus topic	Covered in Unit No	✓	Syllabus topic	Covered in Unit No	✓
Unit P1 Mechanics and waves			Capacitors in dc circuits	7.6	
Vectors and scalars	1.5		Parallel plate capacitor	2.9	
Statics	1.5		Magnetic flux density B	2.10, 2.11	
Dynamics	1.1, 1.2, 1.3		Force on a moving charge	2.10	
Conservation of momentum	1.3		Electromagnetic induction	2.12	
Work, energy and power	1.4		Unit 3 Microscopic explanations of macroscopic phenomena		
Efficiency	1.4		Structure of solids	5.1, 5.2	
Uniform circular motion	1.7		Elasticity	5.2	
Simple harmonic motion	1.6		Internal energy	6.2	
Free and forced vibrations	3.1, 3.2		First law of thermodynamics	6.7	
Wave properties	3.4, 3.5		Thermal energy transfer	6.4, 6.5	
Interference and diffraction	3.5		Thermal conductivity	6.4	
Stationary waves	3.6		Ideal gas law	6.6	
Reflection and refraction	4.3		Kinetic theory of gases	6.6	
Unit P2 Fields and electromagnetism			Photoelectric effect	8.2	
Newton's law of gravitation	2.1–2.3, 2.5		Spectra	8.3, 8.4	
Coulomb's law	2.1–2.3, 2.5		Wave–particle duality	1.10	
Uniform and radial fields	2.5, 2.6, 2.8		Nuclear model of the atom	8.5	
Gravitational field strength	2.1, 2.6		Radioactivity	8.6	
Gravitational potential	2.2, 2.5		Half-life	8.6	
Electric field strength	2.1, 2.5		Binding energy	8.7	
Electric potential	2.2, 2.5		Nuclear fission and fusion	8.7	
Current, charge and pd	7.1, 7.2		Unit 4 Medical physics		
Resistance	7.3		Body mechanics		
Resistivity	7.3		The eye		
dc circuits	7.3		X-rays		
Potential divider	7.4		Radioactive isotopes		
Capacitance	2.9		Radiation exposure and detection		

Syllabus topic	Covered in Unit No	✔
Ultrasonics		
Fibre optics		
Unit 5 Alternating current and electronics		
Capacitor charging and discharging	7.6	
Self-inductance	2.14	
Transformer	2.14	
ac rectifiers	7.7	
Series LR and CR circuits	7.8	
Resonance	7.8	
Electrical communication	7.11	
Radio receiver		
Operational amplifiers	7.10	
Unit 6 Mechanics and properties of matter with engineering relevance		
Torque, work and power		
Machines	1.4	
Rotational kinetic energy	1.8	
Angular momentum	1.8	
Bulk and shear stress	5.1	
Material properties	5.1, 5.2	
Forces and simple structures	1.5	
Pressure in a liquid	5.4	
Archimedes' principle	5.4	
Viscosity	5.6	
Bernoulli's equation	5.5	
Unit 7 Fields and electromagnetic waves		
Kepler's laws	2.7	

Syllabus topic	Covered in Unit No	✔
Satellite motion	2.7	
Magnetic field formulae	2.11	
Charged particles in magnetic fields	8.1	
Electromagnetic radiation	2.15, 3.7	
Interference and diffraction	4.1, 4.2	
Polarisation	3.5	
Unit 8 Particle physics		
Relativity	8.8	
Accelerators and detectors	8.8	
Beta decay	8.6, 8.8	
Leptons and hadrons	8.8	
Quarks	8.8	
Types of interaction	8.8	
Exchange particles	8.8	
Hubble's law		
Expansion of the universe		
Unit 9 Physics in the environment		
Solar energy		
Electromagnetic radiation		
Laws of radiation		
Simple model of the atmosphere		
Coriolis forces		
Saturated vapour pressure		
Physical properties of water		
Hydrological cycle		
Geosphere; radiation balance		
Insulation of buildings		

A-Level Syllabus 9685 (modular)
Modules P1, P2 and P3 are compulsory. Candidates must study three further modules from P4–P9 or module S1 (Science and technology in society) or module S2 (Energy).

AS-Level Syllabus 8385 (modular)
This consists of the three modules P1, P2 and P3.

Paper analysis

All modules are equally weighted. Each module is assessed by means of:

Written paper — *1½ hours* — Structured compulsory questions. Each paper will include 1 question based on data analysis or comprehension.
$13\frac{1}{3}\%$ of total mark ($26\frac{2}{3}\%$ at AS level)

Investigative skills and abilities — Internal coursework assessment
$3\frac{1}{3}\%$ of total mark each module ($6\frac{2}{3}\%$ at AS level)

SCOTTISH EXAMINATION BOARD
Higher Grade

Syllabus topic	Covered in Unit No	✔	Syllabus topic	Covered in Unit No	✔
1.1 Kinematics			2.3 Capacitance		
Dynamics, equations and graphs	1.1		Capacitors in dc circuits	7.6	
Projectile motion	1.2		Capacitors in ac circuits	7.8	
1.2 Dynamics			Uses of capacitors		
Newton's laws of motion	1.3		2.4 Analogue electronics		
Vector nature of force	1.5		Inverting mode	7.10	
Conservation of momentum	1.3		Differential mode	7.10	
Impulse	1.3		Monitoring and control	7.10	
1.3 Properties of matter			3.1 Waves and light		
Gas pressure	6.6		Interference and diffraction	3.5, 4.1, 4.2	
Density	5.4		Refraction, total internal reflection, dispersion	4.3	
Pressure in fluids	5.4		3.2 Optoelectronics		
2.1 Resistors in circuits			Intensity of radiation	3.4	
Electric fields	2.1		Photoelectricity	8.2	
Potential difference	7.1, 7.2		Spectra	8.3, 8.4	
emf and internal resistance	7.3		The laser		
Resistor combinations	7.3		Optoelectronic devices		
Wheatstone bridge circuit	7.5		3.3 Radioactivity		
2.2 Alternating current and voltage			Rutherford scattering	8.5, 8.6	
Frequency, peak and rms values	7.7		Nuclear reactions	8.6	
Resistive circuits	7.7		Dose equivalent, safety limits		
			Fission and fusion	8.7	

All three units are compulsory.

Paper analysis

Paper 1   *1½ hours*   Multiple-choice questions
Short questions
         36% of total mark

Paper 2   *2½ hours*   Questions requiring extended answers
         64% of total mark

WELSH JOINT EDUCATION COMMITTEE
A-level (terminal), A-level (modular)
AS-level (modular)

The linear A-level syllabus is available for first examination in 1996. A modular route, based on compulsory modules P1–P4 is also available at A-level, for first award in 1997 (1996 for the modular AS-level course). The content is the same for both the terminal and modular syllabuses.

Syllabus topic	Covered in Unit No	✓
Module P1 Mechanics and electricity		
1. Basic physics		
Units and dimensions	9.1	
Scalar and vector quantities	1.5	
2. Kinematics		
Rectilinear motion	1.1, 1.2	
3. Dynamics		
Linear momentum	1.3	
Newton's laws of motion	1.3	
Conservation of momentum	1.3	
4. Statistics		
Moment, torque and couple	1.5	
Centre of mass, centre of gravity	1.5	
Equilibrium of rigid bodies	1.5	
5. Work, energy and power		
Work, pe, ke, power	1.4	
Energy conservation	1.4	
6. Rotational dynamics		
Uniform circular motion	1.7	
Centripetal force and acceleration	1.7	
Satellite motion	2.5	
Couple and torque	1.5	
7. Solids under stress		
Stress and strain	5.1, 5.2	
Elastic and plastic behaviour	5.2	
8. Capacitance		
Capacitance	7.6	
Capacitor combinations	2.9	
Energy stored	2.9	
9. Current electricity		
Current, pd, resistance	7.1–7.3	
Nature of conduction	7.1, 7.9	
Resistivity	7.3	
Heating effect	7.2	
emf and internal resistance	7.2	
10. dc circuits		
Series and parallel circuits	7.2, 7.3	
Resistor combinations	7.2, 7.3	
Kirchhoff's Laws	7.3	
Potential divider	7.4	
Potentiometer	7.4	

Syllabus topic	Covered in Unit No	✓
Module P2 Vibrations, waves and atomic physics		
1. Vibrations		
Simple harmonic motion	1.6	
Energy changes	1.6, 3.2	
Damping	3.2	
Forced oscillations and resonance	3.2	
2. Waves		
Progressive waves	3.4	
Transverse and longitudinal waves	3.4	
Wave characteristics	3.4	
3. Superposition		
Interference	3.5	
Diffraction	3.5	
Stationary waves	3.6	
Beats	3.6	
4. Light waves		
Reflection and refraction	4.3	
Interference and diffraction	4.1, 4.2	
Polarisation	3.5	
5. Quantum physics		
The electromagnetic spectrum	3.7	
Spectra	8.3	
Photoelectric effect	8.2	
Wave–particle duality	1.10	
6. Nuclear physics		
Nuclear atom, the nucleus	8.5	
Isotopes	8.5	
Binding energy	8.7	
Fission and fusion	8.7	
Nuclear reactors	8.7	
7. Radioactivity		
Ionising radiation	8.6	
Nature of radiations	8.6	
Radioactive decay	8.6	
Hazards and safety precautions	8.6	
Module P3 Fields and ac		
1. Uniform and radial fields of force		
Electrostatic and gravitational fields	2.1–2.9	
Field strength	2.1	
Parallel plate capacitor	2.9	

Syllabus topic	Covered in Unit No	✓
Electrical and gravitational inverse square law	2.5	
Potential	2.2	
Field strength and potential gradient	2.1, 2.2	
Vector addition of electric fields	2.1	
pe for a system of charges	2.2	
2. Electromagnetism		
Magnetic flux and flux density	2.10, 2.13	
Magnetic fields due to currents	2.11	
Relative permeability	2.13	
Force on a conductor and on a moving charge	2.10	
Measurement of B	2.10	
Deflection of charged particles in fields	8.1	
3. Electromagnetic induction		
Laws of electromagnetic induction	2.12	
Induced emf and charge	2.12	
Self induction	2.14	
4. Alternative currents		
Peak and rms values	7.7	
Transformer	2.14	
Series ac circuits	7.7, 7.8	
Rectification and smoothing	7.7	
Use of cro	7.7	
Module P4 Materials and energy		
1. Ideal gases		
Ideal gas law	6.6	

Syllabus topic	Covered in Unit No	✓
Kinetic theory of gases	6.6	
Kinetic energy of a molecule	6.6	
2. Phases of matter		
Solids, liquids and gases	5.1	
Bonding	5.3	
Intermolecular force curves	2.4	
Explanation of Hooke's Law and expansion	2.4	
3. Solids under stress		
Elastic and plastic behaviour	5.2	
4. Electromagnetic radiation		
The electromagnetic spectrum	3.7	
Optical and x-ray spectra	8.3	
5. Electrical properties		
Nature of conduction in conductors and semiconductors	7.1–7.9	
Variation of resistance with temperature	7.1	
6. Nuclear changes		
Detection of ionising radiations	8.6	
Radioactive decay	8.6	
Nuclear reactors	8.7	
7. Energy and the environment		
Internal energy	6.7	
1st law of thermodynamics	6.7	
Energy transfer	6.7	
Energy resources		
Environmental effects		

All topics are compulsory.

Paper analysis

A-level

Paper 1   *2½ hours*   Based on fields, forces and energy, electricity and magnetism (P1, topics 1–6, 8–10; P3; P4, topic 7)

Section A   *1¼ hours*   6 compulsory short structured questions
17% of total mark

Section B   *1¼ hours*   3 compulsory long questions
17% of total mark

Paper 2   *2½ hours*   Based on molecules and matter, oscillations and waves, atomic physics (P1, topic 7; P2; P4, topics 1–6)

Section A   *1¼ hours*   6 compulsory short structured questions
17% of total mark

Section B   *1¼ hours*   3 compulsory long questions
17% of total mark

Paper 3   *1¾ hours*   2 compulsory comprehension/data analysis questions
17% of total mark

Paper 4   Internal assessment of experimental work
15% of total mark

A-level (modular)
P1–P4 module tests

$1^1/_3$ *hours*	16% each	Each test will consist of 6 short questions and two longer questions.

Synoptic paper

$1^1/_3$ *hours*	16%	3 long questions
Teacher-assessed practical work	20%	

AS-level (modular)
P1 and P2 module tests

$1^1/_3$ *hours*	40% each	Each test will consist of 6 short questions and two longer questions.
Teacher-assessed practical work	20%	

EXAMINATION BOARD ADDRESSES

AEB The Associated Examining Board
Stag Hill House, Guildford, Surrey GU2 5XJ
Tel: 01483 506506

Cambridge University of Cambridge Local Examinations Syndicate
Syndicate Buildings, 1 Hills Road, Cambridge CB1 2EU
Tel: 01223 553311

NEAB Northern Examinations and Assessment Board
12 Harter Street, Manchester M1 6HL
Tel: 0161 953 1180

NICCEA Northern Ireland Council for the Curriculum Examinations and Assessment
Beechill House, 42 Beechill Road, Belfast BT8 4RS
Tel: 01232 704666

Oxford University of Oxford Delegacy of Local Examinations
Ewert House, Ewert Place, Summertown, Oxford OX2 7BZ
Tel: 01865 54291

Oxford and Cambridge (*including Nuffield*) Oxford and Cambridge Schools Examination Board
(a) Purbeck House, Purbeck Road, Cambridge CB2 1PU
   Tel: 01223 611211
(b) Elsfield Way, Oxford OX2 8EP
   Tel: 01865 54421

SEB Scottish Examination Board
Ironmills Road, Dalkeith, Midlothian EH22 1LE
Tel: 0131 663 6601

ULEAC University of London Examinations and Assessment Council
Stewart House, 32 Russell Square, London WC1 5DN
Tel: 0171 331 4000

WJEC Welsh Joint Education Committee
245 Western Avenue, Cardiff CF5 2YX
Tel: 01222 265000

STUDYING AND REVISING PHYSICS

Too many students sit down to a session of work having given little thought as to *how* they will carry out their study. This section is designed to guide you in this area. If you read the information supplied carefully and develop your own study skills you will learn to work effectively; not only will you learn better, but you will also learn faster. Students who have taken courses in 'How to study' unanimously point out that acquiring a good technique has not only improved their understanding, but has also helped to increase their leisure time. There are many good books on studying that are recommended to those who find this necessarily concise section too brief for their needs.

THE DIFFERENCE BETWEEN GCSE AND A/AS LEVEL

When you were studying for GCSE you may have thought that A-level was rather similar but more difficult. This is an easy trap to fall into since at first sight there are some similarities in the syllabuses. These are, however, superficial and your A-level studies will have a greater significance if you fully appreciate what is expected of you and the approach which is required at this level.

The significance of facts

'In GCSE you mainly need to know the facts and remember them, but at A level you need to know much more.'

'At A level you are encouraged to think much more deeply.'

The volume of knowledge required at A level is very much greater than at GCSE and the depth of treatment much more profound – you must be prepared to do plenty of background reading and make your own notes (more of that later). You must get to grips with the basic concepts and have a thorough understanding of the fundamental principles and laws, so that you can apply them to familiar and unfamiliar situations. In order to explain the behaviour of matter you will need to acquire the ability to develop conceptual and mathematical models, modifying and shaping them as you find out more about the physical nature of the world around you. You will discover, and have to learn to accept, that there is not always a 'correct' answer to every question and you must be prepared to amend, or even reject, previously held theories in the light of further evidence. Only in such a way does science move forward.

Mathematical skills

GCSE requires little more than a basic grasp of mathematics and a certain degree of numeracy. You begin to see at A level that the universe appears to be governed by physical laws that have a mathematical basis. Consequently, you need to develop your mathematical abilities in order to understand more fully the simplistic beauty of these laws. You must be able to carry out complex computations quickly and accurately with the help of an electronic calculator and have some idea of the order of magnitude of quantities (the table of values at the end of this book is designed to help you with this). This ability to make sensible approximations should be developed and the significance of experimental errors must be appreciated. A basic working knowledge of calculus is advantageous, although not essential.

Practical skills

Throughout the A-level course you should be involved in practical work. It is much easier to understand ideas that relate to your own experience, and so experiments relating closely to the concepts that you are learning will help reinforce the learning process. Some experiments will lead you to find out and question discoveries for yourself, and others will help you to develop relevant skills in manipulating apparatus, making careful observations and interpreting data.

MODULAR COURSES

Most examining boards now offer modular courses as a flexible scheme which allows a variety of different routes to be followed in order to achieve an AS or A-level. The syllabus is divided into small units (modules), each of which is examined by an end-of-module test. These tests may be taken at appropriate times throughout the course or, if preferred, at the end of the course. The advantages of taking the tests throughout the course are that there is a regular feedback on how well you are doing and there is the possibility of retaking modules in order to improve your grade. Furthermore, the results of modular tests may be 'banked', and then 'cashed-in' at some time within four years to obtain an AS or A level. This allows for a break in a course of study without loss of the credit already obtained. Some modules are compulsory, others may be selected from a list of optional ones. At the end of a modular course there may be a **synoptic paper**. This is a test made up of questions which are taken from across the compulsory modules.

By varying the number and type of modules studied you can obtain an AS or A level in Physics or, by combining Physics modules with those from other sciences, an AS or A level in Science. Currently, an AS level is either two or three modules and an A level four or six modules, depending on the examination board.

STUDY STRATEGIES AND TECHNIQUES

When and where to study

Whether you are studying for your A-level Physics at college, school, evening class or at home, you will need to spend a considerable amount of time working on your own, possibly without a great deal of guidance. Quite often students find this transition from the more formal pattern of GCSE teaching a difficult step to make:

'A level requires you to work more by yourself and also to think for yourself.'

'The emphasis on pupils learning by themselves is greater.'

It is therefore essential that you plan ahead, perhaps as much as a week in advance. For GCSE, it is often sensible to schedule your work so that it is done the day before it is due to be handed in. At A level such an approach is disastrous. Two problems arise; firstly, the nature of the work means that you often encounter difficulties that will need advice from your teacher or the aid of books unobtainable until the following day; secondly, the demands of A level are unpredictable and you may find that you have also been given some other tasks to complete overnight. In all sciences, the need frequently arises to analyse results and plot graphs from one day's practical/demonstration in time for the next lesson; the student who is always unable to carry out such work (because he has too much other work to complete by the following morning) will struggle with his A-level course.

For successful planning you must decide how much work you need to do and how much time you intend to spend on it. You will not study effectively if you work for long periods without a break. After about 40 minutes concentration begins to wane. A break of a few minutes, in which you do something completely different, can soon recharge your batteries and enable you to continue your studying to greater effect. So, try to plan your work so that you fit in these breaks at convenient intervals. *When* you work is very much a personal matter. Some people find that they work best early in the morning, others seem to work better late at night – find out when suits you best and plan accordingly.

Where you work is also important. It is helpful if you can have somewhere that you can call 'your own', even if it is only a table in the corner of your bedroom, where you can work without being disturbed. Make the area welcoming. You may find music helpful – some people do actually work better accompanied by background music, despite popular opinion to the contrary. Your surroundings should be comfortable, warm and, most important, well lit. Make sure that you have essential materials readily at hand: writing paper, pens (several colours are helpful), drawing instruments, calculator and books should be kept conveniently nearby.

Reading for A level

'… more reading is necessary.'

'At GCSE you were given more information. At A level you have to find out more for yourself.'

Since you have been able to read for as long as you can remember, it probably does not occur to you to question your ability to read *effectively*. You can almost certainly improve your reading skills by developing some fairly simple techniques that can be found in specialist books on the subject.

You will read more purposefully if you are:

❶ **motivated** – without a target, reading lacks drive;

❷ **interested** – concentration relates to degree of interest;

❸ **active** – think for yourself; don't expect the author to think for you. Read critically what is said.

During your A-level course you will encounter many references to books and magazines and other resource material. Some of it will only be 'background reading' and the temptation is great to ignore such texts and only concentrate on the 'necessary'. However, modern examinations strongly favour the student who has developed his reading skills by doing such work. The comprehension papers set by some Boards are handled far better by students who are accustomed to appraising scientific writing critically. Further, modern exam questions in Physics tend to be wide ranging and often take a considerable amount of time and skill to comprehend fully. Students who have read little frequently run into time trouble in written examinations, and have an unfortunate habit of misinterpreting questions.

Note taking

'Teaching methods are different … making notes for your own use is essential.'

'More responsibility on pupil to make own notes.'

The amount of note taking that you will be required to do will very much depend on the particular methods adopted by your teacher. However, you should remember that your notes are for *your* benefit, and should eventually form the basis of your revision plan. Bearing this in mind, the way in which you present your notes is going to have a considerable effect on their efficiency in helping you to understand and learn the work.

Notes should be visually outstanding, achieved by the use of CAPITALS, underlining, *symbols* (the √ and ⌇◯), by using a box and by the use of colour.

Your notes should be succinct; remember that you should already have a textbook, so there is no point trying to write another one. Wherever possible try to illustrate your notes with diagrams (a good diagram can save volumes of words), charts and tables. You should always write up your notes, with the help of books if necessary, as soon after the lesson as possible while the subject matter is still fresh in your mind. Make sure you *understand* the work at this stage, or else further development and, later on, revision will be difficult. Seek help from other books or your teacher if necessary.

Learning

'There is a greater emphasis on learning rather than being told in A level.'

'… more to memorise.'

·In GCSE you mainly need to know the facts and remember them, but in A level you need to know much more … you need to know the basic principles.'

Physics is a demanding discipline. There are certain basic laws and principles that must be memorised. Key equations and formulae should be known off by heart, even if a data sheet or booklet is supplied to you during your examination. Although a data booklet is helpful if you happen to forget a particular formula which you need to be able to start answering a question, searching frantically through a data booklet to find a likely formula is a waste of valuable time in an examination. The data booklet should not really be needed for the key questions.

Keep to a minimum the basic laws and equations that you need to memorise. Highlight them in your notes by boxing-in or colouring; or even by making a separate list of basic facts that must be known.

You must *thoroughly* understand fundamental principles and concepts before your attempt to commit them to memory. Certain mathematical and analytical skills are needed. These can only be acquired through *practice*, particularly by answering questions, which is why a large part of this book is devoted to just that.

The important thing is to have the right attitude to learning. If you *want* to learn you *can* learn. To this end you should set yourself goals at which to aim so that the satisfaction of attaining them acts as a spur for the next task ahead.

COURSEWORK

Physics is a practical-based subject and you should have lots of opportunities to develop your practical skills during your course. All A-level and AS-level syllabuses assess practical work, either by means of a practical examination set by your examination board or through internal assessment of practical skills. Whichever of these two assessment routes applies to you, practical work during your course will contribute either directly (i.e. if assessed internally) or indirectly (i.e. if assessed by examination) to your final grade.

The extent and nature of your practical work in Physics will depend on the course you are following and the facilities available to you. There is no typical activity in Physics but your activities will probably fall into one of several categories, as follows.

Short experiments

These are set with a specified aim in mind, perhaps to verify a principle or to measure a specific property. This type of experiment, usually lasting no more than about an hour, may involve testing a suspected link between two physical quantities, measuring a physical property or practising a particular skill.

Usually, a short experiment involves carrying out specific tasks by following a set of instructions. You may need to select suitable instruments to make the measurements. It is often necessary to plot a graph and/or carry out calculations using the measurements.

Some short experiments are used to develop data handling skills and laboratory techniques which then can be put into practice in longer investigations and projects.

Investigations

These usually have a specific aim and you will have to make decisions to reach that aim. In general, investigations offer the opportunity to develop your knowledge of Physics through practical activities.

You must use your time effectively in planning and carrying out an investigation, keeping the aim of the investigation in mind throughout. Once you have made your plans and they have been checked by your teacher, carry out the investigation leaving sufficient time to analyse the measurements, evaluate the results and write your report.

Projects

These are open-ended longer investigations which demand a wide range of skills, from initial planning and design to final evaluation. This type of practical work offers you the opportunity to show that you can use your skills and knowledge of Physics in practical situations.

If your syllabus involves internal assessment of practical skills, you may be expected to carry out several investigations or one or more projects as assessment activities. The skills and techniques used in short experiments and investigations are important in any project so you will need to have made reasonable progress in these shorter activities before attempting a project. In addition, you are expected to exercise planning and organisational skills, as well as an ability for evaluating experimental evidence, and you will have to judge for yourself when to make decisions.

Internal assessment of practical work

Your first coursework assessment(s) might be in the third term and could involve short investigations or a project, depending on your exact course. Subsequent assessment involving either short investigations and/or projects would then be carried out in the second year. Your teacher will tell you at the start of your course what the planned timetable for coursework assessments is.

Schemes of internal assessment of practical skills differ from one Board to another in detail, although most boards assess a common range of skills. These skills appear in each scheme but they are organised and marked in a different way. For example, a particular skill area might be assessed only once on a four-point scale on one syllabus, whereas on a different syllabus it might be assessed twice on a six-point scale. Below is a numbered list of most of the skill areas that feature in the assessment schemes of the examination boards, with the skills in that skill area also given. You may not be assessed on all of these skill areas but you are likely to be assessed on most of them. Before you carry out an assessment exercise, your teacher will tell you which skill areas are being assessed.

The list below is *not* a substitute for a scheme of internal assessment. It shows you what skill areas are in most schemes, but you will need to obtain a copy of your syllabus if you want to see how the skill areas are organised and marked.

❶ **Project specification**
   identify problem
   select aims
   conduct initial tests
   use knowledge and test results to refine aims

❷ **Planning an investigation**
   decide what to measure
   decide what to use
   decide on procedure
   select apparatus
   Check reliability and modify if necessary

❸ **Carrying out an investigation**
   set up apparatus correctly
   use apparatus safely
   follow instructions correctly

❹ **Observing and measuring**
   make precise measurements
   take steps to ensure accuracy

check reliability
check unexpected results
estimate errors

❺ Analysing the data
decide on appropriate graphs
plot graphs accurately
carry out calculations
use result with appropriate theory
analyse errors, represent errors on graphs

❻ Evaluating the results
draw valid conclusions
assess strength of evidence
evaluate method used
make further suggestions

❼ Recording and reporting
record measurements effectively
use correct units, symbols, etc.
keep a diary of progress, etc.
write a report of your work

Project work

Because the nature of a project is open ended, you have many decisions and choices to make. The first of these may be to decide on suitable aims for your project, perhaps given some initial ideas to think about. Don't be too ambitious in the aims of your project, otherwise you might find there are too many variables to measure, monitor and control.

The most difficult choice may well be deciding when to stop your investigations. There will be a limit on the time available and you must leave sufficient time to finalise your project report. However, a good project often has no end point and you simply make as much progress as possible in the time allowed.

At the end of a project, you need to evaluate your work in term of your aims and methods. How far did you achieve your aims? Did you change your aims? If so, why? What improvements could you have made in your methods? The evaluation might include an outline of further possible investigations stemming from the project, and such possibilities could be explored by other students on later projects.

Writing your report

No matter how good a scientist is at experimental work, unless he or she can communicate effectively, the work will be of little benefit since no one is likely to find out about it. For the same reasons, your experimental skills will be wasted unless you can write effectively about your work.

The assessment of your practical work, whether by examination or coursework, depends largely on your written practical reports. Some of your practical marks may be awarded by your teacher observing you at work, but most will depend on what you write down.

In practical examinations, your written report may be tightly defined by the structure of the question. However, if this is not so or if your practical work is assessed through coursework, you need to present your report in a format that is accessible and readable.

In your previous studies of science at GCSE, you ought to have written reports on a range of short experiments. This type of report generally has a standard format, as follows:

❶ Title, e.g. 'measurement of the resistance of a wire-wound resistor by the ammeter-voltmeter method'.

❷ Diagram, e.g. circuit diagram, labelled, with the range of the meters shown.

❸ Method, i.e. what measurements are made and in what order, how the measurements are made.

❹ Results, e.g. a table of measurements.

⑤Treatment of results, e.g. a brief explanation of relevant theory, leading to the chosen graph.

⑥Graph, e.g. voltage (on the vertical axis) against the current.

⑦Calculations, e.g. using the graph gradient to work out the resistance.

⑧Conclusions, e.g. the value of the resistance and comments on the accuracy of the result.

You can use the same format to write about short investigations although, in addition, you will need to explain your reasons for making decisions and choices.

Project reports can be very time consuming and you will need to plan your report before you begin detailed writing – otherwise, you might find some vital aspect of the project has been omitted. Confine your plan to a single sheet of paper so you can spot anything left out. Your report should include the following features:

❶Project outline This should state the aim of the project and the general method used to achieve your aim. You could outline any initial investigations that helped you to define your aim. Also, if your project included more than one investigation, state the purpose of each investigation.

❷Theory Explain what quantities you controlled and measured, and outline the theory involved, showing how your measurements related to your aim.

❸Diagrams Number and label your diagrams of the apparatus so you can refer to them in the report. Diagrams should be straightforward and easy to understand. Confine freehand sketches to your diary and avoid them in your report.

❹Procedure Describe how you used the apparatus to control and measure the variable quantities and what the sequence of making the measurements was. List the precautions taken to ensure consistent and accurate readings and explain how you estimated errors in your measurements. If you changed the direction of the project at any stage, outline the reasons for making the change, taking care not to describe procedures which did not lead to useful results.

❺Measurements Always give units and probable errors where possible. Tabulate your measurements where appropriate, giving your table a numbered heading which you can refer to elsewhere in your report if necessary. If there are a large number of measurements, list them in an appendix to your report and present only a sample set in the main part of your report.

❻Data analysis The aim of this section is to use your measurements and appropriate theory to establish, confirm or use links between measured quantities. Techniques of data analysis include graph work, calculations and analysis of errors. All these techniques are described in detail in Chapter 9.

❼Conclusions In the final section of your report, you will need to use your measurements and data analysis to draw valid conclusions. You will have to consider the strength of evidence behind your conclusions, if possible using your analysis of errors. You could also comment on the methods and techniques used, and make suggestions for further work if more time had been available.

REVISION TECHNIQUES

'Dividing the course up into topics makes revision easier.'

'Topic tests (however irksome!) are a good idea because they let you know what you do or don't understand.'

'I like the idea of frequent tests … makes revision easier.'

Revision is not something that should be left until a few weeks before you sit your A-level examination or module test. Rather, it should be an ongoing process that actually begins from your very first Physics lesson. The amount of knowledge to be accumulated is large and must therefore be taken in and gradually reinforced as you go along. Indeed, revision is really part of the learning process, although it will have to become more concentrated as the examination approaches.

Recall improves immediately after learning, when you have had time to process the information properly, then declines sharply. By use of reviews, memory can be maintained at the original high level. Of course, these times are not absolute, and some reviewing will take place automatically, for example when doing homework questions or class tests. The important principle is *regular* reviews, which means a continuous, planned revision programme.

First of all get *organised*. Devise some form of revision timetable, bearing in mind that learning periods of about 30–40 min produce the best recall. Plan your time carefully, giving yourself definite goals to aim at (e.g. completion of a specific topic), the attainment of which provides you with satisfaction and encouragement.

Next be *active*. Simply reading a text book in a comfortable armchair with your feet up is not conducive to successful learning. You should write out important definitions, formulae and equations; draw diagrams and graphs checking them against your textbook or notes; describe experiments in note form as though you were performing them. Above all practise lots of questions (repetition!).

Hopefully by now you can begin to see the importance of the section on note taking, and how well-organised notes can help during your revision. You can, at this stage, make special revision notes in a form suitable for quick reference, perhaps colourful charts to hang on the wall of the room where you work, or postcards with essential information that you can glance at on the bus or walking to school.

Finally, mention should be made of a valuable learning aid. Whenever you find it particularly difficult to remember equations or information you can invent codes to assist you; these are called 'mnemonics'. Some examples that have been used by A-level Physics students are ROY G BIV (Rainbows), CIVIL (Reactance), Talking Of Animals Some Ostriches Have Curiously Adjusted Heads (Trigonometry), non-listeners go to the sign theatre ($n\lambda = 2d \sin \theta$), Blue Bends Best (Dispersion), I am a knave ($I = nAve$).

THE EXAMINATION

QUESTION STYLES

Before reading on, check your syllabus to confirm the question styles you will meet. Terminal examinations include most of the following question styles. Module tests are usually more limited in the range of styles.

Multiple choice

These papers are sometimes called multiple selection, coded answer or objective tests. Do *not* read through the paper first; you have to attempt all the questions and there are too many to make a quick scan through the entire paper worthwhile. Start at the beginning and work steadily through, but leave out any questions that appear difficult at first sight. On completion go back and attempt the problems that you missed out initially. Just before time is up, make a guess at all the remaining questions. *This is vital*, as marks are not deducted for wrong answers. Above all, do not panic and spend a long time on one question. Just keep going, you are bound to find some that you can answer easily, and these will restore your confidence.

A unique feature of all multiple-choice papers is the fact that you are provided with several answers, only one of which is right. Many students fail to appreciate that instead of looking for the right answer, it can sometimes be much easier to identify the wrong ones! This approach will often enable a guess at an unknown answer to be 'one from two' instead of 'one from five'. The technique is demonstrated in some of the question practice commentaries, but one example may help.

The moment of inertia of a sphere about an axis though its centre of mass is $2mr^2/5$, where m is its mass and r its radius. When on a flat surface the sphere rolls without sliding with a horizontal velocity v, the ratio of its translational kinetic energy to its rotational kinetic energy is:

A $2r:5$.   B $5r:2$.   C $2m:5$.   D $5:2$.   E $2:5$.

The ratio is of two energies, so the answer must have no units. Hence answers A, B and C must be wrong.

Short answers

You are usually expected to answer all the questions, in which case there is no need to read right through the paper; this advice can be important to the many who have time troubles on short-answer papers. However, you should read through each question completely. Frequently the questions are in several parts that are linked. Ideas from later parts of the question may help you follow the earlier parts. You may avoid the pitfall, for example, of giving an answer in part (a) that is expected in part (b). Again move on quickly to the next question as soon as you get bogged down. Leave space on your answer paper/book to enable you to return and complete the previous question later. It is amazing how the subconscious will unravel a tricky problem for you as soon as you have set to work on something else.

Long answers

Most long-answer papers offer you a choice of questions. It is therefore important that you read through the whole paper before choosing which questions you intend to attempt. If the question itself is long it can often be tackled without much planning; the structure of the question will, in effect, generate the plan for you. However, in general on long answers a plan is needed, which means that a significant amount of the total examination time (maybe up to one-third) is spent in thinking rather than writing an answer. This time is allowed for when setting the questions. Time troubles are invariably due to either failure to stick to the terms of reference of the question, failure to use diagrams effectively to save many words, or failure to be concise.

Once again you should read the whole question before you start so that you have an overall picture of the complete question. Calculations are often based on earlier descriptive work, so look for clues to help you tackle the problem. Information on planning long answers appears under Examination Techniques.

Comprehension

Acquire a general idea of the content of the passage by reading through the text at your normal reading pace, avoiding any backtracking (i.e. going back and rereading sections). *If* you have time, it is a good idea to read all the questions before you start writing answers; the content of the questions can often add to your understanding of the passage and earlier questions. As you do a question, reread the section of the text that is relevant before you write your answer. Always answer in your own words; only quote from the passage when asked to do so. As in many other papers the structuring of questions can help you. For example, assume that the questions are laid out as follows:

Question 1.
Question 2. (a)
             (b)
Question 3. (a)
             (b) (i)
                  (ii)
                  (iii)

There will usually be some connection between 2(a) and 2(b), and 3(a) and 3(b); but there can be very strong links between 3(b)(i), 3(b)(ii) and 3(b)(iii).

Data analysis

The questions are normally all compulsory but as they follow the same theme it is worthwhile reading them all before you start. The skills needed for data analysis are outlined in Chapter 9. You are particularly advised to practise if you are slow at graph drawing or analysing results. Candidates without a *scientific calculator* or the ability to operate it expertly can be at the greatest disadvantage on this paper. It is foolish to wait until the examinations to buy your new calculator as practice with it is essential.

Practical problems

This examination is unique to Nuffield, although some of the NEAB short experiments in their practical examination are comparable. However, several boards set similar questions in their short- and long-answer papers, but naturally they supply the results to the experiments rather than asking the candidates to do them.

Most of the marks are available for the 'theory' part of the questions, and Nuffield candidates should complete the experiments quickly and with an accuracy appropriate to the apparatus with which they have been supplied. If you finish a question early, read up about the next one so that when you move on to the apparatus you are already well prepared, or you can use the time to complete some earlier questions. If short of time, always try and complete the new question before finishing off the old one; this is more sensible and better for your morale.

Practical examinations

Read the whole question before you start so that you can plan what you are going to do. A few minutes spent thinking about the best way to arrange your apparatus can save considerable time when you start to take readings. The experiments usually require readings to be taken and then processed, normally with the aid of a graph.

The amount of 'writing up' you are required to do varies according to the board; however, the question will clearly indicate what is needed. Sadly, many candidates write a full 'report' on their experiments when this is not asked for, and vice versa. Frequently a fully labelled diagram of your apparatus will be necessary, and virtually all boards (in their rubric) ask for details of special precautions or experimental techniques that you may use.

Investigate the range over which observations can be made and then take readings at convenient intervals over the whole of that range. Take as many readings as you can in the time available, but always leave sufficient time to process them. Repeat readings if possible, particularly where considerable judgement has to be exercised (as in timing and focusing experiments). Leave your apparatus set up; this will enable you to check any 'strange' points after the graph has been plotted or to take further points in 'important' regions (e.g. when trying to locate a maximum or when drawing a tangent to a curve to find the slope).

In all practical work, check the zero error of all instruments (e.g. meters, micrometers, stopwatches) and record that you have done this. Record *all* the observations that you make, in tabulated form if possible, so that it is easy to check back for any mistakes in the analysis of your results. Ensure that all results of data processing (and the initial readings) are quoted to a reasonable number of significant figures. Finally, if the results of your experiment are not those expected, comment on this and offer some sensible explanation if possible.

EXAMINATION TECHNIQUES

A very common complaint of examiners and teachers is about the student who fails to 'answer the question'. All the following misdemeanours can be classified under this heading: answer too long, answer too short, irrelevant material included, relevant material excluded, absence of diagrams/calculations/graphs that are clearly asked for, entire parts of questions missing, etc. 'Answering the question' is a skill that will only gradually be perfected as you practise throughout your course, and the question practice in this book has many commentaries that are specifically designed to help you develop this technique.

Remember:

'A-level questions are a lot more difficult to answer … they are not so straightforward: i.e. you have to work out what the question is asking before answering it.'

'Lots more depth is required in the answers.'

State, define, explain, discuss, compare, describe, comment

Questions on all types of examination paper (except multiple choice) may use any of these key words in a question. They all have different meanings and interpreting them wrongly can be very expensive in terms of marks or time lost.

State The briefest possible answer will suffice. It may only be one word!

Define A full statement is needed, together with the appropriate equation if possible. Give the meaning of the symbols used in the equation and the units of the quantity to be defined. Beware of padding your precise definition with vague statements that might contradict your definition and lose marks.

Explain You are given a piece of information that must be described in detail. Normally you must use basic physical principles to justify or clarify what you are required to explain. A graph or diagram will often be invaluable.

Discuss Usually you are given something to consider that may or may not be true. You need to indicate the various possibilities, giving evidence and arguments for and against each proposition.

Compare Two or more items have to be compared with each other. You must describe similarities and differences between them. It will not be necessary to describe each item in detail unless the question asks you to 'compare and describe' or 'compare and discuss'.

Describe Most often you are asked to describe experiments. Full details of method, measurements, apparatus and processing of results are needed. Be careful to cut back on detail if the question asks you to 'describe briefly' or 'describe concisely'; here your ability to isolate the key features of the experiment/topic is being tested.

Comment Usually you are being asked to think about something unusual or slightly strange. You must relate the information supplied for comment to your own knowledge of the topic and draw some conclusion. Answers to these questions are rarely longer than a sentence or two.

A common difficulty amongst candidates is the judgement of the length required for an answer. Help can be provided by the amount of space supplied if you are to write your answer on the question paper, or if the mark scheme is supplied you can relate this to the length of the question (normally marks are related to the time needed to complete an answer). You must always use the wording of the question (e.g. state, explain) to assist you in determining the length of your answer.

The following example question and answer may help you to understand the 'jargon'.

Question

(a) **State** one example each of an electrical insulator, semiconductor and conductor.
(b) **Define** resistivity.
(c) **Explain** how a material can be classified as an insulator, conductor or semiconductor.
(d) **Compare** the terms 'resistance' and 'resistivity'. **Discuss** whether either of these quantities can have a negative value.
(e) **Briefly describe** how you would carry out an experiment to measure the resistivity of copper. **Comment** on the fact that a student doing such an experiment finds that

if he leaves the apparatus switched on the resistivity of copper seems gradually to increase.

Answers

(a) Insulator, glass; semiconductor, silicon; conductor, copper.

(b) The resistivity ρ (measured in Ω m) of a material is defined as the electrical resistance per unit length of a sample of the material with unit cross-sectional area. Hence, if a sample has resistance R, length L and cross-sectional area A, then $R = \rho L / A$.

(c) Classification of a material as an insulator, semiconductor or conductor is based upon the value of its electrical resistivity at room temperature. Materials with resistivities of the order of 10^{11} Ω m or larger are called insulators, those with a value of 10^{-6} Ω m or smaller are termed conductors. If the value lies between these limits then the material is called a semiconductor.

(d) 'Resistance' and 'resistivity' are both terms used to describe the opposition offered to the passage of electrical current. However, resistance can only be defined for a specifically sized sample of a given material, whereas the resistivity is a property of the material in general and will have the same value for any size of sample (this is the same comparison as can be drawn between mass and density, or spring constant and Young's modulus). If a negative value of resistance could occur, this would imply that the resistivity of the material is negative. But if the resistance were negative then either the current would reduce as the pd across the sample were increased, or the current would flow in the opposite direction to the applied pd (the two alternatives can arise from different interpretations of the term 'resistance'); such behaviour is not encountered in A-level Physics.

(e) It is left to the reader to find out how this may be done, but even in a 'brief' description the difficulty of obtaining a copper specimen of large enough resistance must be identified. You are expected to comment that the rise of resistivity of the copper suggests that its temperature is rising during the experiment, so the student is probably using a large enough current to heat up his copper specimen.

Planning long answers

Many long-answer questions are structured (i.e. in several parts) and will need little or no planning, but others may require you to produce an essay style answer that will need a plan. Some of the questions of this nature in this book have commentaries designed to assist you with your planning. However, it is worthwhile outlining the method of producing a plan and illustrating it with an example.

After reading the question make the plan as follows:

❶ Jot down broad headings for the main topic areas of the question.

❷ Under each of these headings list the points to be made (e.g. facts, formulae, definitions, units, key experiments).

❸ Arrange your information so that your answer can be written in a well-organised, concise and coherent manner. Make sure that you are answering the 'actual' question, as opposed to the one you had 'hoped for'.

Some long answers involve a wide range of topics that need to be linked together. It can be a great help to put your notes in diagrammatic form with such links; it is possible to plan long answers in a similar fashion as shown in the following example.

Question
Describe how you would explain electromagnetic induction to a friend studying A-level Physics who had missed the teaching of this particular subject.

First, note the topics to be explained: the laws, self-induction and mutual induction, transformers.

Suitable brief notes under these headings can be made. Alternatively, the plan can be constructed in diagram form. More ideas can be added by including extra branches to the diagram; it is easy to write a logically sequenced answer from such a plan.

Remember that your examiner will have to read through many scripts. If you give a poorly planned, rambling answer, any valuable points may prove difficult for the examiner to find; a well-structured answer will not present such problems. Avoid the tragic mistake of giving several pages of largely irrelevant information which will demonstrate your knowledge, waste your time and score no marks. Never 'write down all you know'. Sadly marks can even be lost by the candidate who spoils a good answer with extra 'padding'; try and train yourself to be relevant in answering all questions.

Some long answers will require descriptions of experiments. Here you should plan to describe the experiment as you would perform it. A well-labelled diagram of the apparatus will usually head the answer (no need to use volumes of words to describe the apparatus). Next you describe what measurements you take and how they are taken (e.g. naming important precautions or procedural points); a table with headings can show how the results can be expressed. Finally demonstrate how the results are processed; note that sketch graphs are quick to produce and can secure valuable marks.

Some modern A-level questions provide considerable amounts of data, and the candidate has to choose which items of information are appropriate to the particular question that is being answered. If the data are scattered throughout the question (and they are sometimes to be found on the front of the exam paper as well, or on a formula sheet), then try and write down a list of all values and equations you may need before you answer the question. As you do each part, consult the 'complete' list; this should enable you to answer the question with greatest accuracy and in the least time. Sometimes this technique can help you to unravel a comprehension paper as well, especially if there is much information hidden in the passage rather than being supplied in tabulated form.

Numerical problems: calculations and estimations

In calculations you are asked to find a numerical value based on data that will be supplied in the question or on the front of the exam paper. In an estimation you have to supply the values from which the estimation will be made using your own knowledge of physical constants and values. Apart from the way in which the data for the question are obtained, the method of procedure for both styles of question is the same. Calculations tend to be easier because the data supplied can provide good clues as to which equations and techniques are needed to produce the answer. In an estimation the candidate has no such data and must think of a technique to carry out the estimation without it. Once a technique has been established, the data must also be estimated. Students learning to estimate will find the table of constants and numerical values at the back of this book very helpful.

In a calculation, start by summarising the data, using the accepted symbols and converting each quantity into its correct units. Sometimes a diagram of the situation will be helpful. From this written information you should be able to spot the principles and equations that relate to the situation; there are marks to be gained from simply noting the right ideas and equations, even if you don't know how to use them. See whether you need to make sensible assumptions or approximations, and indicate clearly where you have done this in your answer. Each line of your answer should be a complete sentence, though it may be in the form of mathematical symbols or an equation. This makes it easy to follow your line of argument and spot any mistakes (considerable credit is given for the right method). Your final answer should be quoted to the same number of significant figures as the data, together with suitable units. Check that your answer is sensible. If you have time then look back for silly mistakes. You must always comment on a wrong answer, even if you cannot identify your error.

When estimating, it is first necessary to decide upon a method of producing an answer. Physical principles, laws and equations must be considered. There are often several methods of carrying out an estimate; spotting the shortest and simplest one is a skill that only comes from experience. Some students read the estimating questions early in the examination, but do not tackle them until the end. Often a good idea arises during the course of doing another question, as your subconscious comes to your rescue! Once the 'method' is found, it is necessary to estimate the quantities that will be needed to carry out the method. These quantities are normally the first material to appear in your written answer. They should be quoted to a sensible degree of accuracy (usually either one significant figure or the nearest order of magnitude). Once the quantities to be used have been written down, proceed as with a normal calculation. Take particular care to quote your answer to the right degree of accuracy (usually one significant figure or the nearest power of ten). If you know what the answer to your estimate should be, then this will occasionally help you to spot mistakes or obtain values (for data quantities) of which you are not sure.

Two further points about calculating and estimating are worth making. Firstly, you should work with symbols as far as possible; only put in numbers when you have an equation linking the quantity you require to those that you have available as data. Secondly, try not to leave your answer in incomplete numerical form (e.g. $2\pi g$) unless you are running short of time.

A sample calculation can illustrate most of the techniques needed to answer these questions.

Question

A standard resistor is made from wire having the resistivity of 2.5×10^{-7} Ω m. In order not to overheat it, the current density (i.e. the current per unit cross-sectional area of the wire) should not exceed 15 mA mm^{-2}. What is the maximum potential gradient that may be safely applied to the wire of the resistor?

Answer

Summarise the information with correct units and usual symbols:

ρ = 2.5×10^{-7} Ω m

I/A = 15 mA mm^{-2} = 15×10^{-3} A mm^{-2}

= $15 \times 10^{-3} \times 10^{6}$ A m^{-2}

= 15×10^{-3} A m^{-2} = maximum current density

Write down any relevant equations:

$R = \rho L/A$

$R = V/I$

Potential gradient is a term you may not recognise. However, you could well imagine that as temperature gradient means temperature per unit length, potential gradient means pd per unit length or V/L. You can certainly work this out from the information in the question, and indeed it is the right answer. So continue:

$V/I = \rho L/A$

so $V/L = \rho I/A = 2.5 \times 10^{-7}.15 \times 10^{3} = 3.75 \times 10^{-3}$

Remembering the units and number of significant figures:

Maximum safe potential gradient = 3.8×10^{-3} V m^{-1}

TAKING MODULE TESTS

If you are following a modular syllabus, you will sit module tests set by the board during your course. Although these tests cover the limited range of content comprising the relevant module, the questions are A-level standard and therefore preparation for a module test must be just as careful as for terminal examinations. Most of the comments made in the section 'Final preparation' therefore apply to module tests as well as to terminal examinations. However, there are some important differences:

- Revision for a module test involves only coverage of topics in the module.

- On-going revision from the start of a module is crucial since extensive revision time is unlikely to be available before a module test.

- Module papers are at A-level standard and may include comprehension/data analysis questions.

- Module results are reported on higher education references and therefore it is important to do as well as possible in a module test, even though you may be allowed a further attempt on a later module test.

- In a modular course, there is unlikely to be much time between a mock module paper and the actual paper. It is therefore important to analyse your mock performance rapidly and to act on gaps in knowledge pinpointed by your answers.

All modular syllabuses are required to incorporate terminal examinations. Some syllabuses use final module tests only to meet this requirement, in which case it may be necessary to sit two or more module tests at the end of your course. Although all modules are equally weighted, option modules studied in the latter part of a course usually build on knowledge and understanding in the compulsory modules studied earlier. Questions in option modules may therefore include synoptic elements testing knowledge and understanding from relevant sections of the compulsory modules. The ULEAC syllabus requires candidates to sit a final 'synoptic' examination paper that tests knowledge and skills from all the compulsory topics in the syllabus as well as a final module test. The section on 'Final preparation' is therefore directly relevant to those preparing for final option modules or synoptic papers.

FINAL PREPARATION

You will need to plan and prepare carefully for your final examinations. Your mock examinations or previous module tests will give you an estimate of where you have reached before you enter the final 'strait'. If you plan this last stage carefully and stick to your plans, you should be able to fit in some social activities to prevent mental fatigue setting in. Examinations are a test of stamina, organisation and resilience, in addition to the obvious test of subject knowledge.

Before your mock examinations

Revise thoroughly all the work you have covered from the start of your course. Use the syllabus checklist as a 'ticklist' for revising work covered and don't attempt to revise topics not yet covered in class. Pay particular attention to the topics you covered in the first year of your course as these form the foundation for the later topics.

After your mock examinations

Go over your mock papers and pinpoint gaps in your knowledge and make sure you fill them. You could analyse your performance against the subject checklist for your syllabus. Also, look for strengths and weaknesses in terms of performance on different sections of your papers. Your teacher should give you feedback on where you can improve and where you are doing well. With careful and thorough revision between the mocks and the final exams, you can improve considerably if your marks were not as good as you'd hoped for. Equally, there are those who have been known to snatch defeat from the jaws of victory as a result of a good mock performance followed by a 'laid-back' attitude to the final stage.

As your syllabus reaches completion

- Make out a revision timetable covering all your subjects. Allocate set times for each subject on a weekly timetable and make sure you allow some time for relaxation.
- Within your overall revision timetable, make up a Physics revision schedule on a topic-by-topic basis, with a full round of detailed revision to be completed by at least two weeks before the examination. This will leave time for a further round of 'target' revision in the final week.
- You may find it helpful to organise your revision schedule in cooperation with your teacher so that you are revising the same topic at home as in class. However, this time in class for revision is limited, especially if you are still finishing the syllabus.
- Stick to your revision schedule and tick off the topics as you revise them. Use the questions in this book to check your grasp of the subject and your answer techniques. The question commentaries should prove useful and the answers to numerical problems are supplied. Your teacher will be able to supply you with past papers and you should use these as you do your topic-by-topic revision, keeping a checklist of the questions you have answered. If possible, get your teacher to check your answers.

- Use Chapter 9 of this book to revise data analysis, graph shapes, common equations, etc. Tackle the questions in Chapter 9 to make sure you have mastered these essential topics and work through multiple-choice questions on graph shapes, relationships and equations.

In the last few weeks

Once you have completed the cycle of detailed revision, try some past papers under exam conditions. Section 3 is designed for this purpose and you can check your performance against the answers. Use the commentaries *immediately* after you have tackled the questions to make sure your approach to each was appropriate.

In the last week

Make out a further revision schedule, taking account of when your exams are and what type of questions are on each paper. There is not much point revising options after the first paper if the options were on the first paper. You should aim to revise as much of the syllabus as possible at a broad level for those papers which are 'compulsory' and then target specific topics with detailed revision for those papers with longer questions where you have a choice of which to answer. This is where your revision charts can be very useful in helping you to connect concepts together rapidly.

The night before the exam

Everyone is agreed that the night before the exam you should be as relaxed as possible, but some people insist that you will be reassured mentally if you glance over a few topics you are still not happy about. We do not recommend any revision, nor do we recommend a late night which will leave you tired and jaded for the exam the next day. Go to bed early in the knowledge that any further revision will only diminish your performance.

The day of the exam

Don't enter the exam room without a plan for tackling that specific examination paper. This will enable you to avoid time trouble or choosing the wrong questions.

Do familiarise yourself thoroughly with any formula sheet that you might be permitted to use.

Do ensure that your calculator batteries will not run out during the exam, that you have spare pens and sharpened pencils (in a range of colours but not red), and a ruler and a rubber. Some boards ban the use of correction fluid; check on this.

Don't enter the exam room at the last possible moment. Early candidates may be able to choose their seats, and there is much to do before the examination starts.

Do have a supply of chewing gum (mints, etc.) if these aid your concentration and are permitted in your examination room.

Don't sit an examination if you are unwell without informing the invigilator *before* the start of the paper.

Don't drink (alcohol) before an exam, or eat too much or too little.

Do use the 'waiting' time sensibly before the 'official start'. You can write your name and examination details onto all the answer books/paper. You can study the front of the exam paper for the expected rubric and any constants and data; recall important equations and principles involving the information on the front of the paper. In a practical examination you can study the apparatus carefully.

Don't sit an examination without a convenient means of keeping track of the time that is elapsing.

During the examination

Do read all the questions. Many students ignore a question after starting to read it, only to realise that had they read it through to the end it would have proved far easier than they had imagined.

Don't spend too much time on one question. It is essential to develop techniques for doing this to avoid panic setting in.

Do call the invigilator *immediately* you have any problem, the worst that he/she can do is refuse to help you. This is particularly vital if you need the toilet.

Don't be too fussy about the neatness of your diagrams. Make sure that they are well labelled and large (large 'freehand' diagrams are much easier to follow or correct than small ones).

Do leave plenty of space to finish questions that you leave when you run into difficulties.

Don't forget your units; it is best to convert all quantities to SI units before you use them.

Do write in short sentences, avoiding too many 'ands' and 'buts' in mid-sentence. It is not easy for you or the examiner to follow an argument written in long sentences.

Don't dive into questions without planning or reading the entire question. This can frequently lead to disasters.

Don't spend time reading answers or checking for mistakes until you have completed the required number of questions.

Do comment on any answer that you suspect is wrong.

Do take great care to make sure that your script is securely tied before you hand it in and that all your sheets/booklets bear your name and all the relevant information.

Do try to teach yourself to enjoy the challenge of the examination. There is much skill involved in doing yourself justice; remember that you can always sit it again if things go wrong. Examinations are important, but they are not 'life and death' matters.

After the examination

Don't fret over the many 'obvious' mistakes you will have made and should have avoided. All the other candidates will have made similar errors, and there is nothing you can do to retrieve the situation.

Do forget the paper and start to prepare for the next one. Think positively; only carry out an inquest on a paper you have just sat if similar material may appear on a paper you are still to sit (i.e. if you can learn from your mistakes).

Do remember that if you have found a paper hard, it's likely that so will the rest of the candidates (whether they are ready to admit it or not); this could simply mean that the 'pass mark' on this paper could be correspondingly lower.

Don't get overconfident if you found a paper easy (for the opposite reasons to the previous piece of advice).

A-LEVEL PHYSICS

In this section:

Each chapter features:

■ *Units in this chapter*: a list of the main topic heads to follow.

■ *Chapter objectives*: key ideas and skills which are covered in the chapter are introduced.

■ *The main text:* this is divided into numbered topic units for ease of reference.

■ *Chapter roundup*: a brief summary of the chapter.

■ *Question bank*: exam questions with tutorial comments on the pitfalls to avoid and points to include in framing your own answers.

FORCE AND MOTION

Units in this chapter

Chapter objectives

After working through the topics appropriate to your syllabus in this chapter, you should be able to:

- define and state the units of acceleration, momentum, impulse, moment, couple, angular frequency and angular speed
- use the appropriate dynamics equations to solve problems involving constant acceleration
- understand and be able to use Newton's laws of motion and the principle of conservation of momentum
- use the equations for ke, pe and power
- resolve and combine vectors
- describe the forces on an object in equilibrium and solve static equilibrium problems
- prove that the oscillations of a loaded spring and a simple pendulum are simple harmonic
- use the equations describing simple harmonic motion
- understand the concept of centripetal acceleration
- use the equations for angular speed and for centripetal acceleration to solve problems
- define moment of inertia, angular momentum and rotational ke
- solve problems in rotational dynamics involving angular momentum and energy

1.1 MOTION IN A STRAIGHT LINE

Displacement is a distance moved in a given direction.

Speed is rate of change of distance.

Velocity is rate of change of displacement.

Acceleration is rate of change of velocity.

If the above definitions are applied to an object with initial velocity u that is **uniformly accelerated** to a final velocity v in time t, the following equations can be derived:

E1.1 $\quad a = \dfrac{(v - u)}{t}$ $\qquad$ E1.2 $\quad s = \dfrac{(u + v)t}{2}$

E1.3 $\quad s = ut + \frac{1}{2}at^2$ $\qquad$ E1.4 $\quad v^2 = u^2 + 2as$

where a = acceleration (m s^{-2}), u = initial velocity (m s^{-1}), v = final velocity (m s^{-1}), s = displacement (m), t = time (s).

Velocity–time graphs and displacement–time graphs are valuable ways of depicting motion in a straight line. For example, the motion of a ball thrown directly upwards is shown by the graphs of Fig. 1.1. Because velocity and displacement are **vectors**, care must be taken with the sign convention for direction; in the diagram, the +ve y-axis is used for upwards direction (and −ve y for downwards).

Consider the velocity–time graph. As the ball rises its velocity becomes less and less +ve, becoming zero at its maximum height; then its velocity becomes more and more −ve because it is now falling at increasing speed until it returns to 'ground level'. Note that for an object acted on by gravity only (i.e. negligible air resistance) its acceleration, g, is equal to the slope of the graph and this is constant. Also, since the area under the line gives the displacement, it is important to remember that area above the time axis gives +ve displacement and area below the time axis gives −ve displacement. In this example, there is as much area above the time axis as there is below; hence the total displacement from 'launch' to 'impact' is zero.

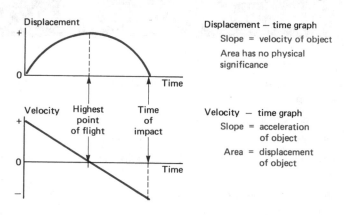

Fig. 1.1 Graphs of displacement and velocity against time for a ball thrown in the air

1.2 PROJECTILE MOTION

The key point in dealing with projectile motion is to remember that gravity always acts downwards only. There is *no* horizontal acceleration, assuming negligible air resistance.

Consider a projectile launched from 0 with initial velocity of magnitude u in a direction at angle A above the horizontal. Let x and y denote horizontal and vertical components of displacement, respectively. The path is shown in Fig. 1.2.

Horizontal motion is analysed separately from vertical motion, as follows:

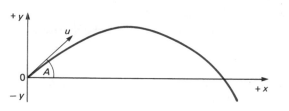

Fig. 1.2 Path of a projectile

Horizontal motion	Vertical motion
Initial velocity = $u \cos A$	Initial velocity = $u \sin A$
Acceleration = 0	Acceleration = $-g$ (– for downwards)
Time taken = t	Time taken = t
Distance moved = x	Distance moved = y
From E1.3, $x = ut \cos A$	From E1.3, $y = (ut \sin A) - \frac{1}{2} gt^2$
From E1.1,	From E1.1,
final speed $v_x = u \cos A$	final speed $v_y = (u \sin A) - gt$

The horizontal velocity remains constant, but the vertical velocity decreases steadily from its initial value at a rate of g. Note that the magnitudes of the displacement and the velocity after time t are given, respectively, by:

$$\text{displacement magnitude (i.e. distance)} = \sqrt{x^2 + y^2}$$

$$\text{velocity magnitude (i.e. speed)} = \sqrt{v_x^2 + v_y^2}$$

1.3 FORCE AND MOMENTUM

Momentum is defined as mass × velocity. Since velocity is a vector, momentum is also a vector. Its unit is kg m s^{-1}.

Newton's laws of motion are:

1. An object stays at rest or at uniform velocity unless acted on by a resultant force.

2. The change of momentum per unit time is proportional to the resultant force, and in the same direction.

3. Action and reaction are equal and opposite.

The first law tells us what a force is (i.e. that which changes an object's state of motion). The second law gives a useful equation:

E1.5 $F = \dfrac{mv - mu}{t} = ma$

where F = force (N), m = mass (kg), u = initial velocity (m s^{-1}), v = final velocity (m s^{-1}), a = acceleration (m s^{-2}), t = time (s).

Note the following points in connection with E1.5:

1. The equation applies only where the mass m remains constant.

2. The unit of force is the **newton** (N), defined as the force that will give a mass of 1 kg an acceleration of 1 m s^{-2}. Always use these units.

3. In general, Newton's second law is represented by the equation $F = \dfrac{d}{dt}(mv)$ where d/dt is the mathematical way of writing 'rate of change of'.

4. Where mass is expelled at a steady rate in a 'jet' at speed v, the general equation reduces to $F = v \dfrac{dm}{dt}$ (= speed × mass loss per second).

5. The **impulse** of a force is defined as force × time. From E1.5, it can be shown that the change of momentum ($mv - mu$) is equal to the impulse Ft.

The **principle of conservation of momentum** states that when no resultant force acts upon a system of bodies, the total momentum of the system remains unchanged. This is a very important principle of physics, and it is closely linked with Newton's third law. This can be seen by considering a collision between mass m, initially moving at velocity u, and mass M, initially at rest. Suppose, after the collision, that the two masses move off in the same direction as that in which mass m was initially moving, as represented in Fig. 1.3. Let the final velocity of mass m be v, and the final velocity of mass M be V. Let t denote the time of duration of the impact.

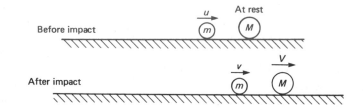

Fig. 1.3 Collision between two masses

For mass m, the change of momentum per unit time is $(mv - mu)/t$; for mass M, the change of momentum per unit time is MV/t. Since change of momentum per unit time is equal to force, then the force acting upon m is $(mv - mu)/t$, and the force acting upon M is MV/t. From Newton's third law, the force acting upon m is equal and **opposite** to the force acting upon M; thus $(mv - mu)/t = -MV/t$ where the $-$ sign represents the opposite directions of the two forces. From this equation, it can be shown that:

total initial momentum (mu) = total final momentum ($mv + MV$)

In other words, the total momentum is unchanged; the principle of conservation of momentum applies to *all* collision and explosion situations, but remember that momentum is a vector and so can either be $+$ or $-$ (according to which direction is defined as $+$) in straight line (i.e. one-dimensional) problems.

Weight is the force of gravity acting upon an object. For a mass m, its weight is mg, since if the object were allowed to fall freely, it would accelerate at g due to the force of gravity only. The correct unit for weight is the newton. Since g can change from one place to another, the weight of a mass can alter from one place to another; however, mass does not vary in this way because it is a measure of the **inertia** (i.e. resistance to change of velocity) of the object.

A person in 'free fall' experiences the sensation of **weightlessness** because that person is without support. The person is still acted upon by the force of gravity, causing an acceleration g. To appreciate this situation, consider a person in a lift that accelerates downwards at a. Fig. 1.4 shows that the person is acted upon by weight mg and by the **reaction** R from the floor of the lift. Thus the force that accelerates that person downwards is $mg - R$; hence, $ma = mg - R$, so giving reaction $R = mg - ma$. Assuming a is less than g, the reaction R will be less than the weight mg, so that the person will feel less support from the floor than usual; in other words, the person feels lighter than usual.

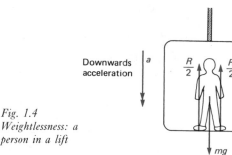

Fig. 1.4 Weightlessness: a person in a lift

1.4 WORK, ENERGY AND POWER

Work done by a force is defined as force × distance moved by the force in the direction of the force. **Energy** is the capacity of a body to do work. When work is done, energy is transformed from one form to another, so energy can be thought of as 'stored work', to be used (or stored) according to whether work is done by (or on) the object in possession of the energy. The **joule** (J) is the unit of work and of energy; 1 J of work is done when 1 N of force moves its point of application by 1 m in the direction of the force.

Power is the rate at which work is done. The unit of power is the **watt** (W) which is equal to a rate of 'doing work' of 1 J s^{-1}. For a constant force F moving at constant velocity v in the same direction as the line of action of the force, the power is equal to the force × velocity; i.e. power = $F \times v$.

The **principle of conservation of energy** states that energy can never be created or destroyed. It is a very important principle of Physics, and there has never been any experimental evidence to suggest it is not true. Energy can be changed from one form to other forms, but the total energy after the change must always be the same as before.

Kinetic energy (ke) is the energy possessed by a mass because of its motion. For a mass m, moving at speed v, its ke = $\frac{1}{2}mv^2$; a simple proof for this can be found in any A-level textbook, and it would be useful to look it up at this stage.

Potential energy (pe) is energy due to position. For a mass m raised through a height h, its gain of pe = mgh provided $h \ll$ earth's radius. See 2.2.

An **elastic** collision is one in which the total ke before the collision is the same as the total ke after the collision. An **inelastic** collision is one in which some or all of the initial ke is transformed into other forms of energy, usually heat or sound or both. Remember that conservation of momentum applies to *both* elastic *and* inelastic collisions.

1.5 STATICS

When an object is in static equilibrium, the forces that act upon it balance one another out, so that there is no resultant force upon the object.

Types of forces

The more common forces met in equilibrium situations include:

❶ **weight**, always considered to act at **the centre of gravity** of an object;

❷ **tension** (or compression);

❸ **friction**, which acts between surfaces (parallel to the surfaces at the point of contact) when the surfaces move (or try to move) relative to one another;

❹ **normal forces**, which act on an object when it is pushing into (or trying to push into) another object. 'Normal' forces act 'perpendicular to' the surface.

The situation in Fig. 1.5 shows the forces acting on a beam resting against a smooth wall on a rough floor. The combination of F_1 and N_1 is the **reaction** of the floor upon the beam.

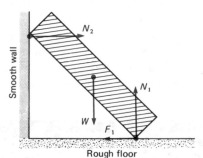

Fig. 1.5 Beam resting against a wall

Vector nature of force

Two important techniques in 'handling' forces are:

❶ **resolving** a force into two perpendicular components, as shown in Fig. 1.6.

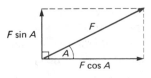

Fig. 1.6 Resolving a force

❷ **combining**, as vectors, several forces. To do this for two forces, either draw a vector diagram and complete the parallelogram, as in Fig. 1.7(i), or resolve one of the forces into two perpendicular components, one component being along the line of action of the other force as in Fig. 1.7(ii); the resultant force R is then given by $R^2 = (F_1 + F_2 \cos A)^2 + (F_2 \sin A)^2$.

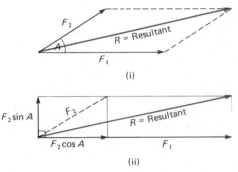

Fig. 1.7 Adding forces:
(i) parallelogram method;
(ii) combining components

Turning forces

Any force that tends to turn an object about a given point may be called a turning force. The **moment** of a turning force about a fixed point is defined as force × perpendicular distance from the line of the force to the point. Note that the unit of moment is N m; the joule is not used because, in a static situation, no work is done.

A **couple** is a pair of equal and opposite forces that do not act along the same line. The moment of a couple is always the same about any fixed point, and is always equal to the product of one of the forces and the perpendicular distance between the lines of action. The moment of the couple shown in Fig. 1.8 is Fd. Note that the term **torque** is usually used for a moment about a fixed axis.

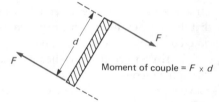

Fig. 1.8 Turning effects

The **principle of moments** states that, for a body in equilibrium, the total clockwise moment about a given point equals the total anticlockwise moment about that same point.

Analysing equilibrium forces

When analysing the forces that act upon a body in equilibrium, always start by making a sketch diagram showing all the forces acting on the object. Then, use of the following three rules should enable you to complete the analysis:

❶ **Resolve** all forces into components in two mutually perpendicular directions. Choose the two perpendicular directions according to the situation. For the beam shown in Fig. 1.5, the best directions are horizontal and vertical; for a mass on a slope, the best directions might be parallel and perpendicular to the slope.

❷ **Balance** the force components for each of these two directions. In other words, for each direction equate the force components acting one way to those acting the opposite way.

❸ **Apply the principle of moments** about the most convenient fixed point. Choose the fixed point as that point through which one (or more if possible) of the 'unknown' forces has its line of action.

1.6 SIMPLE HARMONIC MOTION

Simple harmonic motion (shm) is defined as oscillating motion of an object about a fixed point, such that the acceleration (a) of the object is:

❶ proportional to the displacement (x) from the fixed point,

❷ always directed towards the fixed point.

The definition can be summarised by the following equation:

E1.6 $a = -\omega^2 x$

where a = acceleration (m s^{-2}), x = displacement (m), ω = angular frequency (rad s^{-1}).

The **angular frequency** (ω) is a constant of the motion.

(Note that the $-$ve sign in E1.6 indicates that the acceleration is always directed towards the fixed point.)

According to Newton's second law, the force producing the motion is proportional to the acceleration (i.e. $F = ma$). Therefore, the restoring force of an shm system can be written as $F = -m\omega^2 x = -$ constant $\times x$. In other words, the **restoring force** must be proportional to the displacement from the fixed point and directed towards the fixed point. Thus, any oscillating system for which the restoring force meets these conditions will move in shm.

The **solution of the shm equation** can be evaluated with the aid of differentiation. The general solution is:

$$x = A\sin(\omega t) + B\cos(\omega t)$$

where A and B are constants determined by the initial values of displacement and velocity (v). Note that v is obtained by differentiating x with respect to time (i.e. $v = dx/dt$), and a is given by differentiating v with respect to time (i.e. $a = dv/dt = d^2x/dt^2$). If the initial conditions are $x = x_0$, $v = 0$ and $t = 0$, then the general solution becomes $x = x_0\cos(\omega t)$.

Consider a mass on a vertical spring, as in Fig. 1.9. If the mass is displaced from equilibrium and then released from rest, the displacement is given by the equation $x = x_0\cos(\omega t)$, and its variation with time is represented by the displacement–time graph of Fig. 1.10. Remember that the slope of the displacement–time curve gives the velocity, and the slope of the velocity–time curve gives the acceleration.

The **time period** (T) is the time taken to move through one complete cycle of oscillation. From the above example, one complete cycle corresponds to $\omega T = 2\pi$ so giving $T = 2\pi/\omega$.

Note that the **frequency** (f), the number of complete cycles per second, is given by $f = 1/T = \omega/2\pi$.

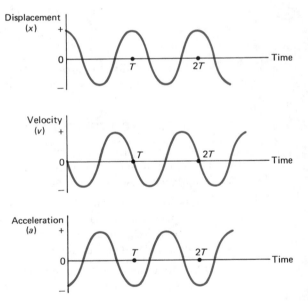

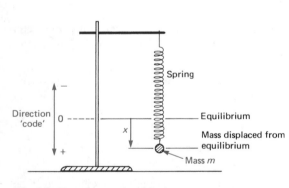

Fig. 1.9 Mass on a spring (shm)

Fig. 1.10 Graphs of simple harmonic motion

Oscillations of a simple pendulum

When a pendulum bob is displaced from equilibrium and then released with the string taut, it oscillates in a vertical plane with its motion along the arc of a circle, as in Fig. 1.11. The diagram shows the pendulum, with angular displacement A from its equilibrium position, as it moves along its path. The restoring force is provided by the tangential component of the

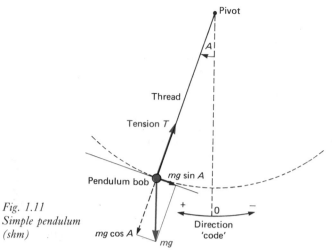

Fig. 1.11
Simple pendulum
(shm)

weight (i.e. $mg \sin A$) so that the acceleration a is given by $a = -g \sin A$, which can be written as $a = -gA$, provided angle A is small enough for $\sin A$ to be approximated to angle A in **radians**.

To show that the motion is shm, angle A must now be linked to the arc length x; in fact $x = LA$ (remember that when $A = 2\pi$ radians, then $x = 2\pi L$ = circumference; a useful check!).

The equation for acceleration can now be written as:

$$a = -\frac{g}{L}x \text{ or as } a = -\omega^2 x \text{ where } \omega^2 = \frac{g}{L}$$

Thus, the acceleration (a) meets the conditions set out in the definition (E1.6), so the simple pendulum moves with shm, provided its angular displacement (A) is small. In practice, this means never greater than about 10°. Finally the time period (T) can be expressed in terms of L and g:

E1.7 $\quad T = 2\pi\sqrt{L/g}$

where L = length of thread from pivot to 'bob' centre (m), g = acceleration due to gravity (m s^{-2}), T = time period (s).

Always remember that the time period T is the time for the bob to pass from one extreme to the opposite extreme and **back again**.

Oscillations of a loaded vertical spring fixed at its upper end

The spring obeys **Hooke's law**; consequently, when the length of the spring changes, there is a change in the tension in the spring **in proportion** to the change of length. When the mass is displaced downwards from its equilibrium position and then released from rest, the mass oscillates about the equilibrium position, as in Fig. 1.10.

To show that the motion is shm, consider the forces on the mass:

❶ In equilibrium, the weight (mg) = equilibrium tension (ke), (e = extension, k = tension per unit extension).

❷ At displacement x from equilibrium:
resultant force = weight (mg) − tension ($ke + kx$)
so   resultant force = $-kx$ since $mg = ke$

The resultant force is in the **opposite** direction to the displacement (i.e. if mass is above equilibrium, force on it is downwards, and vice versa) and acts to try to **restore** the mass to equilibrium. By Newton's second law, the acceleration a is then given by:

$$a = -\frac{k}{m}x \text{ or by } a = -\omega^2 x \text{ where } \omega^2 = \frac{k}{m}$$

The time period is given by:

E1.8 $T = 2\pi\sqrt{m/k}$

where T = time period, m = mass (kg), k = spring constant (N m^{-1}).

For both the loaded spring and the simple pendulum, the displacement x varies with time as given by $x = x_0\cos(\omega t)$, assuming an initial displacement x_0 and zero initial velocity. The amplitude of oscillations is x_0 (i.e. the maximum displacement).

Energy in shm

For a mass moving with shm, there is a continual interchange between its ke and the pe associated with the restoring force. If the mass is initially released at rest from a nonzero displacement (measured from equilibrium), then the initial pe stored in the system due to displacement of the mass from equilibrium changes to ke and back to pe after one half-cycle. At any instant during the motion, **ke + pe = initial pe**, as indicated by Fig. 3.2.

Since the restoring force F is given by $F = -$ constant $\times x$, the **pe is given by** $\frac{1}{2}kx^2$ where k is the force constant; this can be understood by considering the area under the force–displacement graph in Fig. 1.12.

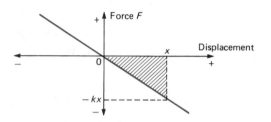

Fig. 1.12 Force–displacement graph (shm)

For displacement x, the area of the shaded triangle $= \frac{1}{2}$ base $(x) \times$ height (kx); the area under a force–displacement curve is equal to the work done, which in this case gives the pe stored.

Since pe + ke = initial pe, it then follows that when the displacement is x and the velocity is v, $\frac{1}{2}kx^2 + \frac{1}{2}mv^2 = \frac{1}{2}kx_0^2$, and because $k = m\omega^2$, then the following useful equation is obtained:

E1.9 $v^2 = \omega^2\left(x_0^2 - x^2\right)$

where $v =$ velocity (m s^{-1}), $\omega =$ angular frequency (rad s^{-1}), $x_0 =$ amplitude (m), $x =$ displacement (m).

When **frictional forces** are present, the total mechanical energy (i.e. pe + ke) of the system is gradually changed to heat energy by the frictional forces. As a result, the amplitude of the oscillations gradually becomes less and less, as shown in Fig. 3.3.

1.7 UNIFORM CIRCULAR MOTION

When a single particle is moving at steady speed (v) on a circular path (radius r), its velocity is constantly changing because its direction is constantly changing. Its time period (T), the time for one complete rotation, is equal to its circumference ($2\pi r$)/speed (v):

E1.10   Angular speed $\omega = \dfrac{2\pi}{T}$ **for steady rotation**

It follows from the above definition that the speed v (i.e. tangential speed) is linked to the angular speed (ω) by:

E1.11 $v = \omega r$

where $v =$ speed (m s^{-1}), $\omega =$ angular speed (rad s^{-1}), $r =$ radius (m).

To cause the continual change of direction of a particle moving in uniform circular motion, a resultant force must act upon the particle; if the force suddenly ceases to act, the particle will move off at a tangent to the circle. Whatever the reason for the resultant force (e.g. tension in a string, magnetic field force on a moving charge), the resultant force is referred to as the **centripetal** force, and its direction is always **towards** the centre of the circle.

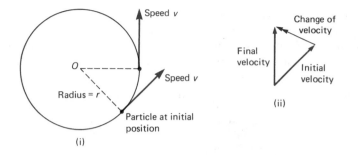

Fig. 1.13 Circular motion: (i) speed at two points in motion; (ii) velocity change between the two points

In Fig 1.13(i), the particle is shown at two positions on its path. The two positions are separated by a short interval of time, so the change of velocity (from one position to the other position) is directed to the centre of the circle, as in the vector diagram in Fig. 1.13(ii). Remember that the centripetal acceleration is equal to the change of velocity per unit time. The centripetal acceleration is given by:

E1.12 $\quad a = -\dfrac{v^2}{r} = -\omega^2 r$

where a = centripetal acceleration (m s⁻²), v = (tangential) speed (m s⁻¹), r = radius of rotation (m), ω = angular speed (rad s⁻¹).

Note that the − sign in the equation indicates that the centripetal acceleration is directed **inwards** (i.e. to the centre of rotation). Also, the angular acceleration is zero because the angular speed is constant. Take care not to confuse angular acceleration with centripetal acceleration.

1.8 ROTATION OF A RIGID BODY

Consider a rigid body that can rotate about a fixed axis. To change the angular speed, a **torque** must be applied to the body; the link between torque applied and angular acceleration produced

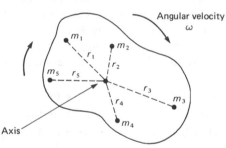

Fig. 1.14 Moment of intertia

is one that can be established by treating the rigid body as a structure of point masses m_1, m_2, m_3,... at distances r_1, r_2, r_3..., respectively, from the axis of rotation, as shown in Fig. 1.14.

Let the angular speed of the body be ω and, since the body is rigid, the angular speed of each point mass will also be ω. The (tangential) speed for each point mass will be ωr_i (using E1.11 where i = 1 or 2 or 3 or...), so that the (tangential) acceleration of each point mass when ω changes will be αr_i where α is the angular acceleration (= dω/ dt). Hence, the force acting upon each point mass must be ($m_i \alpha r_i$), requiring an individual torque (i.e. couple) of moment ($m_i \alpha r_i$)r_i for each point mass. This 'individual moment' can be written as ($m_i r_i^2$)α, so that the total torque required to produce angular acceleration of *all* the particles of the body is $\sum_i (m_i r_i^2)\alpha$. The symbol $\sum_i$ is the summation symbol, so that $\sum_i (m_i r_i^2)\alpha$ is a shorthand expression for $m_1 r_1^2 + m_2 r_2^2 + m_3 r_3^2 + \ldots$

The **moment of inertia** (I) about a given axis of a rigid body is defined as $\sum_i (m_i r_i^2)$. The unit of I is kg m².

Therefore, the moment of inertia is equal to the torque required to produce unit angular acceleration, giving the following equation:

E1.13 $\quad T = I\alpha$

where T = torque (N m), I = moment of inertia (kg m²), α = angular acceleration (rad s⁻²).

This is a useful equation when a constant torque is applied to a rigid body of known moment of inertia. The body will then experience a constant angular acceleration, which will either increase or decrease its angular velocity at a steady rate.

Since I is defined as $\sum_i (m_i r_i^2)$, then the further away from the axis the mass is, the greater will be the moment of inertia. This is illustrated in Fig. 1.15, which shows a hoop of mass M and a uniform disc of the *same* mass. Because the distribution of mass of the disc is closer to the axis than the mass of the hoop, then the moment of inertia of the disc must be smaller than that of the hoop. In fact, the hoop's moment of inertia is simply given by MR^2 because each part of the hoop's mass (M) is at a distance R from the axis.

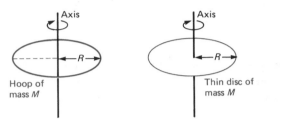

Fig. 1.15 Moments of inertia: hoops and discs

The **angular momentum** of a single particle that rotates about a given point is defined as momentum (mv) × radius of rotation (r). Since $v = \omega r$, then the angular momentum is equal to $m\omega r^2$. From this expression, the angular momentum of a rigid rotating body can be written as $\sum_i m_i \omega r_i^2$ or $\sum_i (m_i r_i^2)\omega$ which then gives:

E1.14 angular momentum = $I\omega$

where I = the moment of inertia (kg m²), ω = angular speed (rad s⁻¹).

The equation $T = I\alpha$ can be written $T = \dfrac{d}{dt}(I\omega)$ since $\alpha = \dfrac{d\omega}{dt}$, so that the link between torque and angular momentum can be established as:

torque = rate of change of angular momentum

If there is no torque acting upon a rigid rotating body, then its angular momentum (and its angular velocity) does not change.

The same rule (i.e. constant angular momentum if no resultant torque acts) applies to a system of rotating bodies in a more subtle form, for even though the total angular momentum must remain the same, one body can lose angular momentum to another body of the system. Thus one part might speed up at the expense of other parts, or a reduction of I for the whole system will cause an increase of ω (but $I\omega$ must remain the same for the whole system). This generalisation of the rule from a single rigid body to a system of rotating bodies is known as the **principle of conservation of angular momentum**, and is usually stated in the following form: if no resultant couple acts upon a system, then its total angular momentum remains constant.

When an object is in rotational motion, it possesses ke on account of that motion. For a rigid body, since the speed of each point mass of the body is given by ωr_i, then the ke of each point mass is $\frac{1}{2}m_i\omega^2 r_i^2$, so that the total rotational ke is given by $\frac{1}{2}\sum_i (m_i r_i^2)\omega^2$ or $\frac{1}{2}I\omega^2$ since $I = \sum_i m_i r_i^2$.

E1.15 Rotational ke = $\frac{1}{2}I\omega^2$

where I = moment of inertia (kg m²), ω = angular speed (rad s⁻¹).

An example of an energy transformation involving rotational ke is provided by a cylinder rolling down a slope. Suppose the cylinder is released from rest at the top of the slope; let v and ω represent, respectively, its speed and angular speed at the base of the slope.

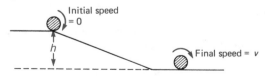

Fig. 1.16 Cylinder rolling down a slope

From Fig. 1.16, loss of pe (mgh) = gain of ke, so that $mgh = \frac{1}{2}mv^2 + \frac{1}{2}I\omega^2$ and, using $v = \omega R$, the speed (v) at the base of the slope can be calculated (given all other values). The final ke is shared between ke of rotation (= $\frac{1}{2}I\omega^2$) and ke of translation (= $\frac{1}{2}mv^2$). For the cylinder in the above example, $I = \frac{1}{2}mr^2$ (assuming it is solid rather than hollow) so the ke of rotation can be shown to be equal to $\frac{1}{4}mv^2$. Hence the final ke is shared between rotation and translation in the ratio 1:2.

In order to change the rotational ke of a rigid body, a couple must be applied; it can be shown that the **work done by a couple = moment of couple × angular displacement**, as shown by the simple arrangement of Fig. 1.17. Thus, work done by a couple = change of rotational ke.

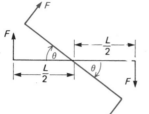

Each force moves a distance of $\left(\frac{L}{2} \times \theta\right)$ and so does work $= \dfrac{FL\theta}{2}$

Hence total work done $= (FL) \times \theta$ = moment of couple × θ

Fig. 1.17 Work done by a couple

1.9 COMPARISON OF LINEAR AND ANGULAR MOTION

A useful comparison can be made by considering:

A mass m moving along a straight line	is equivalent to	A rigid body rotating about a fixed axis
Displacement (s)		Angular displacement (θ)
Velocity (v)		Angular velocity (ω)
Acceleration (a)		Angular acceleration (α)
Mass (m)		Moment of inertia (I)
Momentum (mv)		Angular momentum ($I\omega$)
Force (F)		Torque (T)
$F = ma$		$T = I\alpha$
ke $= \frac{1}{2}mv^2$		ke $= \frac{1}{2}I\omega^2$
Work $= Fs$		Work $= T\theta$

The above comparisons can be extended to the equations of dynamics. For example, the equation $v = u + at$ (E1.1) can be 'translated' into angular motion as $\omega = \omega_0 + \alpha t$, where ω_0 and ω are the initial and final values of angular velocity.

1.10 BEYOND NEWTON'S LAWS

Classical mechanics

Sometimes called Newtonian mechanics, this is based upon Newton's laws as applied to particles. All the preceding topics are based upon classical mechanics. However, there are many situations in which Newton's laws are inadequate, and new theories are needed. The following are some of these situations met in your studies.

Quantum theory

Quantities in classical mechanics can take any value (e.g. any value of ke may be obtained by putting suitable values of m and v into $\frac{1}{2}mv^2$). However, it was first suggested by Planck (in attempting to explain black-body radiation curves, see 6.5) that, in some systems, quantities could only take certain specific values. Such systems are said to be **quantised**, an individual value being called one quantum (plural, quanta). Planck proposed that an oscillator can only gain or lose energy by discrete amounts (quanta) given by:

E1.16 $E = hf$

where E = one quantum of oscillator energy (J), f = oscillator frequency (Hz), h = Planck's constant (Js).

This formula started a whole new branch of physics called **quantum mechanics** and touches your studies through radiation, photoelectricity (see 8.2), energy levels and spectra (see 8.3, 8.4) and heat distribution in solids (see 6.10). Other quantum effects include Millikan's discovery that electric charge is quantised (in units of e, the charge of an electron). Beyond your present studies, quantum mechanics opens up a range of theories and equations far more complex and diverse than classical mechanics.

Relativistic effects

Newton's laws have to be modified for speeds approaching the speed of light, c (at speeds greater then $0.9c$ is the usual criterion). Einstein developed the theory of special relativity and showed, amongst other things, that the two classical conservation laws for mass and energy need to be handled with care. This is because energy has mass according to this conversion formula:

E1.17 $E = mc^2$

where E = energy (J) equivalent to mass m, m = mass (kg) equivalent to energy E, c = speed of light in a vacuum (m s^{-1}).

In small-scale events involving comparatively large energies the 'mass' of the energy involved becomes significant. In your studies, the main use of E1.17 is in explaining mass defects in nuclear physics (see 8.7) and the extraordinary quantities of energy that can be released by nuclear fission and fusion.

The starting point of special relativity is the experimental result that the speed of light (c) is the same for all observers (with constant relative velocities). Many important classical equations can be adapted to deal with relativistic effects (e.g. the Doppler effect, see 3.6). The full consequences and theory of the principles of relativity lie outside the scope of your present studies.

Wave mechanics

Classical mechanics is based upon the study of particles, each particle having nonzero mass. De Broglie showed that, under the right experimental conditions, all particles could be shown to behave like waves (and conversely all waves could be shown to behave like particles). He expressed this link by the following equation:

E1.18 $\lambda = h/mv$ or $\lambda = h/p$

where λ = wavelength (m), h = Planck's constant (J s), m = mass of particle (kg), v = velocity of particle (m s^{-1}), p = momentum associated with the wave (kg m s^{-1}).

This is a fundamental equation of wave mechanics, and is used to explain the particle properties of waves (e.g. photon nature of light), the wave properties of electrons and the impossibility of finding electrons in nuclei. This wave–particle duality, as expressed by **de Broglie's equation**, is an important concept because we tend to explain effects involving submicroscopic particles by large-scale models (e.g. billiard ball models for atomic collisions, wave model for electron diffraction). The concept of particles having wave properties is perhaps easier to accept than the reverse. Fig. 1.18 can help although it must be emphasised that it is intended as a 'thinking model' and *not* as an explanation of wave–particle duality.

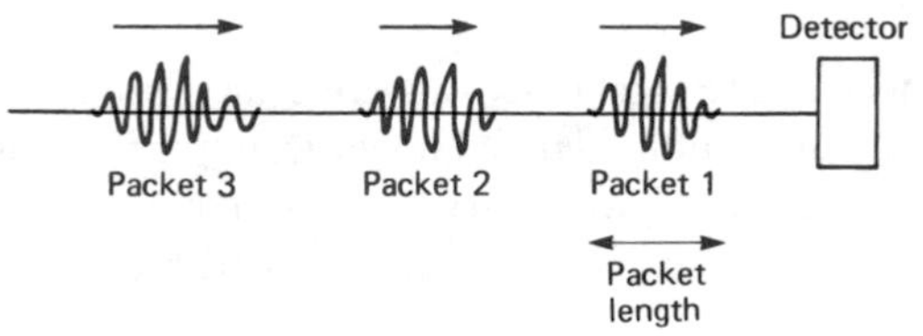

Fig. 1.18 Packets of wave–photons

In the diagram, the wave detector will receive three separate 'packets' of waves in turn. With very little energy arriving between the 'packets', the detector receives three separate impulses of wave energy, as if 'hit' by three particles. If the waves were 'quantised' into packets like this, then a wave-particle combined model makes sense. In electromagnetic radiation, the packets of waves are called **photons**, each with energy given by E1.16, and with packet lengths up to 0.4 m for light (see 4.1)

Statistical mechanics

Classical physics studies individual particles or bodies. The study of systems containing many bodies often requires the mathematical laws of statistics to predict macroscopic (i.e. large-scale) behaviour. Where probability and chance are involved, the type of treatment is referred to as statistical mechanics. Examples in your studies may include the kinetic theory of gases (see 6.6), heat flow and distribution of quanta (see 6.10) and radioactive decay (see 8.6). Normally a model of the system is set up that involves random behaviour; this results in many of the equations of statistical mechanics involving exponential functions.

Into the future

Theoretical physicists have already found it helpful to combine these theories to probe beyond classical mechanics. For example, we find books on relativistic quantum mechanics. New concepts are being proposed all the time.

Chapter roundup

Make sure you have a good grasp of linear mechanics before moving to any other chapter. In addition, statics is used in the topics on solids in Chapter 5, shm links closely with Chapter 3 and circular motion links with Chapter 2 (gravitation) and Chapter 8 (electron dynamics).

Question bank

1 The magnitude of the acceleration of a moving object is equal to the:
   A gradient of a displacement–time graph.
   B gradient of a velocity–time graph.
   C area below a force–time graph.
   D area below a displacement–time graph.
   E area below a velocity–time graph.

(**London**: *all other Boards*)

Points

See 1.1 if necessary.

2 A stone is thrown from *P* and follows a parabolic path. The highest point reached is *T*. The vertical component of acceleration of the stone:
   A is zero at *T*.
   B is greatest at *T*.
   C is greatest at *P*.
   D is the same at *P* and *T*.
   E decreases at a constant rate.

(**Cambridge**: *all other Boards*)

Points

Assume air resistance is negligible and that the Earth's gravitational field is uniform over the range of the flight path. See 1.2 if necessary.

3 Which of the graphs below correctly shows how the acceleration and velocity of a perfectly elastic ball bouncing on a horizontal surface vary with time?

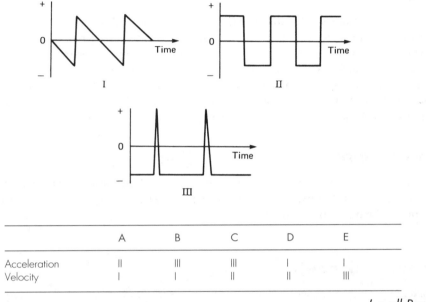

	A	B	C	D	E
Acceleration	II	III	III	I	I
Velocity	I	I	II	II	III

(–: *all Boards*)

Points

Remember that between impacts the acceleration is due to gravity only. Also, the slope of the velocity graph gives the acceleration.

4 The displacement–time graph for a moving body is shown right.

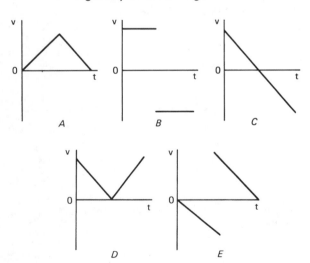

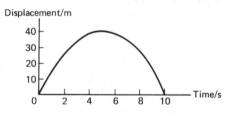

Which of the velocity–time graphs A–E could represent the motion of the same body?

(**SEB**: *all other Boards*)

Points

Remember that the slope of a displacement–time graph gives the velocity.

5 Force is applied to an object of mass 2 kg at rest on a friction-free horizontal surface as indicated on the graph.
After 1 second, the speed of the object in m s^{-1} is:

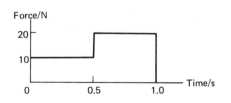

A 7.5   B 12.5   C 15   D 25   E 30

(**SEB**: *all other Boards*)

Points

This question is based on the idea that the area under a force–time graph is equal to the change in momentum.

6 Four identical railway trucks, each of mass m, are coupled together and rest on a smooth horizontal track. A fifth truck of mass $2m$ and moving at 5 m s^{-1} collides and couples with the stationary trucks. After impact the speed of the trucks is v, where v equals:

A $5/6$ m s^{-1}.   B 1 m s^{-1}.   C $5/4$m s^{-1}.   D $5/3$ m s^{-1}.   E $5/2$ m s^{-1}.

(**London**: *all other Boards*)

Points

Draw a diagram similar to Fig. 1.3 and then equate the initial momentum to the momentum after impact and solve to find v.

7 The following are quantities associated with a body performing shm:
1   the velocity of the body
2   the accelerating force acting on the body
3   the acceleration of the body
Which of these quantities are exactly in phase with each other?
A   None of these   B   1 and 2 only
C   1 and 3 only   D   2 and 3 only
E   1, 2 and 3

(**NICCEA**: *all other Boards except SEB*)

Points

See 1.6. The force producing motion is proportional to the acceleration, so acceleration and force must change together.

8 A bead, X, resting on a smooth horizontal surface, is connected to two identical springs and is made to oscillate to and fro along the line of springs.
When the bead passes through the central position, its energy is:
A   zero.
B   mostly potential energy.
C   all potential energy.
D   half potential and half kinetic energy.
E   all kinetic energy.

(**London**: *all other Boards*)

Points

See 3.2 and Fig. 3.2.

9 For a simple pendulum undergoing shm with small oscillations, which of the following correctly describes the ke of the bob and the tension of the thread supporting the bob at zero displacement (i.e. passing through the equilibrium position)?

	A	B	C	D	E
ke =	Max	0	Max	0	Min
Tension =	Min	Max	Max	Min	0

(–: *all Boards except SEB*)

Points

See Fig. 1.11 if necessary. Start by considering the ke at zero displacement – is the object moving fastest or slowest at this point? Thus choose from A–E on the basis of ke, so eliminating some alternatives. Then consider the tension and think of the 'strain' on the thread as the bob moves through a half cycle.

10 A stone of mass 120 g is released from rest from the top of a vertical cliff. After falling for 2.5 s, it hits the beach and penetrates 80 mm into the sand. Calculate (a) the maximum speed of the stone, (b) the height of the cliff, (c) the average force resisting the stone as it penetrates the ground.

(–: all Boards)

Points

Draw a simple sketch and show the points where free fall ends and penetration into the ground starts. For parts (a) and (b), use E1.1 and E1.3 with initial velocity of zero and constant acceleration g.

For part (c), the maximum speed is the initial speed when the stone enters the sand. Its final speed is zero and the distance moved is 80 mm. Hence calculate the deceleration and the force.

11 A trolley of mass 0.80 kg is held in equilibrium between two fixed supports by identical springs (S_1 and S_2) as shown in the diagram; each spring has an extension of 0.10 m. In the second diagram, the trolley is shown moving to the right a distance of 0.05 m. The relationship between the force (F) in newtons and the extension (x) in metres for each spring is given by $F = 20x$.

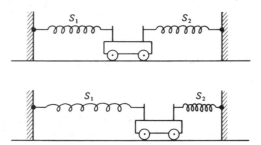

(a) What is the *change* in the force exerted by spring S_1 caused by moving the trolley to the right as in the second diagram?

(b) If the trolley is now released what will be the magnitudes of (i) the resultant force acting upon the trolley at the moment of release, and (ii) the initial acceleration of the trolley?

(c) Showing the steps in your calculation, determine the total energy stored in S_1 and S_2 when the springs are stretched (i) as in diagram 1, (ii) as in diagram 2.

(d) What is the kinetic energy of the trolley as it passes through the equilibrium position?

(O and C Nuffield: *all other Boards)*

Points

(a) The *change* in the force F exerted by S_1 is wanted, corresponding to a *change* of extension of S_1 from 0.10 m to 0.15 m. Use the given equation $F = 20x$.

(b) When the trolley was in the centre, as in diagram 1, the force in S_1 was equal and opposite to the force in S_2. With the trolley moved to the side, S_1 exerts an extra force, as calculated above; S_2 exerts a reduced force, and the reduction in S_2 is equal to the increase of force in S_1. For (i), calculate the force in S_1 and then in S_2, then determine their resultant. For (ii), use E1.5.

(c) (i) Use energy stored = average force × extension; remember that the average force is $\frac{1}{2}$ × force value which gives that extension. (ii) The same method can be used, but remember that both force and extension for S_1 are greater than for S_2, so you will have to calculate energy stored in S_1 first, then repeat the method with different figures for S_2.

(d) At any point after release, ke + pe = initial pe; remember that the pe at equilibrium is nonzero, given by (c)(i) above.

12 A child's toy boat is built with two floats and is propelled by water draining out from a high tank through a hole, as indicated in the perspective and sectional drawings.

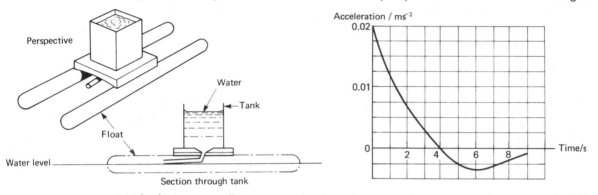

(a) The boat is initially at rest. Why does the water draining out of the tank cause it to accelerate?

(b) The graph shows how the boat's acceleration varies with time.
(i) Find from the graph an approximate value for the maximum speed reached by the boat. Give your reasoning and show clearly how you arrive at your answer. (ii) It can be deduced from the graph that the tank takes at least 6 s to empty. How is this deduction made?

(**O and C Nuffield**: *all other Boards*)

Points

(a) Relate the mass flow at the outlet to the change of momentum, and thus explain why a force acts to accelerate the boat.

(b) (i) Just as displacement is given by the area under a velocity–time graph, so velocity is given by the area under acceleration–time graph.
Remember that area under the time axis counts as –ve, so the maximum velocity is reached at 4.0 s. (ii) When the tank becomes empty, the 'thrust' will become zero, so that the boat will gradually be brought to rest by the viscous drag of the water.

13 (a) The law of conservation of momentum is conserved in any collision. A tennis ball dropped on to a hard floor rebounds to about 60% of its initial height. State how momentum is conserved in this event.

(b) (i) A top class tennis player can serve the ball, of mass 57 g, at an initial horizontal speed of 50 m s⁻¹. The ball remains in contact with the racket for 0.050 s. Calculate the average force exerted on the ball during the serve.

(ii) Sketch a graph showing how the horizontal acceleration of the ball might possibly vary with time during the serve, giving the axes suitable scales.

(iii) Explain how this graph would be used to show that the speed of the ball on leaving the racket is 50 m s⁻¹.

(**AEB** June 90: *all Boards*)

Points

(a) Momentum is conserved provided no external forces act. The ball and the Earth exert equal and opposite forces on each other. Hence the change of momentum of the Earth is equal and opposite to the change of momentum of the ball. Explain why this results in no overall change of momentum.

(b) (i) Use E1.5.

(ii) The acceleration is proportional to the force. Think about how the force changes with time.

(iii) What does the area under an acceleration v. time graph represent?

14 A cable-operated lift of total mass 500 kg moves upwards from rest in a vertical shaft. The graph shows how its velocity varies with time.

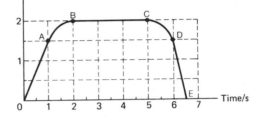

(a) For the period of time indicated by DE, determine (i) the distance travelled, (ii) the acceleration of the lift.

(b) Calculate the tension in the cable during the interval (i) 0A, (ii) BC. Assume that the cable has negligible mass compared with that of the lift, and that friction between the lift and the shaft can be ignored.

(**NEAB**: *all other Boards*)

Points

(a) Use either graphical methods, as described in 1.1, or use the dynamics equations E1.2 and E1.1.

(b) (i) Express the resultant force (assume it acts upwards) in terms of the tension and the weight. Since the resultant force = mass × acceleration, then you ought to be able to calculate the tension, after calculating the acceleration from 0 to A from the graph.

(ii) Start by calculating the acceleration for B to C. Then use the same method as in (i).

15 A particle rests on a horizontal platform which is moving vertically in simple harmonic motion with an amplitude of 50 mm. Above a certain frequency, the particle ceases to remain in contact with the platform throughout the motion.

(a) Find the lowest frequency at which this occurs.

(b) At this minimum frequency, at which point in the motion does contact cease? (Take the acceleration of free fall, g as 10 m s^{-2}.)

(**Cambridge**: *all other Boards except SEB*)

Points

Remember that for shm, the resultant force is always in proportion to the displacement from equilibrium; using the defining equation for shm, E1.6, the resultant force F can be written as $F = -m\omega^2 x$ where ω is the angular frequency, m is the particle mass and x is the displacement from equilibrium. When the mass first leaves the platform during shm, only the weight mg provides the restoring force since the support of the platform is nonexistent at that point; with the sign convention of '+ is downwards', the restoring force is thus $+mg$ at that point.

By considering part (b) first, you should be able to write down the displacement in terms of the amplitude (x_0) when contact ceases. Remember the sign convention: − is up. Then you should be able to deal with (a).

16 A compressed spring is used to propel a ball bearing along a track which contains a circular loop of radius 0.10 m in a vertical plane. The spring obeys Hooke's law and requires a force of 0.20 N to compress it 1.0 mm.

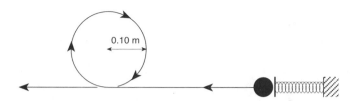

(a) The spring is compressed by 30 mm. Calculate the energy stored in the spring.
(b) A ball bearing of mass 0.025 kg is placed against the end of the spring which is then released. Calculate
  (i) the speed with which the ball bearing leaves the loop,
  (ii) the speed of the ball at the top of the loop,
  (iii) the force exerted on the ball by the track at the top of the loop.
Assume that the effects of friction can be ignored.

(NEAB: *all Boards*)

Points

(a) From Hooke's law, the force for a 30 mm compression is 30 × the force for a 1 mm compression. Work out the force and then work out the energy stored from energy stored = $\frac{1}{2}$ × force × distance the spring is compressed by.
(b) (i) Assume all the energy from (a) is converted to kinetic energy. Given values for the mass and kinetic energy, you can then work out the speed. (ii) When the ball rises from the bottom to the top of the loop, it gains potential energy and loses kinetic energy. Work out the gain of pe from weight × height gain and therefore work out how much ke it still has. Hence calculate the speed from the values of its ke and mass. (iii) The centripetal force at the top is provided by the ball's weight and the force of the track on the ball. Use the speed to work out the centripetal acceleration and hence the centripetal force. Hence work out the force of the track on the ball.

17 A man stands at the Earth's equator. Find (a) his angular velocity, (b) his linear speed, (c) his acceleration, due to the rotation of the earth about its axis.
(1 day = 8.6×10^4 s; radius of Earth = 6.4×10^6 m.)

(Cambridge: *all other Boards except SEB*)

Points

(a) Use E1.10. Remember to give your answer in rad s^{-1}.
(b) Use E1.11.
(c) Use E1.12 for centripetal acceleration, with the radius of rotation equal to the Earth's radius and with the value of angular velocity as in (a).

18 This question is about measuring the acceleration, and so the velocity and displacement, of a moving vehicle, by making observations on masses carried within the vehicle.

  The diagram shows the principle of one sort of accelerometer (device for measuring acceleration). A mass *m* is free to move horizontally within a case, but is restrained by springs fixed to the case. A pointer on the mass can move over a scale fixed to the case. When the case and mass are at rest, the pointer is opposite the zero mark on the scale. When the pointer shows a displacement (*x*) from zero, the net force exerted by the springs is *kx*.

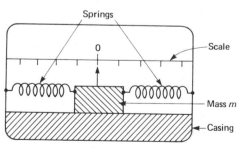

(a) Explain why, when the mass and case are moving at constant velocity in the horizontal direction, the pointer still reads zero. Assume that the velocity has been constant for a long time.

(b) If the casing is in a state of steady acceleration (a) to the left, explain carefully in words why the pointer now has a fixed displacement, saying whether the displacement is to the left or to the right and explaining why. Assume that the acceleration has been constant for a long time.

(c) Give an expression for the magnitude of the displacement in (b).

(d) If the casing were to be suddenly displaced from rest by a sharp blow from a hammer, for example, and then held at rest, describe the subsequent motion of the mass if there is a small amount of friction between it and the casing.

(e) It is suggested that, in use to measure varying accelerations, it would be a good idea to have zero friction between the mass and the casing. Argue briefly for or against this idea.

(f) Suppose that, in use, appreciable changes of acceleration are expected to occur over times not exceeding time t. Give an argument to help decide whether the period T of natural oscillations of the mass and springs should be large, or should be small, compared with t.

(g) In designing an accelerometer for use in a car, a period T of $\pi/5$ seconds was chosen and it was assumed that accelerations up to 2 m s^{-2} should be measured. What would be the displacement at an acceleration of 2 m s^{-2}? (The values of m and k are not needed.)

(h) Suppose that it is decided that an accelerometer for use in a car accelerating at up to 2 m s^{-2} should have a period of 2π seconds. What problems would arise in designing this accelerometer?

(O and C Nuffield: *all other Boards except SEB)*

Points

(a) Zero reading of the pointer on the scale means no resultant force on the mass. Consideration of Newton's first law (see 1.3) should enable you to explain.

(b) Steady acceleration of the mass to the left means that there must be a steady force on the mass to the left. Since the force is provided by the springs there must be a steady displacement from zero; since displacement one way involves a resultant force due to the springs in the other direction, you should be able to explain (i) the steady displacement, (ii) its direction.

(c) Since the net force is given as kx, you should obtain an expression for the magnitude of x in terms of m, k and a.

(d) The friction would cause the subsequent oscillations to be 'damped'. See 3.2 if necessary. With a small amount of friction, the damping is 'light'.

(e) With zero friction and steady acceleration, the mass would oscillate with shm freely, assuming acceleration from rest or constant velocity. The mean position of the pointer would be in proportion to the acceleration. However, here you must consider the further complication of *varying* accelerations, involving a varying 'mean' (i.e. the midpoint between the extreme positions at either end varies).

(f) If the period T is large compared with t, then the pointer will not have reached its correct position by the time the next change of acceleration takes place. The speed of response to applied changes should be discussed in terms of first T large then T small compared with t.

(g) Use E1.6. The value of angular frequency ω is calculated from the given value of T.

(h) Repeat the procedure in (g) to obtain the displacement corresponding to a period of 2π seconds. The problems arise from the displacement value obtained.

19 State the principle of conservation of momentum. A particle of mass m moving with speed v makes a head-on collision with an identical particle which is initially at rest. How would you tell from the subsequent motion of the particles whether they had made (a) an elastic, (b) a completely inelastic, collision? In each case, work out how (if at all) the kinetic energy of the first particle and the kinetic energy of the system as a whole are affected by the collision.

The neutrons in a beam from a reactor have an average energy of 6.0×10^{-13} J. This is reduced to 6.0×10^{-21} J by causing the neutrons to make a series of collisions with carbon nuclei in a moderator. On average, the fractional loss of kinetic energy of a neutron at each collision in the moderator is 0.14. About how many collisions must a neutron make in this process?

(**Cambridge**: *all other Boards*)

Points

See 1.3 for the principle of conservation of momentum.

A head-on collision means that both particles move along the same line after the collision as before. Remember that in an elastic collision between two particles, they 'rebound' off one another without loss of *total* kinetic energy; for a completely inelastic collision, there is no 'rebound' at all, and the two particles stick together.

Consider a series of 'average' collisions for one neutron with initial energy as given. Each collision reduces its energy to 0.86 of its energy before the collision; hence, after two 'average' collisions, the energy becomes 0.86×0.86 of its energy before the first collision. Deal with the problem neatly by considering a neutron making n 'average' collisions; given the initial energy and the final energy after n collisions, you ought to be able to make up an equation for n.

20 (a) Define 'simple harmonic motion'. Give three examples of systems which vibrate with approximately simple harmonic motion. How does the displacement of a simple harmonic oscillator vary with time? What is meant by 'the phase difference' between two simple harmonic motions of the same frequency? Illustrate your answer graphically, by considering the variation of the displacement with time of two motions vibrating with simple harmonic motion of the same frequency but which have phase differences of (i) 90° and (ii) 180°.

(b) At what points in a simple harmonic motion are (i) the acceleration, (ii) the kinetic energy and (iii) the potential energy of the system each at (1) a maximum and (2) a minimum? Sketch graphs showing how (iv) the kinetic energy, (v) the potential energy and (vi) the sum of the kinetic and potential energies for a simple harmonic oscillator each vary with displacement.

(c) Calculate the period of oscillation of a pendulum of length 1.8 m, with a bob of mass 2.2 kg. What assumption is made in this calculation? If the bob of the pendulum is pulled aside a horizontal distance of 20 cm and then released, what will be the values of (i) the kinetic energy and (ii) the velocity of the bob at the lowest point of its swing?

(**London**: *all other Boards except SEB*)

Points

(a) See 1.6 and Fig. 1.10. Approximate shm means a system in which there are frictional forces so giving lightly damped oscillations. For the meaning of phase difference, give a brief written explanation followed by the two graphs. See 3.1 if necessary.

(b) (i), (ii) and (iii) See Fig. 3.2. Give your answers in terms of the displacement, e.g. at the point of extreme displacement. (iv), (v) and (vi) see 3.2. Remember that the total energy remains constant, and that the pe is proportional to (displacement)^2.

(c) See 1.6. Use $g = 10$ m s^{-2}. The assumption is to do with the approximation for sin A. For (i) and (ii), draw a simple sketch showing the pendulum at its lowest point and at the moment of release. Then calculate the vertical displacement of the bob between the two positions using Pythagoras' theorem. You can then calculate the pe difference of the bob between the two positions. The energy changes as outlined in (b) will then enable you to determine the ke and hence the velocity at its lowest point.

21 The motion of a body is said to be 'simple harmonic' if the acceleration of the body is directly proportional to its distance from a fixed point and is always directed towards that point.

(a) Show that the vertical motion of a mass on the end of a helical spring is a case of simple harmonic motion.

(b) A body of mass 0.20 kg hangs from the end of a helical spring, the axis of which is vertical, as shown in the sketch. When the body is raised 0.10 m above its equilibrium position and is then released, it executes shm with a period of 1.0 s. Calculate (i) the force constant of the spring (i.e. the force required to produce unit extension); (ii) the speed of the body as it passes through the equilibrium position; (iii) the maximum kinetic energy of the body; (iv) the magnitude and direction of the acceleration of the body at the lower extremity of its first oscillation; (v) the maximum value of the upward force which the spring exerts on the mass. Find also the distance by which the equilibrium position is lowered if the mass is increased to 0.30 kg. (Acceleration of free fall, $g = 9.8$ m s^{-2}.)

S

0.20 kg

(c) Oscillations of a mass and spring system, as in (b), could be induced if the support S is caused to vibrate vertically through a very small constant distance. For a given frequency of the driving force the amplitude of the resulting oscillations will attain a steady value after some time. (i) Sketch a graph to show how this amplitude may be expected to change if the frequency of vibration of S is varied over a wide range. (ii) On the same axes sketch a second graph to indicate the results if the mass had been immersed in a liquid. (iii) Explain carefully the reasons for the shapes of the graphs.

(**NICCEA**: all other Boards except SEB)

Points

(a) See 1.6.

(b) (i) Use E1.8. (ii) and (iii) Use E1.9 to calculate speed v, then calculate ke. (iv) Use E1.6 with ω calculated from T. (v) When the mass is at its lowest position, the restoring force (i.e. tension − weight) is equal to the mass × acceleration. Hence calculate tension.

(c) (i) This is a resonance situation. See 3.2 and your textbook. (ii) Increased damping will give a less sharp resonance curve. (iii) See 3.2. Give your answer in descriptive terms by explaining why the amplitude builds up at the resonant frequency.

22 (a) A steady stream of balls each of mass 0.2 kg hits a vertical wall at right angles. If the speed of the balls is 15 m s^{-1} and 600 hit the wall in 12 s and rebound at the same speed, what is the average force acting on the wall? Sketch a graph to show how the actual force on the wall varies with time over a period of 0.10 s.

Explain how the average force on the wall could be obtained from this graph. Explain briefly how the above problem can lead to an understanding of how a gas exerts a pressure on the walls of this container.

(b) A car travelling along a level road at a speed of 10 m s^{-1} crashes head-on into a wall. If the mass of the car is 1000 kg, calculate the kinetic energy and momentum of the car just before the collision.

If the impact time (the time taken for the car to come to rest) is 0.2 s, calculate the average force acting on the wall and explain why it is an average. Why is it advantageous for a passenger if the impact time is increased? Make a calculation to support your point. How in practice could car design be improved to achieve an increase of impact time?

(**London**: all other Boards)

Points

(a) Use 'force = change of momentum per second', remembering that the *velocity* will change by *2v* as it is a *vector* quantity. As the stream is regular, you can work out how many balls hit the wall in 0.1 s, assuming the force acts for a very short time interval. You do not have sufficient information to find out the force of each impact, but the average force will be the sum of all the impacts in 0.1 s divided by 0.1 s (what does 'sum of' mean for a graph?). Compare the above situation with the kinetic theory of gases; see 6.6.

(b) Simply substitute the data into the basic equations for kinetic energy and momentum, remembering to give the units of each. Think what you would *see* and *hear* when the car crashes!

  You have already found the momentum, so the average force will be the change in momentum (i.e. to zero) divided by the time taken. What will happen to the magnitude of the force of impact if the momentum can somehow be made to change over a longer time interval?

23 State Newton's laws of motion and outline an experimental method of verifying the second law.

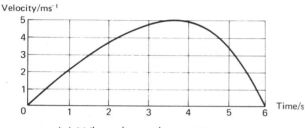

A bead of mass 5 kg is free to slide along a perfectly smooth rigid wire which lies in a vertical plane. The wire is curved, and when the bead is released from rest under the influence of gravity, its speed (*v*) varies with time (*t*) as shown in the graph.

(a) What physical quantities are represented by (i) the total area under the graph, (ii) the slope of the graph at a given point?

(b) At what approximate angles to the vertical is the bead moving at times *t* = 0, *t* = 3.75 s and *t* = 6 s?

(c) What is the difference between the potential energy of the bead at its highest and lowest points?

(d) Give a rough sketch of the shape of the wire, indicating the point from which the bead is released, and its position at times *t* = 3.75 s and *t* = 6 s.

(**O and C**: *all other Boards*)

Points

For Newton's laws of motion, see 1.3 if necessary. Refer to your textbook for experimental verification.

(a) The graph is a speed–time graph. See 1.1.

(b) Since the only forces on the bead are its weight and the normal reaction from the wire, then consideration of this sketch will show you that the force which accelerates the bead is *mg* cos*A* where *m* is the bead mass and *A* its direction to the vertical. Thus, its acceleration is *g* cos *A*; since acceleration can be found from the graph, then angle *A* can be found. Assume *g* = 9.8 m s^{-2}.

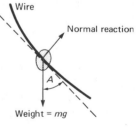

(c) At the highest point its speed is zero. At the lowest point, its speed is greatest. Calculate the pe difference from the ke difference since no friction acts.

(d) From (b), you should appreciate that the wire is steepest at the final part. The lowest point is where the speed is greatest. At *t* = 6 s, the bead should be shown having regained its initial height since its speed is then zero (i.e. all its initial pe at *t* = 0 has been recovered as pe at *t* = 6 s).

Answers: 1 B; **2** D; **3** B; **4** C; **5** A; **6** D; **7** D; **8** E; **9** C; **10**(a) 24.5 m s^{-1} (b) 30.6 m (c) 450 N; **11**(a) 1N (b)(i) 2 N (ii) 2.5 m s^{-2} (c)(i) 0.2 J (ii) 0.25 J (d) 0.05 J; **12**(b)(i) 0.035 to 0.040 m s^{-1}; **13**(b)(i) 57 N; **14**(a)(i) 0.38 m (ii) 3 m s^{-2} (b)(i) 5750 N (ii) 5000 N; **15**(a) 2.3 Hz (b) Highest; **16**(a) 90 mJ (b)(i) 2.68 m s^{-1} (ii) 1.79 m s^{-1} (iii) 0.55 N; **17**(a) 7.3 × 10^{-5} rad s^{-1} (b) 470 m s^{-1} (c) 0.034 m s^{-2}; **18**(g) 0.02 m (h) 2.00 m; **19** 122; **20**(c) 2.7 s (i) 0.25 J (ii) 0.47 m s^{-1}; **21**(b)(i) 7.9 N m^{-1} (ii) 0.63 m s^{-1} (iii) 0.039 J (iv) 4.0 m s^{-2} (v) 2.8 N, 0.12 m; **22**(a) 300 N (b) 50 000 J, 10 000 kg m s^{-1}, 50 000 N; **23**(b) 75.5°, 90°, 60° (c) 62.5 J

CHAPTER 2

FIELDS AND THEIR EFFECTS

Units in this chapter

Chapter objectives

After working through the topics appropriate to your syllabus in this chapter, you should be able to:

- understand the concept of a force field
- define and state the units of electric and gravitational field strength and potential
- relate field strength and potential for uniform and nonuniform fields
- describe the variation of force and pe with separation for two molecules
- use Newton's law of gravitation to explain escape velocity and satellite motion
- use Coulomb's law to work out the electric field strength and electric potential near one or more point charges or near a charged sphere
- use the equation for the capacitance of a pair of parallel plates
- define and state the unit of magnetic field strength (i.e. magnetic flux density)
- solve problems involving current-carrying conductors or moving charges in magnetic fields
- work out the magnetic field strength near a current-carrying wire or a solenoid
- understand the definition of the ampere
- describe and understand the Hall effect
- describe and explain the principle of operation of the moving-coil meter
- understand and apply the laws of electromagnetic induction

2.1 FIELDS: THEIR NATURE, STRENGTH AND FORCES

Objects can exert forces upon each other without touching. Scientists call this 'field' behaviour, and talk about each of the objects producing a field; it is the interaction between the fields that creates the force. The three most important fields for A level are gravitational, electric and magnetic fields; the objects that produce these fields are masses (gravitational fields), charges (electric fields) and objects carrying electric currents (magnetic fields).

A **field** is said to exist around one object if a second similar object placed into that field experiences a force due to the presence of the first object.

Forces between objects can be attractive or repulsive. In field physics, **attractive forces** are given negative values and **repulsive forces** positive values, e.g. +3 N would represent a repulsive force and –20 N would represent an attractive force.

The **strength of a field** is measured by determining the size of the force upon a 'test object' placed in the field.

The 'test object' is always 'one unit' of whatever produces that particular field; a **gravitational field** is tested with a unit mass (1 kg), an **electric field** is tested with a unit charge (1 C) and a **magnetic field** is tested with a wire 1 m long carrying an electric current of 1 A. A further point only concerns magnetic fields: the direction of the field relative to the current–carrying wire is important.

The **force exerted by a field** upon an object can be found from a knowledge of the strength of the field and the nature of the object placed in it. If a charge of +3 C were placed in an electric field of strength 4 N C^{-1}, the charge would experience a force of +12 N (a repulsive force indicated by the positive sign).

2.2 POTENTIAL AND POTENTIAL ENERGY IN FIELDS

It is a familiar experience that objects placed in fields can acquire energy, e.g. two repelling magnets placed close together acquire ke as they spring apart. At first energy is used to place the objects into the fields, e.g. energy has to be supplied to push the two repelling magnets so that they are close to one another. This energy is then stored in the field and the amount of energy stored depends upon the position of the object in the field (as well as upon the strength of the field and the nature of the object involved). For example, if the repelling magnets are pushed even closer together more energy has clearly been stored as work is done in pushing them closer together (and they 'spring' apart faster when released). Energy that is stored according to an object's position is referred to as **potential energy** (pe). There are other forms of pe apart from that stored in fields, e.g. energy stored according to the length (extension or compression) of a spring.

Potential energy is measured by comparing the **pe** of the object at its chosen position with its pe at a reference position. The reference position is usually chosen to be 'infinity' where the pe is defined to be zero.

There are cases when it is more convenient to choose a different reference position, e.g. zero extension of a spring (rather than infinite extension) and, on the Earth, gravitational potential energy is often measured (using *mgh*) relative to the surface of the Earth.

The **potential** at a point in a field is the pe that a suitable 'test object' would have at that point.

The nature of the 'test objects' is explained in 2.1, e.g. if a mass of 2 kg were found to have a gravitational pe of –8 J at a point in a field, then the gravitational potential at that point would be –4 J kg^{-1} (which is the pe that the 'test object', 1 kg, would acquire at that point). In an electric field, the potential would be measured in joules per coulomb (or volts). The difference in potential between two points is referred to as the **potential difference** (pd) between them; this term can be applied to any type of field, not just to electric fields.

The pe of an object placed at a point in a field can be found from a knowledge of the potential at that point and the nature of the object placed there. Thus a charge of +0.1 C placed at a point of potential +300 V (+300 J C^{-1}) in an electric field will have a pe of +30 J. The sign is again important as it still indicates the nature of the force (repulsive in this case) and the direction in which an object will move if allowed to lose pe, converting it into ke.

It is important to note that as soon as a field exists we can talk about the field strength or potential at any point. However, it is necessary to place a suitable object into the field before we can start talking about forces or potential energies.

There is one important general equation of field physics that links field strength and potential (or force and pe):

E2.1 $E = -\dfrac{\mathrm{d}V}{\mathrm{d}r}$ or $F = -\dfrac{\mathrm{d(pe)}}{\mathrm{d}r}$

So $V = -\int E \, \mathrm{d}r$ or pe $= -\int F \, \mathrm{d}r$

where E = strength of field (various units), V = field potential (various units), F = force exerted by a field (N), pe = potential energy in a field (J), r = distance (m).

The distance r is normally measured to the centre of the object producing the field. The negative sign is needed to maintain the sign convention that negative forces are attractive. Beware of the trap of assuming that a field with a negative potential must always produce an attractive force on an object placed in that field. For example, a negative point charge creates a field with negative potential values, but another *negative* charge placed in that field will experience a repulsive force.

At A level the main use of E2.1 comes not in manipulating the calculus integrals and differentials but in being able to interpret graphs according to the equation. The negative value of the slope at any point of a graph of potential (V on y-axis) plotted against distance (r on x-axis) gives the strength of the field at that point (or the actual force is given by the slope of the pe–distance graph). If the area beneath a graph of field strength (E on y-axis) plotted against distance (r on x-axis) is measured between two points, then the pd between those points has been found (or the pe difference comes from a graph of force against distance).

2.3 GRAPHS AND DIAGRAMS OF FIELD STRENGTH AND POTENTIAL

Fig. 2.1 is a diagram of the electric field that exists between the parallel plates of a capacitor charged up to 400 V. The field is illustrated by the solid black lines which show the path of travel of a positive charge released from rest within the field. By definition, the arrows show the direction in which positive charge would move. The strength of the field is indicated by the distance between these '**field lines**'. As the lines are the same distance apart, this shows that the field strength is constant (a **uniform field**), except at the edges where the lines are further apart and hence the field is weaker. A charge placed anywhere in the uniform field will experience a constant force, despite the fact that it may be put at points of different potential. It is possible to represent a field by lines

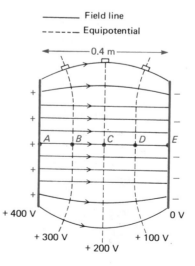

Fig. 2.1 Field lines and equipotentials for a parallel-plate capacitor

(**equipotentials**) joining points in the field that are at the same potential; these are the dotted black lines. The equipotentials are always at right angles to the field lines; this makes it easy to draw an equipotential diagram if the field 'shape' is known. There are further ways of drawing field diagrams, e.g. by using depth of shading to illustrate the strength of the field. In Fig. 2.1 path $ABCDE$ (0.4 m long) has been marked and Fig. 2.2(a) and Fig. 2.2(b) show the variation of field strength (E_E) and potential (V_E), respectively, plotted along that path with the distance (r) measured from the positive capacitor plate.

In Fig. 2.2(a): slope, no great physical significance apart from the fact that zero slope means a uniform field; area, the negative value of the area under the graph between any two points measures the pd between them (E2.1); intercept, the field strength is 1000 N C^{-1} (or 1000 V m^{-1}) (E2.9).

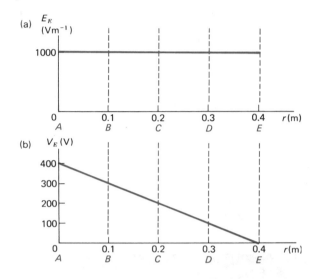

Fig. 2.2
(a) Field strength plotted against distance for Fig. 2.1
(b) Potential plotted against distance for Fig. 2.1

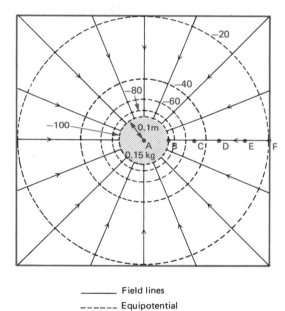

Fig. 2.3 Field lines and equipotentials around a sphere

——— Field lines
- - - - - Equipotential

In Fig. 2.2(b): slope, the negative value of the slope gives the field strength, which is a constant value (1000 V m^{-1}) as the graph is a straight line (E2.1); area, no great physical significance; intercepts, the right-hand plate has been chosen as the zero of potential (0 V) and the left-hand plate is at 400 V.

Note that measuring the slope of Fig. 2.2(b) and taking its negative value leads to Fig. 2.2(a), and measuring the area under the graph of Fig. 2.2(a) and taking its negative value leads to Fig.2.2(b).

Fig. 2.3 shows the field lines and equipotentials around a sphere of mass 0.15 kg due to its gravitational field. Note that the field lines get further apart the greater the distance from the centre of the sphere, indicating that the field strength is getting weaker. The equipotentials are marked in equal steps (20, 40, 60, 80, 100 × 10^{-12} J kg^{-1}) but they are not equal distances apart on the diagram (as they would be in a uniform field). They are of course still at right angles to the field lines. The arrows show the direction of the force on a mass placed into the field. The force is always attractive (unlike electric and magnetic fields when both attractive and repulsive forces are possible). The variation in field strength and potential with distance is shown in Fig. 2.4.

In Fig. 2.4(a): slope, no great physical significance; area, the area under the graph between any two points measures the gravitational pd between them (E2.1); intercepts, etc., the graph starts from the field strength at the surface of the sphere (1000 pN kg^{-1}; p = pico = 10^{-12}) and asymptotically approaches zero field when $r \rightarrow \infty$.

In Fig. 2.4(b): slope, the negative value of the slope gives the field strength at any point (E2.1); area, no great physical significance; intercepts, etc., the graph starts from the surface of the sphere (100 pJ kg^{-1}) and asymptotically approaches zero potential as $r \rightarrow \infty$ (the usual definition of zero potential).

In Fig. 2.3 and Fig. 2.4 the negative values of field and potential are important as they result in negative values for forces and potential energy (as mass values are always positive), correctly indicating an attractive force field. The largest value that the potential can have is zero (at $r = \infty$) as all negative values are less than zero. This is correct as to gain gravitational pe a mass has to be taken further away from the centre of the field; just think of a mass in the Earth's gravitational field gaining pe as it is taken further away from the surface of the Earth. Ultimately the mass will end up at infinity, when it will have its maximum pe; but if the value at infinity is only zero then all other values must be less than zero, i.e. negative! Fig. 2.4(a) and Fig. 2.4(b) can be produced from E2.4 and E2.5, respectively.

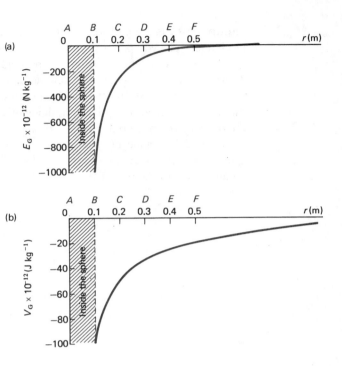

Fig. 2.4 (a) Field strength plotted against distance for Fig. 2.3; (b) potential plotted against distance for Fig. 2.3

2.4 INTERMOLECULAR FIELDS

In many systems an object may be influenced by more than one field, e.g. a charged mass can experience forces due to both electric and gravitational fields. It is further found that a molecule can experience both short-range repulsive forces and long-range attractive forces when near to a single neighbouring molecule. Much can be learnt from studying the force–separation and pe–separation curves for two neighbouring molecules. These are shown as Fig. 2.5(a) and Fig. 2.5(b).

In Fig. 2.5(a): slope, of no great physical significance except for the part between $r = B$ and $r = D$ where the graph is roughly a 'straight line' indicates Hooke's law behaviour; area, the area under the graph between any two points indicates the pe difference between them (E2.1); intercepts, etc., there is zero force between the molecules at $r = \infty$, the molecules cannot get so close that $r = A$, equilibrium occurs when the force between the molecules is zero at $r = C$, the force is repulsive if $r < C$ and attractive if $r > C$, with maximum attractive force at $r = E$.

In Fig. 2.5(b): slope, the negative value of the slope at any separation is the force between the molecules at that separation (E2.1), if $r < C$ then the slope is always negative and the force positive (repulsive), if $r > C$ then the slope is always positive and the force negative (attractive), at $r = C$ the slope and force are zero, the pe is at a minimum and the two molecules are in equilibrium (no net force between them) and at $r = E$ the slope of the graph, although still positive, stops increasing and starts decreasing (point of inflexion) which means that the attractive force reaches its maximum value at this point; area, no great physical significance; intercepts, etc., pe is zero at $r = \infty$, molecules cannot get so close that $r = A$ (infinite energy required).

Fig. 2.5(a) and Fig. 2.5(b) only represent the forces between *two* neighbouring molecules and do not take account of the fact that a molecule may be influenced by many neighbours. However, this simple microscopic model makes several useful predictions. The equilibrium separation occurs when $r = C$ (about 0.2 nm) and a pair of molecules with no thermal energy

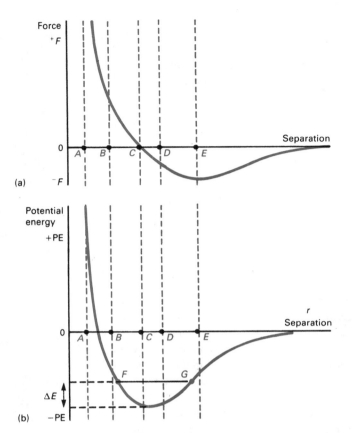

(a)

(b)

Fig. 2.5 (a) Force plotted against separation for two molecules;
(b) potential energy plotted against separation for two molecules

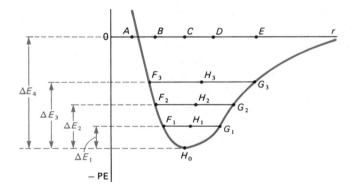

Fig. 2.6
Potential and
thermal energies
of two molecules

(i.e. not vibrating and consequently at **absolute zero**) should be this far apart. Beware of the mistake of confusing the two curves and saying that equilibrium occurs on the force curve at $r = E$ because it is a minimum here, or that it occurs between $r = A$ and $r = B$ on the pe curve because the pe is zero there! We can allow for **thermal energy** by giving the two molecules some extra energy ΔE as shown on Fig. 2.5(b). As in any vibrating system the energy is continually being converted from kinetic to potential and vice versa (see 3.2). The extra energy ΔE allows the molecules to approach as close as F or to get as far apart as G. This is rather like a mass hanging on a spring being given extra energy when it becomes capable of both compressing the spring and extending it; the amount of extension and compression depends on how much extra energy it is given. Fig. 2.6 shows the situation for several different thermal energies given to the two-molecule system.

It is assumed that the average separation of two oscillating molecules is halfway between their distance of closest approach ($r = F$) and their distance of greatest separation ($r = G$); the halfway points ($r = H$) have been marked on the diagram. It can be seen that as ΔE gets larger so the average distance apart of the molecules (H) gets larger; or, macroscopically, as the temperature rises a substance **expands**. Another macroscopic prediction is of **Hooke's law** from Fig. 2.5(a) between $r = B$ and $r = D$, where the straight line indicates force is proportional to separation. The thermal energies of molecules are very small compared with their potential energies so a thermal energy corresponding to ΔE_3 in Fig. 2.6 corresponds to a very high temperature. At this sort of temperature the maximum separation allowed ($r = G_3$) is so great that it is possible in the bulk of a substance for the molecules to change places with other molecules; this type of exchange is thought to be typical of the behaviour that occurs in a **liquid**. Increase the thermal energy to ΔE_4 and the molecules have enough energy for $r = \infty$ so that they can break completely free of one another, forming a **gas**. See 6.2 for the molecular interpretation of latent heat.

2.5 ELECTRIC AND GRAVITATIONAL FIELDS FROM SPHERES AND POINTS

The basic shape of an electric or gravitational field from a sphere or pivot is that of Fig. 2.3, although the diagram is two dimensional and it must be remembered that the fields are three dimensional. The 'field line' diagram is correct for an electric field in which the charge on the sphere or at the centre point is negative; the direction of the arrows would have to be reversed for a positive charge. The equipotentials would have to be renumbered and the sign changed to positive if the field was being generated by a positive charge. The following equations provide mathematical models for these fields and the field and potential graphs in Fig. 2.4.

E2.2 $E_E = \dfrac{1}{4\pi\varepsilon_0} \cdot \dfrac{Q}{r^2}$ (or $F_E = \dfrac{1}{4\pi\varepsilon_0} \cdot \dfrac{Q_1 Q_2}{r^2}$)

E2.3 $V_E = \dfrac{1}{4\pi\varepsilon_0} \cdot \dfrac{Q}{r}$ (or $\text{pe}_E = \dfrac{1}{4\pi\varepsilon_0} \cdot \dfrac{Q_1 Q_2}{r}$)

E2.4 $E_G = -G \cdot \dfrac{m}{r^2}$ (or $F_G = -G \cdot \dfrac{m_1 m_2}{r^2}$)

E2.5 $V_G = -G \cdot \dfrac{m}{r}$ (or $\text{pe}_G = -G \cdot \dfrac{m_1 m_2}{r}$)

where E_E = electric field strength (N C^{-1}) due to a charge Q (C), ε_0 = permittivity of free space (C V^{-1} m^{-1}), F_E = force (N) exerted by the interaction of the electric fields of two charges Q_1 and Q_2, V_E = electric field potential (V or J C^{-1}) due to a charge Q (C), pe_E = electric field pe (J), E_G = gravitational field strength (N kg^{-1}) due to a mass m (kg), G = gravitational constant (N m^2 kg^{-2}), F_G = force (N) exerted by the interaction of the gravitational fields of the two masses m_1 and m_2, V_G = gravitational field potential (J kg^{-1}) due to a mass m (kg), pe_G = gravitational field pe (J).

The distance r (m) in the case of field strengths and potentials is simply the distance from the centre of the field to the point at which the measurement is to be made. When two objects are involved, as when measuring pe and forces, r (m) is the distance between their centres. The constant G is universal, but the value of ε_0 is valid only for fields in vacuums and, although there is virtually no change for its value in air, it must be modified by using the appropriate ε_r for experiments in any other medium (see E2.12). Be careful to watch the sign conventions which define whether forces are attractive or repulsive, and whether the pe increases or decreases as $r \to \infty$ (see 2.1 and 2.2). Note that fields and forces follow a $1/r^2$ (inverse square law) whereas potentials and potential energies obey a $1/r$ rule. You can use E2.4 and E2.5 to check Fig. 2.3 and Fig. 2.4. Remember when calculating forces that Newton's third law (action and reaction are equal and opposite) applies and, although you may only be concerned with the force on an object placed in a field, the object creating the field must experience an equal and opposite force. Experiments to determine the values of G and ε_0 are important and will be found in your textbook. Note that E2.2 can be derived from E2.3 (using E2.1 and a knowledge of how to differentiate); similarly E2.4 can be derived from E2.5.

2.6 EARTH'S 'UNIFORM' GRAVITATIONAL FIELD

If the graph axes and equipotentials were renumbered, Fig. 2.3 and Fig. 2.4 would illustrate the Earth's gravitational field. The following information (and E2.4 and E2.5) is needed:

Radius of the Earth (R) = 6.4×10^6 m

Mass of the Earth (M) = 6.0×10^{24} kg

Fig. 2.3 shows that the field is clearly not uniform, but if a section is examined close to the surface of the Earth it looks as shown in Fig. 2.7. To a good approximation the field appears uniform, and its strength (g) is the force on 1 kg at the surface of the Earth so using E2.4:

E2.6 $g = -\dfrac{GM}{R^2}$

where G = gravitational constant (N m^2 kg^{-2}), M = mass of the Earth (kg), R = radius of the Earth (m).

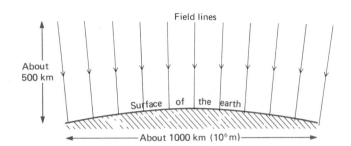

Fig. 2.7 Gravitational field close to the surface of the Earth

From $F = ma$ (E1.5) $F/m = a$, so any value of force per unit mass (newtons per kilogram, the units of gravitational field) is also a measure of acceleration. So g can be regarded as the gravitational field strength at (and just above) the surface of the Earth, or as the acceleration of a mass falling in the uniform gravitational field. The weight (mg) of an object (mass m) is the force of gravity upon it.

The value of g varies slightly owing to the Earth not being a perfect sphere and more importantly, because of its rotation.

2.7 LAUNCHING SATELLITES FROM THE EARTH

What goes up will not come down, if we throw it up fast enough! As something leaves the surface of the Earth it gains gravitational pe as it loses ke. Fig. 2.4(b) can be used to study this effect: if an object is given ke per kg of 80 pJ (p = pico = 10^{-12}) when it leaves the surface of the sphere, eventually its gravitational potential will increase by this amount. This will change its gravitational potential from −100 to −20 pJ kg^{-1} and the object will stop rising 0.5 m away from the *centre* of the sphere and 0.4 m away from the *surface* of the sphere. However, if *at least* 100 pJ kg^{-1} of ke per kg are given to the object, it will have enough energy to escape to $r = \infty$. For the Earth's gravitational field this produces E2.7.

E2.7 $$v = \sqrt{\frac{2GM}{R}}$$

where v = escape velocity (m s^{-1}), G = gravitational constant (N m^2 kg^{-2}), M = mass of the Earth (kg), R = radius of the Earth (m).

The equation is derived from taking the ke ($\frac{1}{2}mv^2$) of a mass m travelling at a speed v and equating it to the pe difference of the mass between being at the surface of the Earth ($-GmM/R$) and at infinity (zero). The escape velocity in the Earth's gravitational field is about 11 km s^{-1}. The escape velocity from the 0.1 m sphere of 0.15 kg mass featured in Fig. 2.3 and Fig. 2.4 can be checked to be about 45×10^{-6} m s^{-1}. The **escape velocity** is the minimum speed away from the centre of a gravitational field that an object must be given to escape completely the influence of the field. The escape velocity may be directed at any angle as long as it is 'away' from the field.

A satellite can be maintained in a stable circular orbit (Fig. 2.8) if a steady force of the right value ($= mv^2/(R + h)$) is always directed towards the

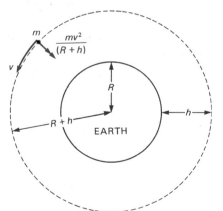

Fig. 2.8 Satellite orbiting the Earth

centre of the circle in which it is travelling. The centripetal force is provided by the gravitational attraction $(GmM/(R + h)^2)$ which is always acting in this direction, so if a satellite can be made to travel at the correct speed then the gravitational force upon it can keep it in orbit, giving

E2.8 $\quad v = \sqrt{\dfrac{GM}{R + h}}$

where v = speed of satellite in orbit (m s^{-1}), G = gravitational constant (N m^2 kg^{-2}), M = mass of the Earth (kg), R = radius of the Earth (m), h = height of orbit above *surface* of the Earth (m).

Note that neither the escape velocity nor the speed of orbit for a particular height depend upon the mass of the projectile involved. It is a popular mistake to confuse the distance of an object from the centre of the Earth $(R + h)$ with its height above the surface of the Earth (h). It is possible to calculate the period (time for 1 revolution) of an orbit from its circumference $(2\pi (R + h))$ and the satellite speed (v). If the period is 24 hours then the Earth rotates with the same period as the satellite; this means that the satellite can stay stationary above a fixed point on the Earth (**parking orbit**). Calculation will show the height of such an orbit to be 36 000 km. Gravitational field equations have further applications in the study of Kepler's laws and planetary astronomy, etc. Textbooks will provide more information if needed.

2.8 UNIFORM ELECTRIC FIELDS

It can be shown that uniform electric fields exist between parallel charged plates or close to large planes with an even distribution of charge over their surface (see your textbook). Consider a charge of $+q$ being moved from R to S through the pd between the plates of V volts. The charge will gain electric pe of qV joules as work is done against the repelling constant force F exerted by the field. The situation is shown in Fig. 2.9. So work done in moving charge from R to S = the constant force $\times$ distance involved, or $qV = Fd$. But in any electric field of strength E, the force F upon a charge q is given by Eq, or $qV = Eqd$, hence

E2.9 $\quad E = V/d$

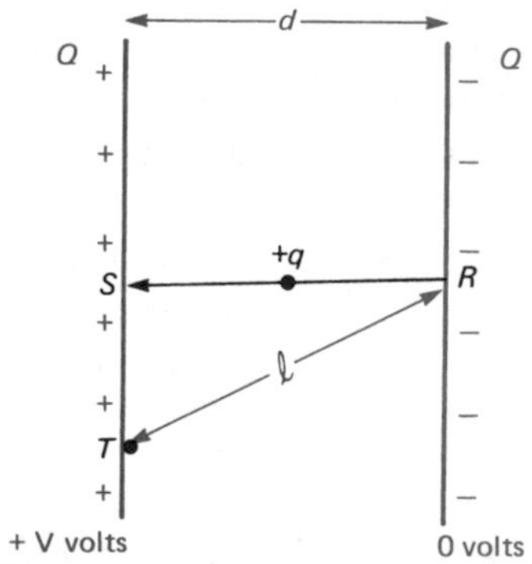

where E = strength of uniform electric field (N C^{-1} or V m^{-1}), V = pd between two points in the field (J C^{-1} or V), d = distance between the two points (m).

It is important to remember that this formula applies only to uniform fields, and the distance d must be measured along a field line that joins two points at the correct pd V. The strength of the field between points R and T (a pd of V volts apart) is *not* V/l, for although l is the distance between these points it is not measured along a field line. Beware also of calculating the pd incorrectly, e.g. if point R were at $-V$ volts and point S at $+V$ volts then the pd involved is $2V$ volts and *not* 0 volts.

If the capacitor plates have surface area A, it would seem sensible to assume that the field generated would depend on the charge per unit area (Q/A) upon the plates; this quantity is frequently referred to as the **charge density** (σ). So we have $E \propto Q/A$. Taking the proportionality we convert it into an equation with the aid of a suitable constant called ε_0, so

E2.10 $\quad \varepsilon_0 E = Q/A$ (or $\varepsilon_r \varepsilon_0 E = Q/A$)

where ε_0 = permittivity of free space (C V^{-1} m^{-1}), E = strength of uniform electric field (N C^{-1} or V m^{-1}), Q = total charge on each capacitor plate (C), A = area of each capacitor plate (m^2), ε_r = relative permittivity of a medium (no units).

This equation *defines* ε_0 which applies to a **vacuum (free space)**. The strength of the field is found to vary if the medium is changed; in this case the value of ε_r for the medium

Fig. 2.9 Moving a charge in a uniform field

concerned must be used. The larger the value of ε_r, the bigger will be the charge stored on a pair of capacitor plates for a given pd; the medium between the plates plays an important role in the storing of energy in the field and charge on the plates. This definition of ε_0 applies to parallel plates and rectangular symmetry, so it is no surprise to find 4π appearing in E2.2 and E2.3 which describe a spherically symmetrical situation (e.g. area of a square of side $r = r^2$, but surface area of a sphere of radius $r = 4\pi r^2$). ε_0 is also known as the **electric constant**.

2.9 CAPACITANCE OF PARALLEL PLATES AND ISOLATED SPHERES

The **capacitance** of a capacitor is a measure of the charge (in C) stored in it per volt of potential difference that is used to charge it giving:

E2.11 $C = Q/V$ (or $Q = CV$)

where C = capacitance in farads (F or C V^{-1}), Q = charge stored (C), V = pd used to charge capacitor (V).

It is possible to combine E2.9 and E2.10 so $\varepsilon_r\varepsilon_0 V/d = Q/A$, and rearranging this we get $Q/V = \varepsilon_r\varepsilon_0 A/d$ which defines the capacitance of a parallel plate capacitor:

E2.12 $C = \dfrac{\varepsilon_r\varepsilon_0 A}{d}$

where C = capacitance of parallel plate capacitor (F), ε_r = relative permittivity of medium between plates (no units), ε_0 = permittivity of free space (C V^{-1} m^{-1}), A = surface area of each plate (m^2), d = separation of the plates (m).

If a conducting sphere with a charge Q residing on its surface is considered, it must obey E2.3 which can be rewritten as $Q/V = 4\pi\varepsilon_0 r$, which defines the capacitance of an isolated conducting sphere:

E2.13 $C = 4\pi\varepsilon_0 r$

where C = capacitance of isolated conducting sphere (F), ε_0 = permittivity of free space (C V^{-1} m^{-1}), r = radius of the sphere (m).

In this case the pd to which the sphere is charged is assumed to be the pd between the surface of the sphere and infinity (zero volts as usual).

In Fig. 2.10: slope, constant, its value (Q/V) is the capacitance; area, electrical energy stored in capacitor; intercepts, etc., graph passes through origin because for zero charging voltage no charge is stored.

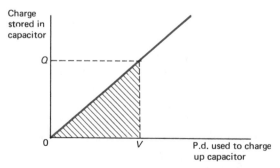

Fig. 2.10 Charge and energy stored in a capacitor

In Fig. 2.10 the area under the graph is the energy stored in the capacitor and as the shaded area is $\frac{1}{2}QV$ and $Q = CV$ (E2.11):

E2.14 Energy stored $= \frac{1}{2}QV = \frac{1}{2}CV^2 = \frac{1}{2}Q^2/C$

where electrical energy stored in capacitor is in joules, C = capacitance of capacitor (F), Q = charge stored in capacitor (C), V = pd used to charge capacitor (V).

2.10 MEASURING MAGNETIC FIELDS AND THEIR FORCES

At A level the concepts of magnetic potential and pe are not required; so only an understanding of magnetic fields and the forces they exert is needed (together with some applications). Magnetic fields are created by electric currents. To understand how this can be applied to permanent magnets it is necessary to examine them microscopically and explain that the currents involved can be linked to the electrons moving (moving charge is electric current) within the atoms. The **strength of a magnetic field** is found by measuring the force on a wire 1m long carrying a current of 1 A at right angles to the field. The magnetic field strength is sometimes called the 'magnetic flux density' or the 'magnetic induction'. So magnetic field strengths (B) are measured in newtons per ampere per metre ($\text{N A}^{-1}\text{m}^{-1}$); this unit is known as the **tesla** (T). However, a major complication arises when dealing with magnetic fields (as opposed to electric or gravitational fields): the direction of the current being used to test the field, relative to the direction of the field itself, is vital. The field and current must be at right angles to one another, otherwise the force involved will be reduced; the force becomes zero if they are parallel to one another.

Fig. 2.11 shows that if the angle between the magnetic field (B) and current (I) is θ, then the vector components of the current at right angles and parallel to the field are $I \sin \theta$ and $I \cos \theta$, respectively. The force on the $I \cos \theta$ component is zero (as it is parallel to the field) and the force on the ('right angle') $I \sin \theta$ component is 1 newton per ampere of current per metre length of wire per tesla of magnetic field giving:

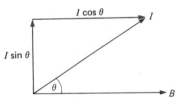

Fig. 2.11 Current (I) and magnetic field (B) at an angle (θ)

E2.15 $F = BIL \sin\theta$

where F = force exerted by magnetic field on wire (N), B = strength of the magnetic field (T or $\text{N A}^{-1}\text{m}^{-1}$), I = current in the wire (A), L = length of wire in the field (m), θ = angle between the current and field (see Fig. 2.11).

This formula is based upon the macroscopic concept of electric current, but it can easily be changed to a microscopic version which deals with moving charges:

E2.16 $F = Bqv \sin\theta$

where F = force exerted by magnetic field on charged particle (N), B = strength of the magnetic field (T or $\text{N A}^{-1} \text{m}^{-1}$), q = size of charge upon particle (C), v = speed of the particle (m s^{-1}), θ = angle between the magnetic field and the direction in which the particle is travelling.

Not only are the current and field at right angles to one another to enable the resulting magnetic force to be measured, but *the resulting force is at right angles to both the current and the field*. To get the correct direction of the force, the **left-hand rule** has to be used (check in your textbook if you need to remember how to use the rule).

The usual way of remembering the left-hand rule is via the **M** in thu**M**b which represents the Motion (direction of the force), the **F** in First finger (Field) and the **C** in seCond finger (Current). Beware of the trap of thinking that the F stands for force. Another popular error occurs when applying the rule to moving charged particles; remember that electric current has its direction defined from the direction in which positive charge moves, so electrons travelling to the 'left' represent a current travelling to the 'right'.

There are three fields created by currents in wires that are particularly important at A level:

❶ **A long straight wire** produces a field of concentric circles getting weaker as you get further away from the wire (direction comes from the corkscrew rule).

❷ **A solenoid** is a long coil filled with air (or a vacuum). The field is like that of a bar magnet except that the inside of the solenoid is accessible and the field here is found to be **uniform**. The direction of the field is found by treating it like a bar magnet and saying that the North-seeking end is where the current flows anticlockwise ⟳ and vice versa.

❸ **A flat circular coil** (i.e. one of zero length) produces a field similar to that of a 'squashed solenoid'. Most importantly these coils are used in pairs (**Helmholtz coils**) when they can produce a uniform field near their axis over a reasonably large volume.

Further information and diagrams of the three fields are available in your textbook. Remember that the field direction is defined so that it fits in with the left-hand rule. The following formulae give means of calculating the field strengths for the three cases:

E2.17 $\quad B = \dfrac{\mu_0 I}{2\pi r}$

where B = field strength at distance r from a long uniform wire (T), μ_0 = permeability of free space (N A^{-2}), I = current in wire (A), r = radial distance from the wire (m).

E2.18 $\quad B = \dfrac{\mu_0 N I}{L}$

where B = strength of uniform field inside a solenoid (T), μ_0 = permeability of free space (N A^{-2}), N = total number of turns of wire in solenoid (no units), L = length of solenoid (m), I = current passing through the solenoid (A).

E2.19 $\quad B = \dfrac{\mu_0 N I}{2r}$

where B = strength of the field at the centre of a flat circular coil (T), μ_0 = permeability of free space (N A^{-2}), N = total number of turns of wire in coil (no units), I = current passing through the coil (A), r = radius of the circular coil (m).

2.11 USING MAGNETIC FIELD FORMULAE

Parallel wires

If two wires are parallel a distance r apart, then the field at one wire (carrying a current I_1) due to the other wire (carrying a current I_2) is given by $\mu_0 I_2 / 2\pi r$ (E2.17). So:

E2.20 $\quad \dfrac{F}{L} = \dfrac{\mu_0 I_1 I_2}{2\pi r}$

where F/L = force per unit length on each wire (N m^{-1}), μ_0 = permeability of free space (N A^{-2}), I_1 = current in one wire (A), I_2 = current in the other wire (A), r = distance between the wires (m).

Two points of importance arise when using this formula. First it must be noted that the formula gives the force per unit length (*not* the force) on the wires. Second both wires experience an equal and opposite force (Newton's third law) and the force between the wires is attractive if the parallel currents run in the same direction (repulsive if the currents run in opposite directions). Students are accustomed to 'like' charges 'repelling', hence they wrongly assume that currents in the same direction also repel.

Electric currents give rise to several effects: magnetic, heating, chemical, dynamic, light production, etc. It could be possible to define the ampere using any of them; however, it is the magnetic effect which has been chosen.

One **ampere** is the constant current that if maintained in two parallel conductors of infinite length, of negligible circular cross-section, and placed 1 m apart in a vacuum, would produce between those conductors a force of 2×10^{-7} N per metre length.

It can be seen from E2.20 that the definition of the ampere also fixes the value of μ_0 (the **permeability of free space**) to be $4\pi \times 10^{-7}$ N A^{-2}. μ_0, although defined for a vacuum, has almost exactly the same value in air; however, the medium can affect the strength of the magnetic field and a term μ_r (**relative permeability**) needs to be introduced to allow for the nature of the medium. Hence to allow for a change of medium $\mu_0\mu_r$ should be put into all the formulae in place of μ_0; note that relative permeability has no units. The similarity between $\mu_r\mu_0$ and $\varepsilon_r\varepsilon_0$ (E2.10) is obvious; both sets of constants help define the strength of field produced appropriate to the medium involved (the larger the value of the constants the larger the field strengths in any given situation). μ_0 is also called the **magnetic constant**.

Hall effect

A piece of conducting material has a current I passing through it and a magnetic field of strength B is applied at right angles to the current. The current is passing to the right and the magnetic field is directed into the paper so the force on the current is 'upwards' (left-hand rule, E2.16). In Fig. 2.12 it has been assumed that the current to the right is being carried by negative charge carriers moving to the left (each carrying $-q$ of charge). Hence, the size of the upward force on them is Bqv (E2.16) where v is the drift velocity of the charge carriers (E7.3) carrying the current.

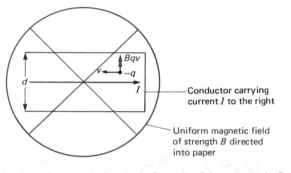

Conductor carrying current I to the right

Uniform magnetic field of strength B directed into paper

Fig. 2.12 Force on an electron carrying a current in a magnetic field

Hence the charge carriers will move under the influence of the magnetic field towards the top of the conductor, creating a negative region at the top of the conductor (surplus negative charge carriers) and a positive region at the bottom (deficiency of negative charge carriers). This means that an electric field is created that tends to attract the charges back to the bottom of the conductor as in Fig. 2.13. With the top of the conductor negative and the bottom positive a pd V_H (called the **Hall voltage**) exists between the top and bottom of the slice. A uniform electric field of strength V_H/d (E2.9) will be created that will result in a downward force Eq on each of the negative charge carriers as shown in Fig. 2.13. As more and more charge carriers move to the top of the conductor, so the electric field (and V_H) increase. Eventually equilibrium is established when the two forces on the charge carriers become equal and cancel each other out; this is when a steady Hall voltage will appear between the top and bottom of the slice. So eventually $Bqv = Eq = V_Hq/d$ or rearranging:

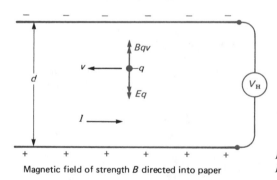

Magnetic field of strength B directed into paper

Fig. 2.13 The Hall effect

E2.21 $\quad V_H = Bvd$

where V_H = Hall voltage (V), B = strength of the magnetic field (T), v = drift velocity of charge carriers (m s^{-1}), d = thickness of the conductor (m).

Using the drift velocity formula (E7.3) the formula can be rewritten:

E2.22 $\quad V_{\mathrm{H}} = \dfrac{BId}{nAq}$

where I = current passing through material (A), n = number of charge carriers per cubic metre (m^{-3}), A = cross-sectional area of material (m^2), q = charge on each charge carrier (C).

 The Hall effect can be observed in metals, but semiconductors provide much larger Hall voltages owing to the higher drift velocities that result from the smaller values of n in these materials (see E2.21/22). The theory is correct for an n-type semiconductor in which most charge carriers are electrons. However, to convert to positive charge carriers (e.g. holes – as in p–type semiconductors) requires little modification; with the current in the same direction it is positive charge carriers that move upwards and so the only modification needed is to the direction of the electric field (and hence the direction of the Hall voltage) which will be reversed in this type of material. Several uses have been found for the Hall effect; they include determining the sign (positive or negative) of charge carriers and (most importantly) the measurement of magnetic field strengths.

Voltage generator

Consider a metal rod of length d being moved to the left at a steady speed v in a uniform

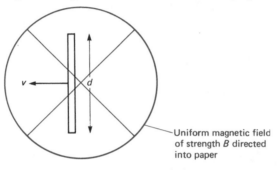

Fig. 2.14 Conductor moving in a uniform magnetic field

Uniform magnetic field of strength B directed into paper

magnetic field of strength B directed into the paper as shown in Fig. 2.14. The only charge carriers free to move anywhere in the metal are negative electrons which are being carried to the left at a steady speed v. But this is exactly the same situation that has been discussed in the Hall effect theory (where the movement was provided by an electric current); so we know that a steady pd of V volts will be set up across the rod with the top of the rod negative and the bottom positive such that:

E2.23 $\quad V = Bvd$

where V = induced voltage across rod (V), B = strength of uniform magnetic field (T), d = length of rod (m), v = speed of rod at right angles to the field ($m\ s^{-1}$).

 Remember that the direction of the voltage can be influenced by the direction of either the magnetic field or the movement, and also by whether the charge carriers are positive or negative.

2.12 PRINCIPLES OF ELECTROMAGNETIC INDUCTION

The voltage generator (2.11) illustrates the basic theme of electromagnetic induction: the creation of an emf (voltage) using a magnetic field, charges free to move *and* relative movement between the field and the charges. In 2.11 the relative movement is created by moving the charges; there is no reason why the magnetic field should not be moved instead (indeed this happens in many electromagnetic devices when the field is often 'moved' by changing its strength, e.g. using ac to generate it). When emfs are induced by these methods, the process is referred to as **electromagnetic induction**. A study of the Hall effect theory (2.11) shows the role played by both electric and magnetic fields in creating the pd; hence the suitability of the term electromagnetic.

It is helpful when studying electromagnetism to introduce the concept of magnetic flux (φ) measured in **webers** (Wb or T m^2). **Magnetic flux** is measured by multiplying an area by the strength of the magnetic field passing at right angles through that area. This gives:

E2.24 $\varphi = BA \sin\theta$

where φ = amount of magnetic flux (Wb), B = strength of the uniform magnetic field (T), A = area through which the field is passing (m^2), θ = angle between the plane of the area and the direction of the magnetic field.

When measuring flux it is important to remember the $\sin\theta$ term, which allows for the fact that the area and field must be at right angles to one another. Note that B (T) could also have its units written as webers per square metre (Wb m^{-2}). To see how using magnetic flux can be useful it is valuable to reconsider the term vd in E2.23, which can be represented as in Fig. 2.15. The shaded area is vd which is the area swept out by the moving conductor per second (as it moves a distance v in one second and has a length d). This means that the flux that the rod moves through in one second (Bvd) is the area that the rod moves through in a second (vd) multiplied by the strength of the field (B) at right angles to that area giving:

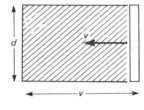

Fig. 2.15 Area swept out per second by a conductor

E2.25 Induced voltage = flux swept out per second

or   Induced voltage = rate of change of flux linkage

or $$V = -\frac{d\varphi}{dt}$$

Care has to be taken when using E2.25 in calculating the flux 'linking' the circuit. In a coil of cross-sectional area A with a field B passing at right angles down the axis of the coil, the flux passing through the coil is BA. However, this flux passes through each turn of the coil, so if there are N turns to the coil then the flux 'linkage' is NBA. It is the change in flux linkage that gives rise to induced emfs.

The size of the **induced voltage** ($d\varphi/dt$) is often referred to as the **Faraday law** (occasionally the **Faraday–Neumann law**), and the direction of the induced voltage (as indicated by the minus sign in E2.25) is a consequence of Lenz's law. **Lenz's law** states that the direction of any induced emf is always such that it opposes the change that created the emf. This result is necessary for the conservation of energy. In the voltage generator/Hall effect (2.11) the direction of the magnetic force that redistributes the charges in the conductor (creating the emf) is upward (Fig. 2.13). However, the emf created generates an electric field that applies a force on the charge carriers in the opposite (downward) direction. Hence, Lenz's law is being obeyed. Thus the direction of the induced emf can be determined either by an application of the left-hand rule (2.10) or by use of Lenz's law.

2.13 CALCULATING FLUX: MAGNETIC/ELECTRIC CIRCUIT ANALOGY

The key to determining induced emfs is to obtain an expression for the magnetic flux (φ) in any situation. The modern approach to calculating flux is to use the magnetic/electric circuit analogy as shown in Table 2.1.

Table 2.1   *Analogous magnetic and electrical quantities*

Magnetic terminology	Analogous electrical quantities
Flux (Wb) φ Magnetic flux is driven around a magnetic 'circuit'; all the flux entering a box must leave that box.	Electrical current (A) I Electric current is 'driven' around an electric circuit; all the current entering a junction in the circuit must leave that junction.
Current-turns (A) NI It is the current-turns of a coil that drive the flux around a magnetic circuit (N = number of turns in coil, I = current in coil); NI is sometimes called the magnetomotive force or mmf.	Potential difference (V) V It is the pd in an electric circuit that drives the electric current around that circuit; V can be referred to as the electromotive force or emf.
Reluctance (H^{-1} or $A V^{-1} s^{-1}$) R_m Increased reluctance causes a reduction in the amount of flux that can be created for a given value of current-turns.	Resistance (Ω) R Increased resistance causes a reduction in the amount of electric current that can be created for a given value of pd.

By definition analogous quantities must obey the same mathematics. So once a good understanding of electrical circuit theory has been acquired, this can then be applied to magnetic circuits. The equations can be adapted simply by replacing the electrical quantities with their analogous magnetic quantities, e.g. $I = V/R$ in an electrical circuit, so in a magnetic circuit:

E2.26 $\varphi = NI/R_m$

where φ = amount of flux created (Wb), N = number of turns in the coil (no units), I = current passing through the coil (A), R_m = reluctance of the circuit (H^{-1}).

A problem still exists in deciding how the reluctance is to be measured. The analogy can be used again as follows: $R = \rho L / A$ in an electrical circuit (E7.8), so in a magnetic circuit:

E2.27 $R_m = \left(\dfrac{1}{\mu_r \mu_0} \right) L / A$

where R_m = reluctance of sample of material (H^{-1}), μ_r = relative permeability of material (no units), μ_0 = permeability of free space ($N\,A^{-2}$), L = total length of sample (m), A = uniform cross-sectional area of sample (m^2).

In electrical circuits $1/\rho$ is known as the **electrical conductivity** of a material: it can be seen from E2.27 to be analogous to $\mu_0\mu_r$ which hence can be regarded as the **'magnetic conductivity'** of a material (i.e. a measure of how easy it is to generate a given flux or field in a material, see 2.11). It is possible to pursue the analogy much further using more basic electrical equations; however, this is not usually done at A level. An important practical example of the use of the analogy may help. All the magnetic flux in Fig. 2.16 will be contained in the iron ring, so the total length (L) of the magnetic circuit is the circumference ($2\pi r$) and

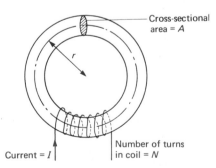

Fig. 2.16 Generating flux in an iron ring

Cross-sectional area = A

Current = I

Number of turns in coil = N

the flux passes through a uniform cross-sectional area (A), so combining E2.26 and E2.27 we get:

E2.28 $\varphi = NI\mu_0\mu_r \dfrac{A}{2\pi r}$

where φ = flux generated in a closed ring (Wb), N = number of turns of wire around ring (no units), I = current passing through wire (A), μ_0 = permeability of free space (N A^{-2}), μ_r = relative permeability of ring material (no units), A = uniform cross-sectional area of ring (m^2), r = average radius of the ring (m).

This formula will hold true for any ring as long as all the flux generated by the coils stays within the ring. We can imagine the equivalent electric circuit as perhaps consisting of a wire (the ring) shorting out two terminals of a battery (the coil and current). Thus the reason the current passes through the wire and not through the surrounding air is because the resistance of the wire is so much lower than that of the air. Hence in the magnetic circuit, for all the flux to remain within the coil, the reluctance of the coil must be far lower than that of the surrounding air. As iron, for example, has a μ_r of about 1000, meaning that it is about 1000 times better at 'conducting' magnetic flux than the air ($\mu_r \approx 1$), it is clear why virtually no flux 'travels outside' an iron ring.

2.14 INDUCTANCE AND TRANSFORMERS

A changing flux within a coil will generate an emf across the ends. But a changing flux ($d\varphi/dt$) is proportional to a changing magnetic field (dB/dt) and a changing magnetic field is created by a changing electric current (dI/dt). So if a steady dc current ($dI/dt = 0$) is passed through a coil of zero resistance there will be no pd across the coil, owing either to $V = IR$ or an emf being induced. However, if an ac current ($dI/dt \neq 0$) is passed through the coil an emf will be induced across the ends of the coil. This voltage (the induced emf) must be proportional to rate of change of current giving:

E2.29 $V_L = -L\dfrac{dI}{dt}$

where V_L = voltage across an inductor (V), L = inductance of the coil in henrys (H or V s A^{-1}), I = current passing through the coil (A), t = time (s).

Because the coil is inducing a voltage in itself, the property L is often referred to as the **self-inductance** of the coil. The minus sign in the formula is an aid to remembering that the induced emf acts in such a way as to oppose the changing current that causes it (Lenz's law, see 2.12). Often dI/dt is measured directly off a graph, or an oscilloscope trace is used to display a plot of current against time.

Proof of the following formula for the energy stored in an inductor can be found in your textbook:

E2.30   Energy stored = $\frac{1}{2}LI^2$

where electrical energy is stored by the magnetic field around the inductor (J), L = inductance of the inductor (H), I = current in inductor (A).

Often a changing current in one coil is used to induce a voltage in another coil. It can be arranged that the same flux can pass through both coils. This can be done by winding the coils directly on top of one another or by using a flux circuit to link the coils as in the **transformer** shown in Fig. 2.17. If a changing current is passed through the primary coil (dI_p/dt) all the flux generated will also link the secondary coil. So, similarly to E2.29, the voltage induced in the secondary is proportional to the changing current in the primary:

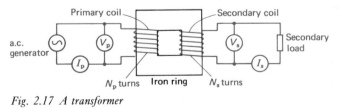

Fig. 2.17 A transformer

E2.31 $V_s = -\dfrac{M dI_p}{dt}$

where V_s = voltage induced in secondary coil (V), M = mutual inductance (H), I_p = current in the primary coil (A), t = time (s).

Remember that this formula does not only apply to transformers but to any pair of coils that share the same flux and can mutually induce voltages across one another; hence the term **'mutual inductance'**.

Transformers are easily the most efficient machines ever built (98% or better), so it is reasonable to say that the power input is equal to the power output giving:

E2.32 $\quad V_p I_p = V_s I_s$

where V_p = rms voltage across primary (V), I_p = rms current through primary (A), V_s = rms voltage across secondary (V), I_s = rms current through secondary (A).

It is important to realise that all electromagnetic effects involve changing voltages and currents (i.e. ac) and this must be recognised when making measurements. The formula E2.32 would still be valid if all the rms values were replaced by peak values (see 7.7).

The other important transformer formula recognises that with the same flux through each coil the voltage across each coil must be proportional to the number of turns:

E2.33 $\quad V_p / V_s = N_p / N_s$

where V_p = rms voltage across primary (V), V_s = rms voltage across secondary (V), N_p = number of turns in primary (no units), N_s = number of turns in secondary (no units).

Again *both* values could be replaced by their peak values. Always check that the bigger voltage appears across the coil with more turns.

2.15 ELECTROMAGNETIC FIELDS

It is unfortunate that permanent magnets were discovered long before the magnetic effect of an electric current, thus 'wrongly' creating two separate subjects of electricity and magnetism. Modern day scientists are left with many awkward legacies arising from the confusion of terminology. It is clear that magnetic fields are created by electric currents which are themselves maintained by electric fields or pds. Hence associated with every changing magnetic field (B) must be a changing electric field (E). Such changing 'joint' fields are called **'electromagnetic fields'**.

Fig. 2.18 shows how the strength of the electric and magnetic fields vary (represented by the lengths of the arrows) with displacement in space. Note that the two fields are at right angles to one another, but their magnitudes are in phase. When an electromagnetic field exists, the fields vary sinusoidally (as shown) and wave energy is found to travel at right angles to both fields at a speed c. This speed (the speed of electromagnetic waves/light) is a constant dependent on the medium and the frequency. There is an equation that not surprisingly links the speed to the electric and magnetic field constants for the medium of propagation:

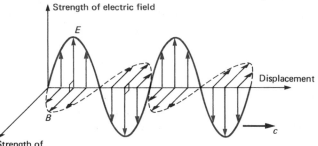

Fig. 2.18 Electric and magnetic fields of a plane polarised electromagnetic wave

E2.34 $\quad c = \dfrac{1}{\sqrt{\varepsilon_0 \mu_0}}$

where c = speed of light in a vacuum (m s^{-1}), ε_0 = permittivity of free space (C V^{-1} m^{-1}), μ_0 = permeability of free space (N A^{-2}).

Note that 'vacuum' and 'free space' mean the same. All electromagnetic waves travel at the same speed in a vacuum, regardless of frequency.

Chapter roundup

You need to use the general concepts of field strength and potential in Chapter 5 (structure of solids), Chapter 7 (electricity) and Chapter 8 (electrons in the atom and radioactivity). The work on magnetic fields including electromagnetic induction is built on in Chapter 7 (alternating currents).

Question bank

1 Suppose a particle *P* is projected in a uniform field, which can be magnetic, electric or gravitational, as shown in the diagram. For the particle to move in the plane of the paper in the *circular* path indicated, the conditions would have to be:

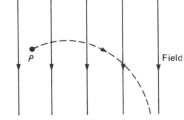

	Particle	**Field**
A	Positively charged	Electric
B	Positively charged	Magnetic
C	Negatively charged	Electric
D	Uncharged	Gravitational
E	There are no conditions which could produce such a motion.	

(**London**: *all other Boards except SEB*)

Points

A charge in a uniform electric field will behave in the same way as a mass in a uniform gravitational field. Is a circular path possible in either of these situations? Also, in the question, the path, field and force are all in the *same plane*, which makes one answer impossible (see note after E2.16).

2 Mars has a diameter of approximately 0.5 that of Earth, and a mass of 0.1 that of Earth. The surface gravitational field strength on Mars compared with that on Earth is greater by a factor of:

A × 0.1 B × 0.2 C × 0.4 D × 2.0 E × 10.0

(–: *all other Boards except SEB*)

Points

Use E2.6.

3 *X* and *Y* are two points at respective distances *R* and 2*R* from the centre of the Earth, where *R* is greater than the radius of the Earth. The gravitational potential at *X* is –800 kJ kg^{-1}. When a 1 kg mass is taken from *X* to *Y*, the work done on the mass is:

A –400 kJ. B –200 kJ. C +200 kJ. D +400 kJ. E +800 kJ.

(**Cambridge**: *all other Boards except SEB*)

Points

Remember that gravitational potential *V* near a planet varies with the inverse of the distance from the planet's centre. Determine *V* at point *Y*, given that *V* = –800 kJ kg^{-1} at point *X*. Then use the fact that the work done per unit mass is equal to the change of potential (i.e. final potential – initial potential).

4 The gravitational pd between a point at the surface of a certain planet and a point 10 m above the surface is 4.0 J kg⁻¹. Assuming the gravitational field is uniform, the work done, in J, in moving a mass of 2.0 kg from the surface to a point 5.0 m above the surface is:
A 0.16   B 1.0   C 2.0   D 4.0   E 16

(**NICCEA**: *all other Boards except SEB*)

Points

See 2.2 if necessary.

5 Suppose that the acceleration of free fall at the surface of a distant planet were found to be equal to that at the surface of the Earth. If the diameter of the planet were twice the diameter of the Earth, then the ratio of the mean density of the planet to that of the Earth would be:
A 4:1   B 2:1   C 1:1   D 1:2   E 1:4

(**London**: *all other Boards except SEB*)

Points

Use E2.6 and the fact that M = density × volume (for a sphere = $4\pi R^3/3$). Equate: g for Earth = g for planet. Then solve to find the ratio of the densities.

6 An isolated hollow pear-shaped conductor is given a charge. Which of the following statements is true?
1   The potential is the same at all points on the conductor.
2   The charge is concentrated at the pointed end of the conductor.
3   No electric field exists inside the conductor.
Answer: A   if 1, 2, 3 correct.
            B   if 1, 2 only.
            C   if 2, 3 only.
            D   if 1 only.
            E   if 3 only.

(**London**: *all other Boards except SEB*)

Points

Details of the distribution of charge on a conductor can be found in most textbooks. Remember that a conductor must be an equipotential surface or else a current would flow until it is.

7 Two point charges X = +2 μC and Y = –3 μC are placed 100 mm apart. The electric potential V due to the two charges (along the line between them) will be zero at a distance, in mm, from X of:
A 30   B 40   C 50   D 60   E 70

(–: *all Boards except SEB*)

Points

Make a simple sketch to start with, then write down the equation for V near a point charge; see E2.3 if necessary. For zero potential due to X and Y, the contribution from X is equal and opposite to that for Y.

8 The equipotentials of a pair of point charges +Q and +Q are shown in the diagram. If an electron moves from X to Y:

A   its electric pe falls.
B   its electric pe is unchanged.
C   it experiences a resultant force pulling it back to X.
D   it experiences a resultant force pushing it away from X.
E   it does not experience a resultant force.

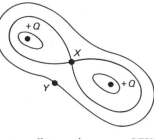

(–: *all Boards except SEB*)

Points

Remember that an equipotential is a line along which a point charge has constant electric pe. Also, the resultant force on the electron is the vector sum (see 1.5 if necessary) of the two individual forces acting on it.

9 X and Y represent two parallel metal plates. X is connected to the cap of a gold-leaf electroscope and Y is connected to earth. The plates are charged and the leaf rises. When an uncharged slab of perspex is inserted between the metal plates, the divergence of the leaf:
A   decreases.
B   increases, then decreases.
C   increases.
D   remains unaffected.
E   drops to zero immediately.

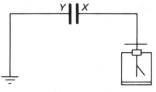

(**SEB**: *all other Boards*)

Points

The electroscope indicates the pd between the plates. The charge on the plates remains unchanged when the perspex is inserted. What happens to the capacitance of the plates when the perspex is inserted? Remember that perspex is a dielectric; see 2.9 if necessary. Knowing what happens to the capacitance, and remembering that the charge is fixed, consider what happens to the pd. Use E2.11 if necessary. Then choose from A–E accordingly.

10 Which of the following statements about the force between the atoms in a diatomic molecule is/are correct?
1   The force obeys the inverse square law for all separations of the atoms.
2   The force changes from an attraction when the separation is large to a repulsion when the separation is small.
3   The force is zero when the potential energy of the system is a minimum.
A None of these.   B 1 only.   C 2 Only.   D 3 only.   E 2 and 3 only.

(**NICCEA**: *all other Boards except SEB and AEB*)

Points

See 2.4 and, in particular, Fig. 2.5(a) and Fig. 2.5(b). Note that the diagram of pe against separation of the molecules provides the answer to 3 above; this should be carefully distinguished from the diagram showing variation of force with separation.

11 Two solenoids P and Q, of equal length but different numbers of turns, are arranged coaxially as shown in the diagram. P has 200 turns and Q has 300 turns.

There is a current of 1 A in Q. What must be the current in P in order that there is no resultant field at X, midway between the coils?

A 2/3 A   B 3/4 A   C 1 A   D 4/3 A   E 3/2A

(**London**: *all other Boards except SEB*)

Points

Refer to E2.18 (although this formula is for the field inside the *centre* of a solenoid, the field at X, along the axis will be proportional to the field at the centre of each of P and Q since the coils are of equal length and X is midway between them). Equate: field due to P = field due to Q, and solve to find the current in P.

12 A long, horizontal, straight wire is placed parallel to the plane of a coil as shown in the diagram. Both the coil and the wire carry currents in the directions indicated.

The force at the point P on the wire due to these currents is:

A   in the direction X.
B   in the direction Y.
C   out of the plane of the paper.
D   into the plane of the paper.

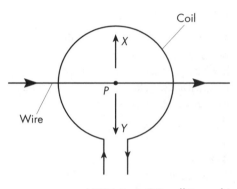

(**AEB** June 90: *all Boards*)

Points

Work out the direction of the magnetic field of the coil at the centre of the coil. Then use the left-hand rule to work out the force direction on the wire. See p.78 if necessary.

13 A simple solenoid X is made by winding a 1 m length of insulated wire around a hollow cardboard tube of diameter 1 cm and length 10 cm so as to produce a uniform spacing of turns in a single layer along the whole length of the tube. A similar solenoid Y is then made by winding a 2 m length of wire around another tube of diameter 2 cm and length 20 cm, again so as to produce a uniform single layer of turns along the whole length. How many times must be the current in Y compared with that in X to produce the same magnetic flux density inside each tube?

A $\frac{1}{2}$ B 1   C 2   D 4   E 8

(–: *all Boards except SEB*)

Points

Remember that the magnetic flux density along the axis of a solenoid is proportional to (i) the number of turns per unit length, (ii) the current.

14 Two long parallel wires X and Y carry currents of 3 A and 5 A, respectively. The force experienced per unit length by X is 5×10^{-5} N to the right as shown in the diagram. The force per unit length experienced by wire Y is:

A 2×10^{-5} N to the left.
B 3×10^{-5} N to the right.
C 3×10^{-5} N to the left.
D 5×10^{-5} N to the right.
E 5×10^{-5} N to the left.

(**Cambridge**: *all other Boards except SEB*)

Points

Remember Newton's third law: see 1.3 if necessary.

15 Two moving-coil galvanometers, 1 and 2, are constructed with exactly similar permanent magnets and springs, but with coils of different numbers of turns (n_1, n_2) effective areas (A_1, A_2) and resistances (R_1, R_2). When they are connected in series in the same circuit, their deflections are, respectively, θ_1 and θ_2. The ratio θ_1/θ_2 equals:

A $A_1 n_2/A_2 n_1$ B $A_1 R_2/A_2 R_1$ C $n_1 R_2/n_2 R_1$ D $A_1 n_1/A_2 n_2$ E $A_2 R_1/A_1 R_2$

(**Cambridge**: and NEAB, AEB and Oxford)

Points

The deflection of the coil is given by $\theta = BAnI/k$ where I = current, k is the 'strength' of the hairsprings and B is the magnetic field strength. You will find a proof of the equation in your textbook. Since the magnets and springs are exactly similar, B and k are the same for the two galvanometers. Also, because they are connected in series, I is the same. Use the above expression to derive an expression for θ_1/θ_2 in terms of the factors of the expression that are not the same.

16 Coil P is connected to a 50 Hz ac source and lies close to a separate coil Q which is connected to the Y-input terminals of an oscilloscope. A wave-shaped trace appears on the screen of the oscilloscope.

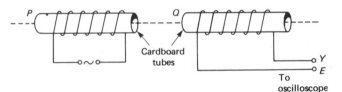

What would be the effect upon the oscilloscope trace of linking the coils by a common soft-iron core

	Height of trace	Number of waves on screen
A	increases	increases
B	decreases	increases
C	stays the same	increases
D	increases	stays the same
E	stays the same	stays the same

(**SEB**: all other Boards)

Points

The introduction of the iron core will have two effects: (i) it will increase the magnetic flux in the primary coil P (see 2.14); (ii) it will increase the flux linkage between coils P and Q, thus increasing the amplitude of the induced emf in coil Q without increasing frequency.

17 Which one of the following effects observed in the study of electromagnetism is a consequence of self-induction?
A   A large voltage may occur when the current in an iron-cored coil is interrupted.
B   A magnet plunged into, or removed from, a solenoid in a circuit encounters forces opposing its motion.
C   Galvanometer coils are often wound on light metal frames.
D   The introduction of an iron core increases the magnetic field of a solenoid.
E   The magnetic flux density well inside a long solenoid is independent of the cross-sectional area of the solenoid.

(**London**: and all other Boards except SEB)

Points

Observe that the question refers to *self*-induction, so consider E2.29.

18 *X* and *Y* are two parallel wires in the plane of the paper.

X is fixed and Y is free to move. When the same current passes through each wire in the same direction, Y will:
A   move towards X in the plane of the paper.
B   move away from X in the plane of the paper.
C   move upwards out of the paper.
D   move downwards into the paper.

(**AEB** June 92: *all Boards*)

Points

See p.79 if necessary.

19 (a) The magnitude of the force on an electron when it is at a point A, 5.0 mm from a point charge of 2.0×10^{-6} C, is 1.2×10^{-10} N. What is the magnitude of the force on an electron when it is at point B, a distance of 15 mm from the same point charge?

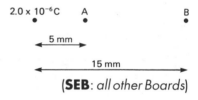

(b) Calculate the electric field strength due to the 2.0×10^{-6} C charge at point A.

(**SEB**: *all other Boards*)

Points

(a) Remember that the force between two point charges is an *inverse square* law, i.e. $F \propto 1/d^2$ which can be written as $F \times d^2$ = constant. Therefore, $F_1 d_1^2 = F_2 d_2^2$ and since F_1, d_1 and d_2 are given, then F_2 can be calculated.
(b) Remember that electric field strength = force per unit (+) charge. Assume that the charge on an electron is 1.6×10^{-19} C.

20 The diagram shows the variation of the Earth's gravitational potential *V* with distance *r* from the centre of the Earth.

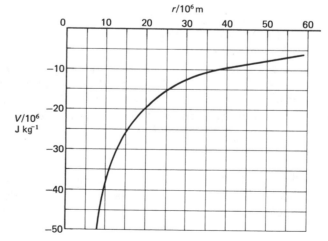

(a) (i) What is the potential at an infinite distance from the Earth? (ii) Why is the potential on the graph negative?

(b) What is the value of the gravitational field at 25×10^6 m from the centre of the Earth? Show how you arrive at your answer.

(c) How much energy is required to raise a mass of 70 kg from 10×10^6 m to 50×10^6 m above the Earth's centre? Show how you arrive at your answer.

(**O and C Nuffield**: *all other Boards except SEB*)

Points

Those in difficulties can refer to 2.2, 2.3 and 2.5. These topics will explain most of this question. Be careful in (b) as the graph shows potential and the 'field' is being asked for; use E2.4 and E2.5 to help solve this. From these formulae it can be seen that it is only necessary to divide the potential by the distance. An alternative method is to use the negative value of the slope of the potential graph at 25×10^6 m from the centre of the Earth (see E2.1). Part (c) is more straightforward; simply read the potential difference between the two points off the graph and multiply by the mass involved to find the energy needed.

21 A closed wire loop in the form of a square of side 4.0 cm is mounted with its plane horizontal. The loop has a resistance of $2.0 \times 10^{-3}\,\Omega$, and negligible self-inductance. The loop is situated in a magnetic field of strength 0.70 T directed vertically downwards. When the field is switched off, it decreases to zero at a uniform rate in 0.80 s. What is (a) the current induced in the loop, (b) the energy dissipated in the loop during the change in the magnetic field? Show on a diagram, justifying your statement, the direction of the induced current.

(**SUJB**: *all Boards except SEB*)

Points

(a) The change of magnetic flux through the loop induces an emf in the loop. The induced emf then drives a current round the loop since the loop is a closed circuit. Calculate the initial flux through the loop; see E2.24 if necessary. Then use E2.25 to calculate the induced emf. Finally, use the value of induced emf and the given value of resistance to calculate the current. Remember that the area of the loop must be in units of m^2.

(b) Since the emf is constant (because the flux changes at a steady rate), the current is constant. To calculate the energy dissipated, calculate the power involved from E7.4, and then calculate the energy dissipated in 0.8 s.

Draw the loop as seen from above, and indicate that the magnetic field lines pass into the plane of the diagram. The decrease of applied flux induces a current in such a direction as to try to maintain the flux through the loop. Use your knowledge of magnetic field patterns to work out the current direction in the loop that would give flux in the same direction as the applied flux.

22 Write equations for Coulomb's law in electrostatics and Newton's law of gravitation. Identify the symbols in the equations.

Explain what is meant by the 'electric potential' of a conductor. Deduce, from first principles, an expression for the potential of an isolated, charged conducting sphere. Electrons 'leak' easily from the surface of many stars so that such stars acquire a positive charge. This charging stops when the charge on the star is so large that protons in the surface also begin to be repelled. This occurs when the sum of the gravitational potential energy and the electrical potential energy of a proton near the surface is zero.

(a) Write down the equation relating these two energies.

(b) Show that, in the steady state, the maximum charge carried by a star of given mass is independent of its radius.

(c) Calculate the maximum charge for the Sun which has a mass of 2.0×10^{30} kg.
(**Cambridge**: *all other Boards except SEB*)

Points

For Coulomb's law of force between two point charges, see E2.2. For Newton's law of gravitation for the force between two masses, see E2.4.

To explain electric potential, see 2.2. E2.3 gives the formula for the electric potential at the surface of an isolated, charged conducting sphere. For the proof of this equation, see your textbook.

(a) The field potential at the surface of an object such as a charged star can be written as k/r; for the gravitational field, $k = -GM$ and for the electric potential $k = Q/4\pi\varepsilon_0$ where M is the mass and Q is the charge of the star. If the mass and charge of a proton are represented by m_p and $+e$, then the pe of the proton due to the gravitational field can be written down, as can the pe of the proton due to the electric field. At the surface, these two expressions are equal in size although of opposite sign (i.e. electric pe = $+X$ joules, gravitational pe = $-X$ joules).

(b) Use your expressions from (a). Make Q the subject of the equation.

(c) Substitute values for G, e, m_p, ε_0 and M.

23 (a) (i) State Newton's Law of gravitation. (ii) Find the dimensions of the gravitational constant G. (iii) Calculate the work done against gravitational attraction when two particles of masses m_1 and m_2, initially at distance r apart, are moved to an infinite separation.

(b) (i) Show that the mass M of the Earth may be expressed in terms of the radius R of the Earth, the gravitational constant G and the acceleration of free fall g, by the formula $M = gR^2/G$. (ii) What approximations, made about the Earth in the foregoing derivation, restrict the validity of the formula? (iii) Calculate the radius of the Earth, assuming the mean density of the Earth's crust, 3.0×10^3 kg m^{-3}, is representative of the Earth as a whole. (iv) The mean radius of the Earth is known to be 6360 km. What does your result suggest about the mean density of the Earth's crust compared with the mean density of the rest of the Earth? Explain your answer.

(c) An astronomer has found the mass of the Earth to be 6.00×10^{24} kg. His observations of the Moon show that its period of rotation round the Earth is 27.3 days. Find the radius of the Moon's orbit, assuming it to be circular and concentric with the Earth.

(Gravitational constant $G = 6.67 \times 10^{-11}$ in SI units; acceleration of free fall = 9.81 m s^{-2}; volume of a sphere = $4/3\pi r^3$.)

(**NICCEA**: *all other Boards except SEB*)

Points

(a) See E2.4. (ii) Make G the subject of the equation of (a)(i), then substitute the dimensions of each term into the rearranged equation. Remember that force can be represented by mass $\times$ acceleration. Thus prove the dimensions of G are M^{-1} L^3 T^{-2}. (iii) A proof of E2.5 is required. Use your textbook.

(b) (i) Use Newton's law of gravitation to show that $g = GM/R^2$. Hence... etc. (ii) What are the assumptions underlying the law of gravitation that do not apply to the Earth? Is g a universal constant? (iii) From mass = volume $\times$ density, obtain an expression for the mass of the Earth in terms of R. Use this in the equation $M = gR^2/G$. (iv) The calculated value of R is too high, so check from your equation in (iii) how the value of density would need to be changed to give a lower value for R.

(c) See 2.7 and E2.8. Since speed $v = 2\pi(R + h)/T$ then combine the two expressions and then solve for $(R + h)$. Take care with units!

24 (a) What characteristics of solid materials suggest that the forces between molecules in such substances have both an attractive and a repulsive component? Explain your answer.

Fig. 1 shows how these components might vary individually with separation for a pair of molecules. Construct on Fig. 1, using the same axes, a graph showing how the *resultant* force between the molecules varies with their separation. Use the graph you have drawn to determine (i) the equilibrium separation of the molecules, and (ii) the separation when the intermolecular attraction is a maximum.

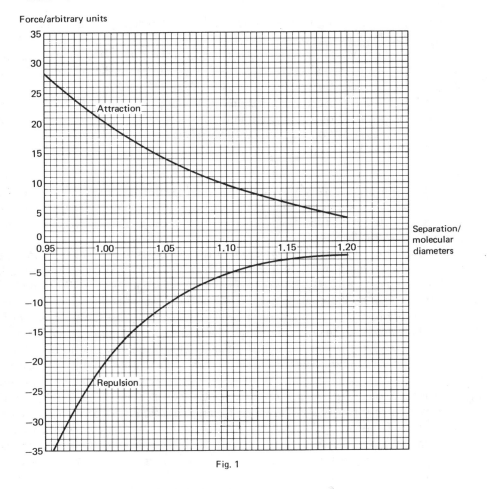

Fig. 1

(b) Fig. 2 shows how the potential energy of this pair of molecules depends on their separation. Account for the shape of the curve by referring to the graph you have drawn on Fig. 1. Use the graph in Fig. 2 to find (i) the energy, in arbitrary units, required to squeeze the molecules together from a separation of 1.0 molecular diameters to a separation of 0.9 molecular diameters, and (ii) the energy, in arbitrary units, required to free the molecules from each other. If the pair of molecules, initially at the equilibrium separation found in (a), are given 10 units of vibrational energy find, using the graph, their new mean separation.

(c) In the gaseous state, the forces between molecules appear to be negligible except at very high pressures. Show that this behaviour could be explained if the forces between the gas molecules were of the form shown in Fig. 1.

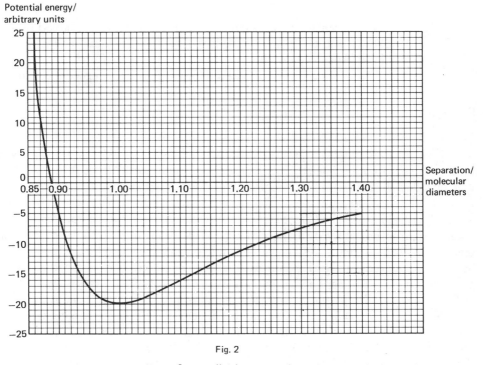

Fig. 2

(**London**: *all other Boards except SEB, AEB and NICCEA*)

Points

(a) Think how solids behave when stretched and compressed. Find the algebraic sum of the two forces at convenient values of molecular separation (e.g. at separation = 0.975, resultant force = (+24) + (−28) = −4) and plot several values. Don't be misled by the fact that the usual convention of making repulsive forces positive has not been adopted.

(b) Refer to Fig. 2.5(a) and Fig. 2.5(b) together with the accompanying explanation. (i) The energy must be increased from its value at 1.00 to the new value at 0.90. (ii) When the molecules are free from each other their pe is defined as zero, so the energy must be increased from the minimum value up to zero. To find the mean separation, take the average of the minimum and maximum displacements at the level which is 10 energy units above minimum pe.

(c) At normal pressures the gas molecules will be far apart (so force = ?) and at very high pressures the molecules will be only a few molecular diameters apart (so force will be ...?).

25 (a) The defining equation for self-inductance is $E = -L dI/dt$. Explain what is meant by 'self-inductance', and define the SI unit of self-inductance.

(b) The apparatus represented in the diagram is to be used to demonstrate self-induction in dc circuits. In circuit 1, the value of the resistor R is 3 Ω, and the lamp (of constant resistance 9 Ω) is correctly lit when the switch is closed. In circuit 2, the resistance of the coil L (of about 2000 turns) is 3 Ω and its self-inductance (when the two parts of the core are together as shown) is 4.0 H, the lamp (of constant resistance 9 Ω) is similar to that of the first circuit.

(i) State and explain what difference would be observed in the behaviour of the lamps when the switches are closed in each of the circuits. (ii) In circuit 2 what is the greatest current and the greatest rate of change of current after the switch is closed? (iii) No difference is observed in the behaviour of the lamps in the two circuits when the switches are opened after having been closed for some time. State any real difference between the behaviour of the circuits, and explain why it is not shown by the

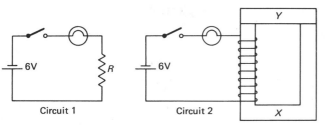

Circuit 1   Circuit 2

lamps. (iv) With the switch in circuit 2 closed, the yoke *Y* is removed and then replaced. State and explain what is observed in the behaviour of the lamp if this is carried out (A) slowly and (B) quickly.

(**Oxford**: *all other Boards except London and SEB*)

Points

(a) See 2.14 and E2.29.

(b) (i) The observed difference in the lamps is required, so you should state if their individual brightnesses grow at the same rate, and if their final brightnesses are the same. Remember that self-inductance opposes the change of current (which is growth here) but does not affect the value of the final steady current. Give your explanation in terms of current, but remember to state that the lamp gets brighter as the current gets larger. (ii) The greatest *current* occurs when the current has grown to its 'final' steady value when $dI/dt = 0$. Then, all the battery emf will be dropped across *R* since the self-induced emf due to *L* will have fallen to 0. So use $I = V/R$. The greatest *rate of change* of current occurs when the self-induced emf is equal and opposite to the battery emf, and there is no pd across *R*. This is the situation when $I = 0$ (i.e. initially). Calculate dI/dt from the defining equation. (iii) In this situation the effect of *L* is to try to maintain the current, so giving a big emf and a spark across the switch gap. (iv) Removal of the yoke causes the flux in the core to drop, and decreasing flux gives an induced emf in the same direction as the battery emf. Remember the *rate* of change determines the size of the induced emf. Answer in terms of lamp brightness.

26 A straight solenoid of length *L* and circular cross-section of radius *r* is uniformly wound with a single layer of *N* turns and carries a current *I*. The solenoid is so long that end-effects may be neglected. Write down expressions for the flux density *B* inside the solenoid and the total flux throughout the solenoid.

(a) The current in the solenoid is increased at a uniform rate from zero to *I* in time *t*. Find an expression for the back-emf induced in the solenoid during the change.

(b) Find the electrical work done against this back-emf in establishing the current.

(c) The work found in (b) is the energy stored in the magnetic field of the solenoid. Show that this energy is $B^2/2\mu_0$ per unit volume.

(d) What becomes of this energy when the circuit is broken?

(**O and C**: *all other Boards except SEB*)

Points

For *B* inside a long solenoid, see E2.18. Magnetic flux is defined in E2.24; remember that $B \times A$ (*A* = area of cross-section) gives the flux per turn, so that total flux is *BAN*. You must give *A* in terms of the radius *r*.

(a) Your expression for the total flux ought to be of the form Φ = constant $\times$ *I* (where Φ = total flux). The back-emf is given by the rate of change of flux which equals constant $\times$ rate of change of current. See E2.25 and E2.29 for further details.

(b) The constant in (a) is the self-inductance of the solenoid. The electrical work done is given by E2.30. You must write the equation in terms of the factors that make up the self-inductance.

(c) Assume the volume of the magnetic field is the same as the volume of the inside of the solenoid (i.e. length $\times$ area of cross-section). You will need to rewrite your expression for (b) in terms of *B* rather than *I*, and then 'divide' that expression by the volume expression.

(d) The sudden collapse of the field of the solenoid induces a large emf across the switch gap, and this emf tries to maintain the current flow. A spark may result.

Answers: 1 E; 2 C; 3 D; 4 D; 5 D; 6 A; 7 B; 8 C; 9 A; 10 E; 11 E; 12 B; 13 C; 14 E; 15 D; 16 D; 17 A; 18 A; 19(a) 1.3×10^{-11} N (b) 7.5×10^8 N C^{-1}; 20(a)(i) 0 (b) 0.61 N kg^{-1} (c) 2.24×10^9 J; 21(a) 0.70 A (b) 7.8×10^{-4} J; 22(c) 155 C; 23(b) (iii) 1.17×10^7 m (c) 3.83×10^8 m; 24(a)(i) 1.00 (ii) 1.10 (b)(i) 15 (ii) 20, 1.07; 25(b) (ii) 0.5 A, 1.5 A s^{-1}

OSCILLATIONS AND WAVES

Units in this chapter

Chapter objectives

After working through the topics appropriate to your syllabus in this chapter, you should be able to:

- describe the differences between longitudinal and transverse waves
- describe progressive and stationary waves in terms of the motion of the particles of the medium
- describe the main properties of waves, including reflection, refraction, interference and diffraction
- explain the formation of stationary waves and relate the wavelength and frequency to the dimensions
- appreciate the nature and properties of electromagnetic waves
- explain the formation of beats
- explain the Doppler effect and solve problems using the appropriate equations

3.1 GENERAL INFORMATION ABOUT OSCILLATIONS

If any quantity changes regularly with time about a fixed value it can be said to be oscillating. At A level most examples feature the change in displacement of an oscillating mass; it should always be remembered that other systems 'oscillate' (e.g. electric and magnetic fields oscillate to create electromagnetic waves, and current can oscillate in electric circuits). It is possible to represent oscillatory motion by plotting graphs of the oscillating quantity against time. Fig. 3.1 shows three different possible modes of oscillation for a body vibrating about a fixed point.

In Fig. 3.1: slope, velocity of oscillator; area, no great physical significance; intercepts, etc., the oscillations pass through zero displacement twice every cycle.

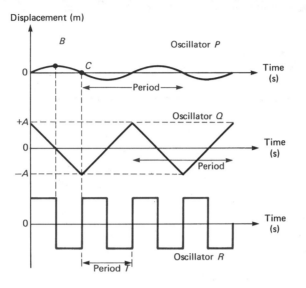

Fig. 3.1
Displacement–time
graphs for three
oscillators

It is important to remember that a vector quantity (displacement) is being plotted on the vertical axis and the purpose of the positive or negative sign is purely to indicate direction; so the 'smallest displacement' occurs at zero displacement, *not* at a negative amplitude ($-A$). Of the three systems shown in Fig. 3.1, oscillator P is by far the most important; any oscillator such as this which has a displacement–time graph whose shape happens to be sinusoidal (sine wave form) is said to be oscillating with shm (see 1.6). Any systems performing shm will also have velocity–time and acceleration–time graphs that are sinusoidal; the three graphs will not be in phase. These three graphs are shown in Fig. 1.10.

The **phase** of a particular point in an oscillation is a measure of what fraction of an oscillation has been completed compared with a chosen 'starting point'. A complete oscillation is normally expressed either as 360° or 2π radians (as in 'once round' a circle). For example, point B on oscillator P is $\frac{1}{4}$ of a cycle *before* point C, so we say in 'phase terms' that B *leads* C by 90° (or $\frac{1}{2}\pi$); alternatively, C *lags* B by 90°.

Two other important quantities can be measured directly from the graphs of Fig. 3.1. The **amplitude** of an oscillation is the largest displacement encountered during the oscillation. The amplitude of oscillator Q is A.

The **period** of an oscillator is the time taken to complete one cycle. The period of oscillator R in Fig. 3.1 is T. The frequency of oscillation is defined by:

E3.1 $f = \dfrac{1}{T}$

where f = frequency of oscillation (Hz or s^{-1}), T = period of oscillation (s).

The **frequency** of an oscillator is the number of cycles completed every second. A comparison of oscillators P and Q in Fig. 3.1 reveals that they have different amplitudes and 'motions', but the same period (and hence the same frequency). As the oscillations are continuous (without any discontinuities, i.e. breaks or jumps, in the 'wave' pattern of the graph), no matter which point on the time axis is used, the same phase relationship between P and Q is always observed. For example, at point B oscillator P is at its positive amplitude whereas, at the same moment in time, oscillator Q is already $\frac{1}{4}$ of a cycle *after* its positive amplitude, so P *lags* Q by 90°. Only phase measurements up to 180° are used, so it is *not* usual to say instead that P leads Q by 270°. Note that there is *not* a constant phase relationship between P and R (or Q and R) as they do not have matching frequencies, e.g. P lags R by $\frac{1}{4}\pi$ at B, P lags R by $\frac{1}{2}\pi$ at C. Note that systems completely out of phase (in antiphase) can be said to lead *or* lag by 180°. There is zero phase difference between systems in phase.

3.2 ENERGY STORED IN OSCILLATORS

Oscillators such as those illustrated in Fig. 3.1 store a fixed amount of total energy in two ways: as ke and as pe. At maximum displacement ($x = \pm A$) there is zero ke (oscillator instantaneously has zero velocity) so the pe stored equals the total energy; at the equilibrium position the pe stored is zero and the ke equals the total energy. For an shm oscillator this can be shown as in Fig. 3.2.

In Fig. 3.2: slope of pe graph, negative value of the force exerted on oscillator (E2.1); slope of ke graph, no physical significance; areas of both, no physical significance; intercepts, etc., ke zero at maximum amplitudes.

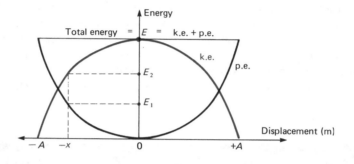

Fig. 3.2 Energy stored in a simple harmonic oscillator

Both ke and pe when added together will always give the total energy stored, e.g. at a displacement of $-x$, pe $= E_1$ and ke $= E_2$ so the total energy $E = E_1 + E_2$ (see 1.6). The total energy stored by a simple harmonic oscillator is given by:

E3.2 $E = \frac{1}{2}m(2\pi f)^2 A^2$

where E = total energy stored by shm oscillator (J), m = mass of oscillator (kg), f = frequency of oscillator (Hz), A = amplitude of oscillations (m).

Note that the pe and ke both depend on the square of the displacement giving the parabolic shape of the curves in Fig. 3.2.

However, oscillators normally lose their stored energy due to damping effects when energy from the oscillating system is passed to the surroundings (e.g. by heating up the air as a result of doing work against air-resistance forces).

Oscillations of constant amplitude as a result of zero damping are called **free oscillations** (as in Fig. 3.1).

Oscillations of reducing amplitude owing to energy from the system being absorbed by the surroundings are called **damped oscillations** (see Fig. 3.3). If a system is so damped that it *just* fails to oscillate it is said to be **critically damped**. More damping than critical damping is referred to as **overdamping** (see Fig. 3.3).

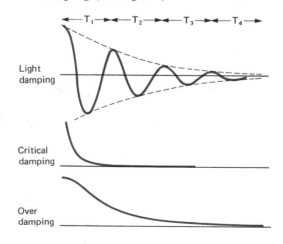

Fig. 3.3 Damped oscillations

It is a common mistake to think that the period changes in a damped system, for until the oscillations have completely died away the period stays the same (e.g. T_1, T_2, T_3 and T_4 in Fig. 3.3 are all the same). In any system of damped oscillations it is possible to draw a locus showing the changing amplitudes (the dotted lines in Fig. 3.3); this locus takes the shape of an exponential decay (or rise).

If a system is made to oscillate by periodic impulses from a **driver** system, then these oscillations are referred to as **forced oscillations**. Forced oscillations will have the same (or almost the same) frequency as the driver system. It is important to realise that forced oscillations can maintain a constant amplitude despite the presence of damping; the energy needed to prevent the amplitude decaying is provided by the driver.

If the **natural frequency** of an oscillator is equal to the frequency of a driver in a forced oscillations system, then **resonance** will occur (see 7.8). The forced oscillations will acquire a large amplitude limited only by the damping losses. The phase of the displacement of the driver will be found to lead that of the oscillator by 90°.

Several useful resonance effects can be encountered in A-level Physics, e.g. tuning circuits for radios, studies of absorption spectra, 3 cm waves produced in cavity resonators, loudspeaker designs that benefit from bass resonant cavities. However, resonance can have negative effects, e.g. causing distortion in hi-fi, unpleasant vibrations in cars at certain engine speeds and even collapsing bridges and buildings purely from the effects of winds. Careful damping of oscillations can be used to control room acoustics, car suspensions, servomechanisms for robots, etc.

3.3 FREQUENCIES OF DIFFERENT SIMPLE HARMONIC OSCILLATORS

An shm can be fully described from two items of information: the amplitude A and the frequency f. How to analyse the motion is explained in 1.6; the following equations define the frequency of different simple harmonic oscillations:

E3.3 $$f = \frac{1}{2\pi}\sqrt{\frac{g}{l}}$$

where f = frequency of oscillation of simple pendulum (Hz), g = acceleration due to gravity (m s^{-2}), l = length of the pendulum (m).

E3.4 $$f = \frac{1}{2\pi}\sqrt{\frac{k}{m}}$$

where f = frequency of oscillation of mass on spring (Hz), k = spring constant of spring (N m^{-1}), m = mass attached to spring (kg).

E3.5 $$f = \frac{1}{2\pi}\sqrt{\frac{1}{LC}}$$

where f = frequency of oscillation of electric current (Hz), L = inductance in circuit (H), C = capacitance in circuit (F).

Proof of these equations may be found in textbooks. It is possible to derive E3.5 from E3.4 using an electrical/mechanical analogy.

3.4 GENERAL INFORMATION ABOUT WAVES

Imagine a line of particles that can oscillate up and down as shown in Fig. 3.4. In Fig. 3.4(a) just the left-hand oscillator (*A*) has energy, the two amplitudes of its oscillation being shown

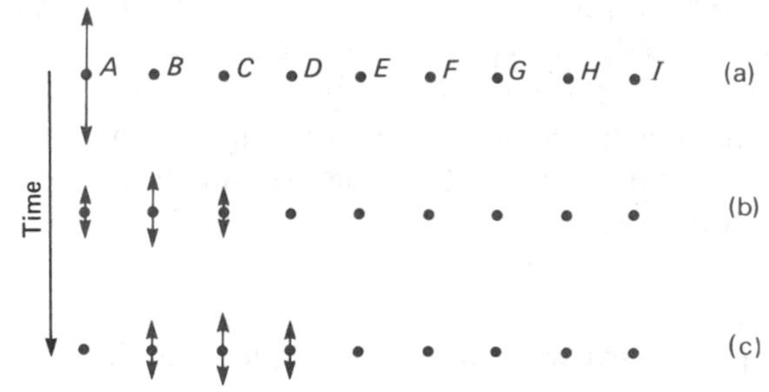

Fig. 3.4 A transverse progressive wave pulse

by the arrows. A little time later *A* has passed some of its energy onto *B* and a little has travelled onto *C* (Fig. 3.4(b)). Later still, *A* has stopped oscillating completely (Fig. 3.4(c)) and the energy has been passed further to the right down the line of oscillators. We have just described what is referred to in Physics as a **transverse wave pulse**. It is possible to repeat this description when talking about oscillations to the left and right (i.e. oscillations in the same direction as the energy travels instead of at right angles to it – or transversely – as before). In this new case we have a **longitudinal wave pulse**. If instead of feeding a limited amount of energy along the line of oscillators we continually feed in energy at the left-hand end, energy will continually emerge at the right-hand end; now we say that a **progressive wave** is travelling from left to right. The progressive wave can be transverse or longitudinal. If the motion of the oscillators is at all damped, the medium will absorb some of the oscillator energy, and hence the wave energy will be gradually absorbed as the progressive waves travel through the medium. Sound wave energy being absorbed as it travels through the air is an obvious example; the oscillator energy (wave energy) will heat the gas.

Fig. 3.5 shows a system in which the amplitude of each oscillator stays the same. Hence no energy is being passed down the line and this situation is referred to as a **standing wave** or a **stationary wave**. The example shown is transverse, but again it could be longitudinal. Points on the standing wave where the amplitude of the oscillations is zero are called **nodes** (no disturbance); points on the standing wave where the oscillations have the greatest amplitude are called **antinodes**. If the oscillator's motion is at all damped (the medium is absorbing energy) then it will be necessary to feed in energy continually to maintain the amplitude of the standing wave (see resonance in 3.2).

Fig. 3.4 and Fig. 3.5 do not show the actual vertical positions of the oscillators; the oscillating particle may be found anywhere along the arrowed line. In nearly all important wave motions the oscillators perform shm and a photograph of the positions of the oscillators carrying a progressive or standing transverse wave might look like Fig. 3.6.

Fig. 3.6 defines the wavelength and amplitude of the wave. Of course if Fig. 3.6 is representing a standing wave then *A*, *E* and *I* will never oscillate, whereas for a progressive wave all particles oscillate. It is interesting to note that on a standing wave *B*, *C* and *D* are in phase with one another, but exactly out of phase with *F*, *G* and *H*. On a progressive wave there will be the same phase difference between each successive oscillator; in Fig. 3.6 this phase difference is 45°. The sinusoidal shape of the wave is important but must not be

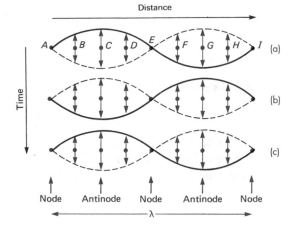

Fig. 3.5 A transverse standing wave

confused with Fig. 3.1 where the horizontal axis represents time and not distance as in Fig. 3.6. It is vital to note that this introduction to waves has talked about particles oscillating in a **medium**; but any oscillating quantity can produce waves and it may not need a medium to propagate (carry) them. An obvious example is the electromagnetic wave which can travel through free space and is carried by oscillating electric (and magnetic) fields.

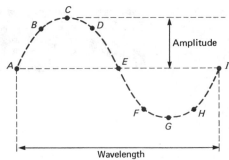

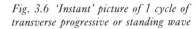

Fig. 3.6 'Instant' picture of 1 cycle of transverse progressive or standing wave

Standing waves are created when two waves of equal frequency and amplitude travelling in opposite directions combine with one another when they are exactly out of phase. This situation occurs when a wave is reflected and the point of reflection serves either as a node or antinode depending on the exact situation. The consequence of this is that it is possible to use the same formulae and measurements for progressive and standing waves:

E3.6 $v = f\lambda$

where v = speed of progressive wave (m s^{-1}), f = frequency of the wave (Hz), λ = wavelength of the wave (m).

When using this formula with standing waves, it is vital to remember that v refers to the speed at which the energy is carried by a progressive wave; so the information refers to the speed of the progressive waves that created the standing wave. A further error can occur when measuring the wavelength of standing waves; looking at Fig. 3.5 it is often wrongly assumed that the wavelength is the distance between two nodes (e.g. from A to E). Fig. 3.6 should make it clear that the wavelength is *twice* the distance between nodes.

The energy carried by oscillators is proportional to the square of their amplitude (E3.2); hence there is a proportionality between the square of the amplitude of a wave and its intensity. The **intensity** at a given point is the energy per second passing through a unit area in a direction along the normal. For a point source of power W emitting waves in all possible directions in a nonabsorbing medium, the intensity at distance r from the source is $W/4\pi r^2$ since the total area which the waves pass through at this distance is $4\pi r^2$ (i.e. the area of a sphere of radius r). Thus, for a point source, the intensity follows the inverse square law:

E3.7 $I = \dfrac{k}{r^2}$

where I = intensity at distance r, k = constant ($= W/4\pi$).

3.5 PROPERTIES OF WAVES

Reflection

The **laws of reflection** that apply to smooth surfaces are:

❶ The **incident ray**, **reflected ray** and **normal** to the reflector at the point of incidence all lie in the same plane.

❷ The **angle of incidence** equals the **angle of reflection**.

Reflection can be explained using a particle model, or using a wave model and Huygens constructions (see 4.3). It can be important to know that waves undergo a 180° phase shift on reflection, e.g. a trough travelling up a slinky spring is reflected as a peak.

Refraction

Refraction means bending; waves are 'bent' (refracted) only if there is a change in the medium through which they travel. This may be a sudden change (e.g. waves travelling from air to

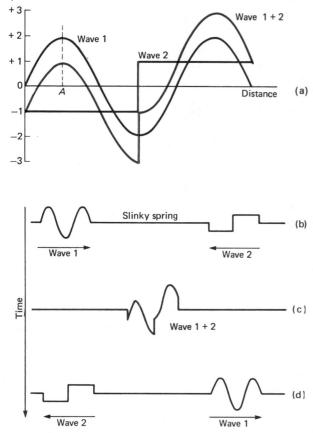

Fig. 3.7 Superposition of waves

glass) or a gradual change (e.g. waves travelling through layers of air at different temperatures). Refraction occurs as a result of the difference in velocity of the wave in different media.

Principle of superposition

The principle of superposition can be applied to any waves that are caused by the same type of oscillation, e.g. to two (or more) sound waves. It cannot be applied to fundamentally different waves, e.g. a water wave and a light wave. The principle is summarised by Fig. 3.7.

The first part of the principle of superposition states that **where two waves of the same type meet in space, the resulting disturbance is the sum of the disturbances due to the individual waves.** This simply means that they can be added together graphically as in Fig. 3.7(a), e.g. at position A, wave 1 has a displacement of +2, wave 2 has a displacement of -1; adding these means that the combined wave has a displacement of +1. Note that in superposition the waves do not need to have a constant phase relationship, or be at the same frequency or amplitude. Thus it would be possible to apply the principle to an intense beam of X-rays mingling with a weak light source.

The second part of the principle is illustrated in Fig. 3.7(b), (c) and (d). These show successive moments in time as wave pulses as in Fig. 3.7(a) pass along a slinky spring, superpose and then continue on their way. Although when superposed a new wave shape is formed, a little time later both pulses emerge unscathed from the encounter (exactly as they were originally before they met). There has been no lasting interaction between the waves.

Although reflection and refraction are properties of both waves and beams of particles, any property that can only be explained by superposition is a 'waves-only' property. Interference is such a property, as is diffraction.

Interference

Interference of waves (longitudinal or transverse) takes place whenever two or more waves meet in space in such a way that there is a constant phase relationship between them (hence they must also have the same frequency, wavelength and speed) and they are of similar amplitudes. Waves that have these properties are said to be **coherent**. Interference results in the amplitude of the waves being constant at the point where they meet. If two waves of identical amplitude interfere, it is possible that zero amplitude will result (**destructive interference**); alternatively it is possible to have a maximum amplitude equal to that of the two waves added together (**constructive interference**). When interference takes place, wave energy does not go missing – it is redistributed. Hence the 'energy missing' at points of destructive interference is found as 'extra energy' at points of constructive interference.

Diffraction

In a uniform medium (no refraction) waves travel in straight lines (**rectilinear propagation**). The arrowed lines (**rays**) in Fig. 3.8 show the direction of travel. The waves in the diagram are **plane waves** (parallel rays), and the waves encounter two obstacles (forming a gap). If there was no diffraction the situation of Fig. 3.8(a) would occur; common knowledge about waves suggests that this situation is impossible. Imagine Fig. 3.8 (a) happening in a ripple tank; along the lines *AB* and *CD* there would be complete calm on one side of the line and large waves on the other side, yet such discontinuities are *not* characteristic of waves. The waves must build up gradually to their amplitude. The principle of superposition and Huygens' principle (see 4.1) show that bending occurs at the edges as in Fig. 3.8(b), with the waves building up from zero amplitude at *F* (or *H*) to their full amplitude at *G* (or *J*). If the points *G* and *J* can be made to meet, then plane waves (parallel rays) have been converted into **circular waves** (rays spreading from a point): this effect occurs when the gap width (d) becomes as small as the wavelength (λ). Hence **diffraction effects** become particularly noticeable when the size of gaps (or obstacles) becomes so small that they are comparable to the wavelength of the waves. Full diffraction theory shows that there must also be places of constructive and destructive interference in the regions of diffraction (i.e. on the right-hand waves this happens between *F* and *G*, and also between *H* and *J*) – see your textbook and 4.2.

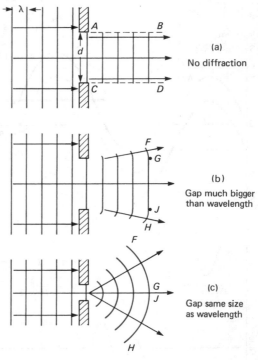

(a)
No diffraction

(b)
Gap much bigger than wavelength

(c)
Gap same size as wavelength

Fig. 3.8 Diffraction by a 'gap'

Polarisation

Waves and the media carrying them are often invisible; this means that the only way a transverse wave can be successfully distinguished from a longitudinal wave is by a demonstration of polarisation. Polarisation is a property of *transverse waves* only. If Fig. 3.4 were viewed from the right-hand side of the page, then you would be looking down the line of oscillators with the wave travelling towards you: the oscillators would be seen vibrating along the line defined in Fig. 3.9 (a). However, the requirement for a transverse wave to travel is that the oscillations should be at right angles to the direction of wave travel. This still happens in Fig. 3.9(b), which would mean in Fig. 3.4 that the oscillators would have to move 'in and out' of the paper rather than 'up and down' as they are shown. If the oscillations only take place in one plane as in Fig. 3.9(a), (b) or (d), then the waves are plane polarised. Other special polarisation effects can be achieved according to specified modes of oscillation. Oscillations may occur in two directions simultaneously (Fig. 3.9(c)), and in unpolarised waves the oscillation

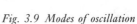

(a)   (b)   (c)   (d)

Fig. 3.9 Modes of oscillation

is random. Surface water waves, most radio waves and 3 cm microwaves from a Klystron are all plane polarised. Reflection can polarise by cutting out much of one mode of oscillation, a fact exploited as the 'glare reducing' property of Polaroid sunglasses. Longitudinal waves are always produced by oscillations along the line of propagation and hence cannot exhibit polarisation effects.

3.6 SOUND

Sound waves are **longitudinal** pressure waves travelling through a medium (i.e. they cannot travel through a vacuum). In air, macroscopically, this can be considered to generate regions of higher pressure (called **compressions**) and regions of lower pressure (called **rarefactions**). In air, microscopically, the air molecules perform shm along the line of travel of the wave.

Fig. 3.10 is an 'instantaneous' picture of a longitudinal wave produced in the same way as Fig. 3.6 was for a transverse wave. Longitudinal waves have all the same properties as transverse waves (reflection, refraction, interference, diffraction, superposition, standing waves, etc.) with the one notable exception of polarisation.

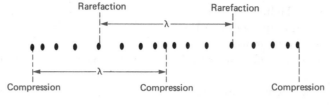

Fig. 3.10 'Instant' picutre of a longitudinal progressive or standing wave

Beats

The effects of beats can occur with *all* types of waves, although it is most often encountered with sound. It occurs when two wave forms of slightly different frequencies (f_P, f_Q) but similar amplitudes meet at a point in space.

Fig. 3.11 has been drawn for transverse waves (the diagram is easier to draw and understand), but the theory is the same for longitudinal waves. It is important to realise that the diagram shows two wave forms, P and Q, meeting at a point, that the horizontal axis represents time and that the waves are added together using the principle of superposition to produce the resultant wave form $P + Q$. At points A and C waves P and Q are in phase, so the resultant wave form has maximum amplitude; at points B and D waves P and Q are exactly out of phase (phase difference of π or 180°), so the resultant wave form has zero amplitude. Thus at the point in space where the wave forms meet, at one instant a 'loud' signal will be heard (at time A), gradually reducing to no signal (at time B), then increasing to another loud signal (at time C), and so on. This 'loud' and 'soft' variation is the effect called **beats**; the period of the beats is T (see Fig. 3.11) and the beat frequency ($1/T$) is given by:

E3.8 $f = f_P - f_Q$

where f = frequency of the beats (Hz), f_P = frequency of one of the wave forms producing the beats (Hz), f_Q = frequency of the other wave form producing the beats (Hz).

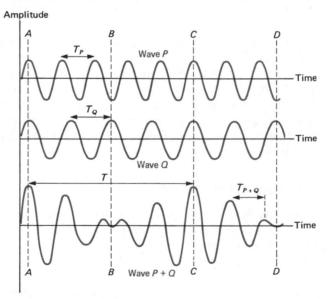

Fig. 3.11 Combining two waves of different frequencies to produce beats

It is important to realise that the beat frequency (defined by T) is not the frequency of wave $P + Q$ (defined by T_{P+Q}) (see Fig. 3.11); the beat frequency is the frequency of the amplitude variation of the wave $P + Q$. Using sound waves, beats become clearly audible when frequencies are nearly matched. This leads to a useful application in the tuning of musical instruments. The 'beating' effect between two electromagnetic waves of slightly different frequency is used in police radar traps (see Doppler effect) and modern (superheterodyne) radio receivers.

Doppler effect

An observer who stands still will receive waves (speed v) from a stationary transmitter at the same frequency at which they were emitted (f). However, motion of either the observer or the transmitter along the line between them will cause the observed frequency (f') to be different from the basic frequency of the transmitter (f). Similar effects can also be observed if the medium is caused to 'move'.

For **sound** suppose the transmitter and observer are moving in the *same* direction at speeds u_s and u_0, respectively.

(a) In 1 s, the transmitter emits f waves which occupy a distance $(v - u_s)$. Hence the wavelength is given by the equation:

E3.9 $\lambda = \dfrac{(v - u_s)}{f}$ where v = wave speed and f = transmitter frequency

(b) The observer receives waves of this wavelength. In 1 s, a stationary observer would receive v/λ waves; since the observer here is moving away from the transmitter, the number of waves received each second is reduced by u_0/λ. Hence the observed frequency is given by:

E3.10 $f' = \dfrac{(v - u_0)}{\lambda} = \dfrac{(v - u_0)}{(v - u_s)} \times f$

Notes:
❶ If either velocity is reversed, change the appropriate sign in the above formula.
❷ If the direction of either velocity is not along the line between the transmitter and the observer, the velocity component along the line should be used.

If **electromagnetic waves** are involved, then E3.9 and E3.10 may not apply. It is simplest to deal with the motion of the transmitter *relative* to the observer; then, since the speed of light is the same for any observer, the observed frequency is given by:

E3.11 $f' = f\dfrac{c}{(c - u_r)}$

where f' = observed frequency (Hz), f = transmitted frequency (Hz), c = speed of light (m s^{-1}), u_r = speed of transmitter relative to the observer (m s^{-1}).

The Doppler effect in sound is well known for altering the pitch (frequency) of train whistles, car horns, etc. In a police radar speed trap the radar beam reflected from the moving vehicle has a frequency shift dependent upon the vehicle velocity. The frequency shift is detected by mixing the 'transmitted' and 'reflected' beams and measuring the 'beat frequency' that results. The Doppler shift of light frequencies (**red shift/blue shift**) has applications for astronomers involved in measuring the motion of stars and planets. Problems arise for spectroscopists from the way in which the Doppler effect (due to the motion of atoms) causes the lines of a line spectrum to be broadened.

Standing waves on strings and in pipes

A stretched string may be made to vibrate transversely or longitudinally but most applications involve transverse waves on the string, which can then set up longitudinal sound waves in the surrounding medium. The ends of the string must be fixed, and reflections at these points (nodes) create standing waves. In a pipe, longitudinal sound waves can be reflected at open (antinode) or closed (node) ends to create standing waves. The lowest frequency that will generate a standing wave (f) is known as the **fundamental** and other modes of vibration that create standing waves at higher frequencies are called **overtones**.

❶ For a stretched string of length L and mass per unit length μ at tension T, the

fundamental frequency f = wave speed/wavelength = $\dfrac{1}{2L}\sqrt{\dfrac{T}{\mu}}$ since the wave speed =

$\sqrt{T/\mu}$ and the fundamental wavelength is $2L$.

❷ For an air column of length L closed at one end, the fundamental wavelength λ is given by $\lambda/4 = L + e$, where e, the end correction, is the distance from the open end of the tube to the antinode (i.e. the length of the standing wave does *not* equal the length of the tube).

❸ For an air column of length L open at both ends, the fundamental wavelength λ is given by $\lambda/2 = L + 2e$, since the tube is open at both ends.

In both 2 and 3 above, the fundamental frequency $f = c/\lambda$, where c is the speed of sound in the tube.

It is important to realise that the standing waves in the pipe in Fig. 3.13 are longitudinal, but diagrams invariably use a transverse standing wave graph to show how the amplitude of the longitudinal wave varies. The transverse standing wave lines drawn in Fig. 3.12 and Fig. 3.13 are the solid and dotted black lines in Fig. 3.5 and they show the maximum displacement (or amplitude) of the standing wave. Although the wavelengths have all been marked as λ, the size of these wavelengths gets smaller as higher overtones are excited and the frequency rises (the speed of the waves staying the same). A study of pipes and stretched strings is obviously of great importance in a study of the behaviour of musical instruments.

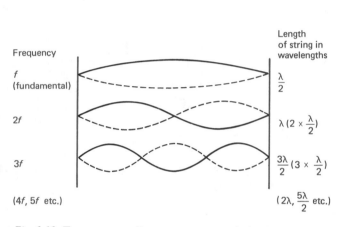

Fig. 3.12 *Transverse standing waves on a stretched string*

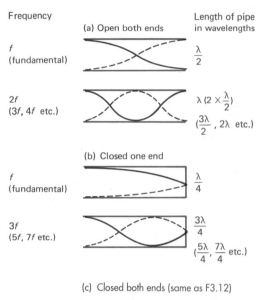

Fig. 3.13 *Standing waves in pipes*

3.7 ELECTROMAGNETIC WAVES

The transverse oscillations that carry these waves are of electric fields and magnetic fields (see 2.15) which need no medium to support them; hence electromagnetic waves can travel through a vacuum. The waves are transverse and always travel at the same speed (c – the speed of light) in a specified medium (regardless of wavelength), though of course if this specified medium is changed then the speed changes. Electromagnetic waves are created by the acceleration of charged particles.

Confusion sometimes arises because the categories of radiation can sometimes overlap, e.g. it is possible to have γ-rays and X-rays at 10^{-11} m wavelength. These radiations are, of course, absolutely identical; they are only given different names because they have been produced in different ways. Quantum theory says that the energy of any oscillator is quantised in multiples of hf (1.10); often hf is such a small quantity of energy compared with the total energy of the oscillator that the quantum effect can be ignored. However, individual quanta of electromagnetic radiation (**photons**) can carry so much energy (because of their high frequencies) that individual photons can become important (particle model of light). See Fig. 3.14.

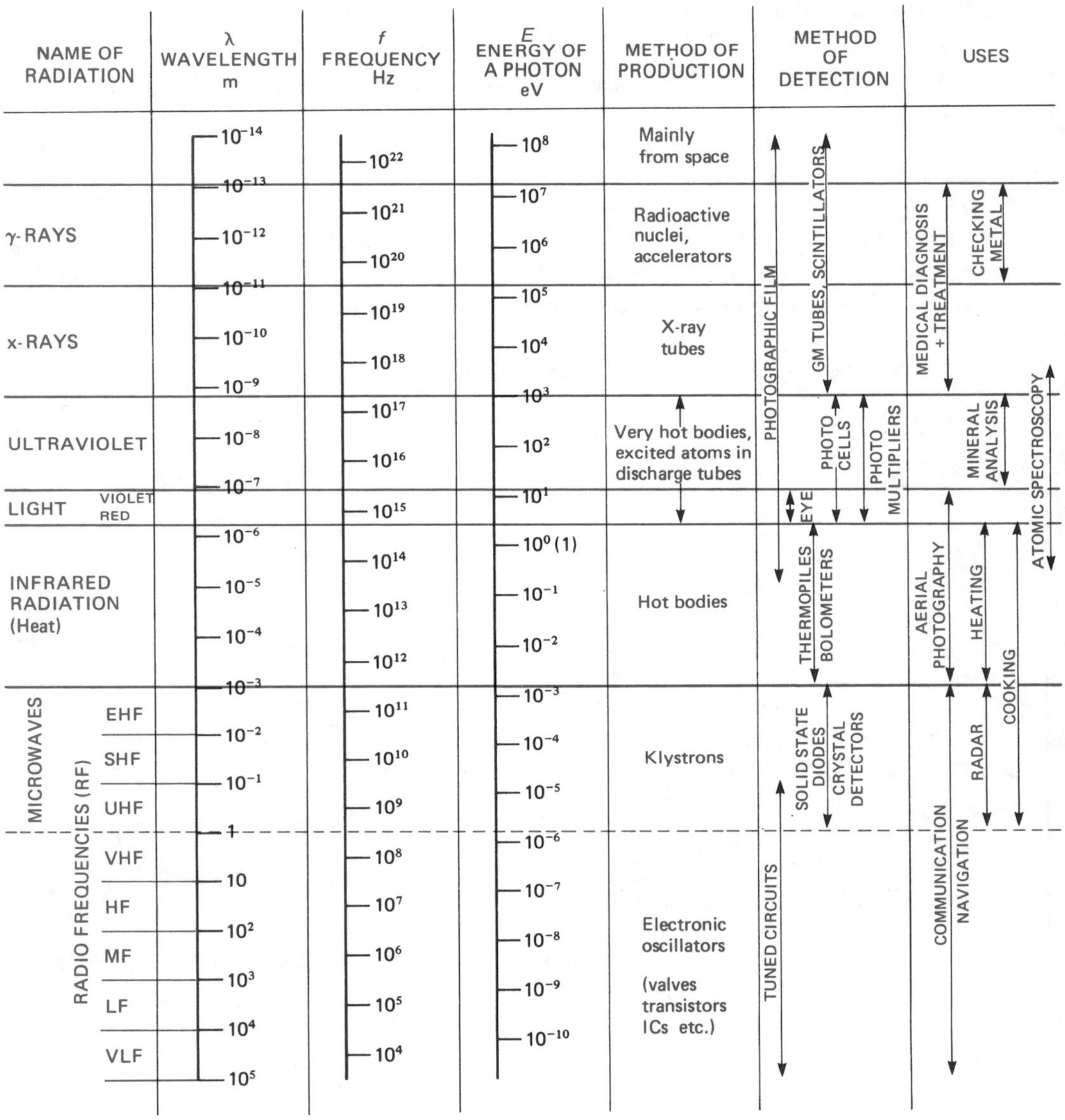

NAME OF RADIATION	λ WAVELENGTH m	f FREQUENCY Hz	E ENERGY OF A PHOTON eV	METHOD OF PRODUCTION	METHOD OF DETECTION	USES
	10^{-14} 10^{-13}	10^{22}	10^{8} 10^{7}	Mainly from space		
γ-RAYS	10^{-12} 10^{-11}	10^{21} 10^{20}	10^{6} 10^{5}	Radioactive nuclei, accelerators	GM TUBES, SCINTILLATORS	MEDICAL DIAGNOSIS + TREATMENT / CHECKING METAL
x-RAYS	10^{-10} 10^{-9}	10^{19} 10^{18}	10^{4} 10^{3}	X-ray tubes	PHOTOGRAPHIC FILM	
ULTRAVIOLET	10^{-8} 10^{-7}	10^{17} 10^{16}	10^{2}	Very hot bodies, excited atoms in discharge tubes	PHOTO CELLS / PHOTO MULTIPLIERS	MINERAL ANALYSIS / ATOMIC SPECTROSCOPY
LIGHT VIOLET RED		10^{15}	10^{1}		EYE	
INFRARED RADIATION (Heat)	10^{-6} 10^{-5} 10^{-4} 10^{-3}	10^{14} 10^{13} 10^{12}	10^{0} (1) 10^{-1} 10^{-2} 10^{-3}	Hot bodies	THERMOPILES BOLOMETERS	AERIAL PHOTOGRAPHY / HEATING / COOKING
MICROWAVES — EHF SHF UHF	10^{-2} 10^{-1} 1	10^{11} 10^{10} 10^{9}	10^{-4} 10^{-5} 10^{-6}	Klystrons	SOLID STATE DIODES / CRYSTAL DETECTORS	COMMUNICATION / NAVIGATION / RADAR
RADIO FREQUENCIES (RF) — VHF HF MF LF VLF	10 10^{2} 10^{3} 10^{4} 10^{5}	10^{8} 10^{7} 10^{6} 10^{5} 10^{4}	10^{-7} 10^{-8} 10^{-9} 10^{-10}	Electronic oscillators (valves transistors ICs etc.)	TUNED CIRCUITS	

Fig. 3.14 The electromagnetic spectrum

Radio and TV broadcasts

These are carried by radio waves and microwaves. Signals carrying information modulate the carrier waves. A radio or TV channel is a band of radio or microwave frequencies allocated to carry signals from a particular station.

The **bandwidth** of a channel is the width of the band of frequencies allocated to that channel. For example, the bandwidth of a medium wave radio channel is 4 kHz which is sufficient to carry audio signals without significant loss of quality. VHF and TV channels have much wider bandwidths because much more information needs to be carried.

The range of a broadcasting station is determined by the wavelength of the carrier waves:

❶ Low-frequency radio waves follow the Earth's curvature and are therefore used for intercontinental broadcasting.

❷ Medium-frequency radio waves (MW) reflect from the ionosphere of the atmosphere in suitable weather conditions. This is why MW programmes can have a long range.

❸ High-frequency (HF), VHF and UHF radio waves carry radio and TV channels and have 'line of sight' range only since they pass straight through the atmosphere.

❹ Microwaves are used for satellite broadcasting. A communications satellite must be in a **geostationary** orbit. Receiver dishes on the ground need to point to the satellite. The **beamwidth** of a dish or aerial is the angle from the direction of maximum signal strength to the direction giving 50% signal strength.

3.8 WAVE SPEED FORMULAE

General formula   E3.6 $v = f\lambda$

where v = speed of wave (m s^{-1}), f = frequency of wave (Hz), λ = wavelength of wave (m).

Water waves   E3.12 $v = \sqrt{gh}$

where v = speed of 'ripple tank type' waves whose $\lambda \gg h$ yet amplitude $\ll h$ (m s^{-1}), g = acceleration due to gravity (m s^{-2}), h = depth of shallow water (m).

Mechanical waves   E3.13 $v = \sqrt{\dfrac{T}{\mu}}$

where v = speed of waves (longitudinal or transverse) along a stretched string or spring (m s^{-1}), T = force used (tension) to stretch the string (N), μ = mass per unit length of string or spring (kg m^{-1}).

E3.14 $v = x\sqrt{\dfrac{k}{m}}$

where v = velocity of wave along a line of masses joined together by springs (m s^{-1}), x = distance between centres of the masses (m), k = spring constant of the springs (N m^{-1}), m = mass of each mass (kg).

Sound waves   E3.15 $v = \sqrt{\dfrac{E}{\rho}}$

where v = velocity of sound in a solid (m s^{-1}), E = Young's modulus for the solid (Pa or N m^{-2}), ρ = density of the solid (kg m^{-3}).

E3.16 $v = \sqrt{\dfrac{\gamma P}{\rho}}$

where v = velocity of sound in a gas (m s^{-1}), γ = ratio of molar heat capacities of the gas (no units), P = pressure of the gas (Pa or N m^{-2}), ρ = density of the gas (kg m^{-3}).

Electromagnetic waves   E2.34 $c = \dfrac{1}{\sqrt{\varepsilon_0 \mu_0}}$

where c = velocity of light in free space (m s^{-1}), ε_0 = permittivity of free space (C V^{-1} m^{-1}), μ_0 = permeability of free space (N A^{-2}).

E3.17 $c_m = \dfrac{c}{n_m}$

where c_m = velocity of light in medium 'm' (m s^{-1}), c = velocity of light in a vacuum or air (m s^{-1}), n_m = refractive index for light passing from air or vacuum into medium 'm' (no units).

In the two electromagnetic wave speed equations the word 'light' can be replaced by 'electromagnetic waves' as, of course, all electromagnetic waves travel at the speed of light.

Chapter roundup

This chapter provides the foundations for concepts in physical optics in Chapter 4 and alternating currents in Chapter 7, and for understanding the interaction of electromagnetic radiation and matter in Chapter 8.

Question bank

1   A wave travelling in the positive direction of the x-axis produces displacements of the particles in its path which at one instant are as shown.

   The wave is reflected from a boundary P which does not permit particle movement. Which graph below best represents the displacement of the particles at some instant during the passage of the reflected wave?

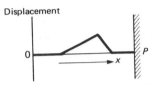

(London: *all other Boards*)

Points

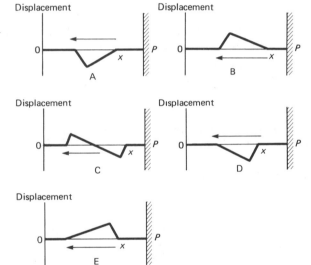

The pulse will undergo a 180° phase change on reflection (see 3.5), which eliminates two answers. It will keep the same shape, so eliminating another answer. Which of the two remaining has the same shape *in the direction of travel* as the original?

2   Physical properties shared by both radio waves and sound waves are:
   1 both can be polarised.
   2 both can be reflected.
   3 both can be diffracted.

   A if 1, 2, 3 all correct.   B if 1, 2 only correct.
   C if 2, 3 only correct.   D if 1 only is correct.
   E if 3 only is correct.

(–: *all Boards except SEB*)

Points

Which of the above properties are common to all types of waves? Remember that only transverse wave forms can be polarised, so are the two wave forms both transverse? See 3.5 if necessary.

3 Here is a list of things involving oscillations:
1 visible light.
2 the 'hum' of a mains transformer.
3 a pendulum of length 10 m.
4 '1500 m wavelength' radio waves.
5 the ebb and flow of the tide.
Assume that the speed of light is 3×10^8 m s^{-1}. Which sequence
A–E correctly places them in order of increasing frequency?
  A 32154.   B 32451.   C 53241.   D 53124.
  E 52143.

(–: *all Boards except SEB*)

Points

Try to pick out the highest and the lowest frequencies, and see if you can make your choice from those two. Assume that the wavelength of visible light is approximately 600 nm.

4 The diagram shows the variation of displacement with time for which *one* of the following objects and situations:

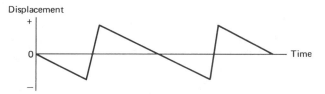

A A rubber ball dropped onto a sponge mat.
B A rubber ball dropped onto a concrete floor.
C The central point of a wire vibrating in its fundamental mode.
D The spot on a CRO screen with a linear time base on the *X*-plates, and no *Y*-input.
E The 'bob' of a simple pendulum.

(–: *all Boards except SEB*)

Points

The graph above is clearly not sinusoidal, so you can eliminate those choices that involve shm and sine waves. Before making your final choice from the remaining alternatives, remember that the graph is of *displacement* against time, not velocity against time.

5 Two sources of waves, S_1 and S_2, a distance a apart, vibrating in phase, produce an interference pattern of nodes and antinodes. This pattern is observed on a line

XY parallel to the line joining the sources a distance d from it. The separation between adjacent nodes increases when:
1 the source separation a is increased.
2 the distance d is increased.

3 the frequency of vibration of the sources is increased.
Which of these statements is/are correct?
A 2 only.   B 1 and 2 only.   C 1 and 3 only.
D 2 and 3 only.   E 1, 2 and 3.

(**NICCEA**: *all other Boards*)

Points

See E4.3, which becomes $\lambda = ya/d$ in the notation of the question. Remember that the separation between nodes is y in the equation.

For 3, assume the wave speed is constant, so a change of frequency changes the wavelength.

6 Which of the following statements about electromagnetic waves is/are correct?
1 X-rays in a vacuum travel faster than light waves in a vacuum.
2 The energy of an X-ray photon is greater than that of a light photon.
3 Light can be polarised but X-rays cannot.
A 1 and 2.   B 2 and 3.   C 1, 2 and 3.   D 2 only.
E 3 only.

(–: *all Boards except SEB and NICCEA*)

Points

Remember that light and X-rays are both electromagnetic radiations. Is electromagnetic radiation a transverse or a longitudinal wave form?

7 A parallel beam of white light is passed through a piece of Polaroid into a transparent specimen. When viewed through a second piece of Polaroid crossed with respect to the first, a number of coloured bands is observed in the specimen. Possible reasons for this are that the specimen is
1 being subjected to a nonuniform stress distribution.
2 causing interference in multiple slits in the Polaroid.
3 rotating the plane of polarisation of the incident light.
Answer:
A if 1, 2, 3 correct.
B if 1, 2 only.
C if 2, 3 only.
D if 1 only.
E if 3 only.

(**London**: *all other Boards except SEB and NICCEA*)

Points

Refer to your textbook to distinguish between photoelasticity (i.e. double refraction under stress) and optical activity (i.e. rotation of the plane of polarisation). Polaroid does not produce interference of light.

8 A transmitter of 3 cm electromagnetic waves and a small metal plate are set up as shown. A receiving aerial is connected to a suitable amplifier and meter. (The speed

of electromagnetic radiation is 3.0×10^8 m s^{-1}.) Which of the following is then true?
1 When the receiving aerial is moved along the line XY, a maximum response is noted every 1.5 cm.

2 The frequency of the waves is 10^{10} Hz.

3 When the receiving aerial is placed at a suitable point behind the plate but on the line *XY* produced, a response can again be noted because diffraction occurs at the edges of the plate.

Answer:

A if 1, 2, 3 correct.

B if 1, 2 only.

C if 2, 3 only.

D if 1 only.

E if 3 only.

(**London**: *all other Boards*)

Points

Refer to Fig. 3.6 and E3.6 (noting the warning regarding standing waves following E3.6). Diffraction is discussed in 3.5; note that the wavelength is 3 cm and the plate is *small*.

9 When a sonometer wire is set up as shown, the length *XY* vibrates with a fundamental frequency of 50 Hz. The fundamental frequency could be doubled by:

A halving the load *M*.   B doubling the load *M*.

C doubling the length *XY*.   D halving the length *XY*.

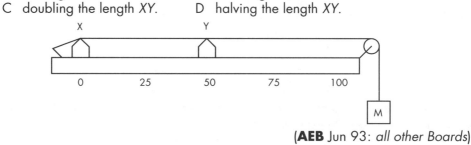

(**AEB** Jun 93: *all other Boards*)

Points

The formula for the fundamental frequency of a standing wave on a string can be derived from E3.6 and E3.13. Remember that the tension is *Mg*, and that the wavelength of the fundamental mode is twice the distance *XY*.

10 Stationary waves are set up in five tubes containing air, three of which are closed at one end. The diagrams represent the displacements of the particles of air from their mean positions at each point along the tube axis. Each tube has a length *L*, in metres, and an end correction *e*, in metres. The speed of sound in each of the air columns is c, in m s^{-1}. Which diagram represents a stationary wave of frequency, in Hz, $c/4(L + e)$.

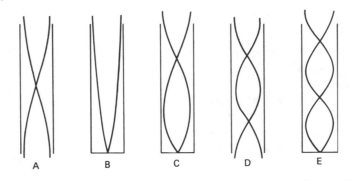

(–: *all Boards except SEB*)

Points

See 3.6 and Fig. 3.13. The diagram clearly illustrates the end errors so you must remember an end correction *e* must be added to *L* for each open end.

11 A vibrator sets up standing waves in a string as shown in the diagram. If the frequency of the vibrator is 20 Hz, then the speed of the waves is:
A 3.0 m s^{-1} B 10.0 m s^{-1} C 15.0 m s^{-1}
D 22.5 m s^{-1} E 43.0 m s^{-1}

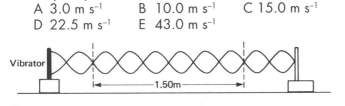

(**SEB**: *all other Boards*)

Points

You must first determine the wavelength by considering the diagram. See Fig. 3.5 if necessary. Then calculate the wave speed using E3.6.

12 A string is stretched under constant tension between fixed points *X* and *Y*. The solid line in the diagram shows a standing (stationary) wave at an instant of greatest

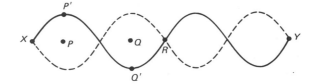

displacement. The broken line shows the other extreme displacement. Which one of the following statements is *correct*?
A The distance between *P* and *Q* is one wavelength.
B A short time later, the string at *R* will be displaced.
C The string at *P'*, and the string at *Q'*, will next move in opposite directions to one another.
D At the moment shown, the energy of the standing wave is all in the form of kinetic energy.
E The standing wave shown has the least possible frequency for this string stretched between *X* and *Y* under this tension.

(**Cambridge**: *all other Boards except SEB*)

Points

See 3.6 and Fig. 3.5. Remember that displacement nodes (points of zero displacement) are spaced out at half-wavelength intervals. At the moment shown, the displacement of each particle is at its maximum value so the velocity of each particle is zero.

13 Which one of the following groups of electromagnetic waves is in order of increasing frequency?
A gamma-rays, ultraviolet rays, radio waves.
B gamma-rays, visible light, ultraviolet rays.
C visible light, infrared radiation, microwaves.
D microwaves, ultraviolet rays, X-rays.
E radiowaves, visible light, infrared radiation.

(**Cambridge**: *all other Boards except SEB*)

Points

See Fig. 3.14.

14 A 3 cm wave transmitter was set up in front of a metal barrier in which there were two narrow, parallel gaps *P* and *Q*.

If the distance *AP* = 88.1 cm and the distance *AQ* = 92.6 cm, comment on the intensity of radiation detected by the probe detector at *A*. Explain clearly your reasoning.

(**SEB**: *all other Boards*)

Points

The path difference *AQ* – *AP* = 4.5 cm. Does this mean that waves from P and Q will interfere constructively or destructively at *A*? See 4.1 if necessary.

15 (a) The sketch shows a long wooden rod clamped at its centre. Longitudinal vibrations of period *T* may be excited in the rod using a resined cloth gripped tightly to the rod and drawn sharply in the direction of the arrow. Sketch, in exaggerated form, the outline of the rod at the times indicated:

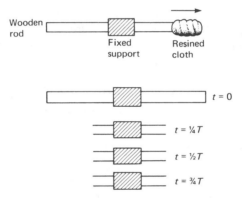

(b) (i) Explain the combination of simple travelling waves which gives the vibration in the rod. (ii) The waves in the gas in a sounding resonance tube have some features in common with the waves in the rod. Name two of them.

(c) The shortest length of the air column in a tube, closed at one end, resonating with a tuning fork of frequency 384 Hz was 206 mm. Another resonance was found at 624 mm. Find the speed of sound in the air inside the tube.

(**NICCEA**: *all other Boards except SEB*)

Points

(a) A stationary wave is set up in the rod as a result of the initial vibration being reflected at each end. Where will the nodes and antinodes occur in the simplest mode of vibration? Look at Fig. 3.6 and consider *C* and *G* as the ends of the rod. Remember that you are dealing with longitudinal vibrations, so that, in a displacement diagram, movement to the right is plotted above the axis and movement to the left below it.

(b) (i) See 3.4 (ii) See Fig. 3.5 and Fig. 3.13 (first part), and 3.6 which deal with standing waves in pipes.

(c) See Fig. 3.13 and use E3.6, remembering to allow for the end correction.

16 (a) The diagram right shows a standing wave in a long narrow spring which is vibrating from side to side on a smooth horizontal surface. What factors together determine the frequency of the vibrations?

(b) The diagram below shows a standing wave in the same spring when it is arranged vertically between two supports; the speed of propagation of the wave, and hence the wavelength, is now decreasing towards the lower end. Why does the speed decrease towards the lower end?

(c) Suggest one similarity and one difference between the standing waves shown in the diagram for part (b) and electron standing waves in an atom.
    (**O and C Nuffield**: *all other Boards (a & b only) except SEB*).

Points

(a) The frequency of any wave form is defined by its wavelength and its speed of propagation. What defines the wavelength in this system, and what controls the velocity of transverse waves down a stretched spring (E3.13)?

(b) E3.13 provides the clue. The role played by the tension in the spring is crucial. If the spring can be regarded as 'heavy', how will the tension towards the bottom of the spring compare with the tension at the top? Does this affect the wave velocity in the right way to explain the diagram?

(c) The electron in an atom can be modelled as a standing wave in a box. The diagram shows a standing wave in a box (the similarity), but does the electron have a wavelength that varies along its path as shown in this diagram? (See Unit 10 of Nuffield Student Books.)

17 (a) A device for measuring the frequency of mechanical vibrations consists of 11 thin steel strips of different lengths each fixed at one end to a block. The frequency of the fundamental transverse oscillations of the strips ranges from 35 Hz to 45 Hz in 1 Hz steps. The device is attached to a machine which is vibrating, and it is observed that the 40 Hz strip is set into strong oscillation while all the others show little movement. (i) Explain the principle on which the device operates. (ii) What conclusions can be drawn about the frequency of vibration of the machine?

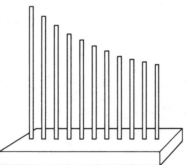

(iii) What should be observed if the experiment were repeated with the strips immersed in a bath of light oil?

(b) A quartz crystal is in the form of a rectangular bar clamped at the midpoint and oscillates longitudinally in its fundamental mode. The crystal is designed to oscillate at a frequency of 100 kHz. Calculate: (i) the length of the crystal;

(ii) the limits between which the length of the crystal must lie if the frequency is to be correct to within ± 10 Hz. (Take the speed of longitudinal waves in quartz to be 5.72×10^3 m s^{-1}.)

(Oxford: *all other Boards except SEB*)

Points

(a) (i) The vibrations of the machine are transmitted through the base to each rod. Each rod undergoes forced oscillations. See 3.2. The key point is resonance. (ii) At resonance, the frequency of natural oscillations (i.e. fundamental) of the resonant system is equal to the frequency of the 'forcing' system (i.e. of the machine vibrations). Use this fact to state the machine vibration frequency. Also, since none of the other rods show much movement, you can comment on the 'spread' of the machine vibration frequency spectrum (i.e. the presence of other frequencies in the range 35–45 Hz in the range of frequencies produced by the machine). (iii) Oil immersion would have a considerable damping effect on the response. See 3.2. You should give your answer in terms of a comparison of observations with and without oil.

(b) (i) At the midpoint there must be a node (i.e. zero disturbance), so in the fundamental mode there would be an antinode at either end with no intervening antinodes or nodes between the centre and the end. You must consider if this means that the length from end to end is $\frac{1}{4} \times$ wavelength or $\frac{1}{2} \times$ wavelength, etc., so that you can write down the length in terms of the wavelength. Then use the given values of wavespeed and frequency to calculate the wavelength. Once you have the wavelength value, then you can calculate the bar length. (ii) The percentage limits for the length must be the same as the percentage limits for the frequency, so work out the percentage limits for length first, then calculate the actual limits, using the length value from (b) (i).

18 (a) Distinguish between transverse and longitudinal waves. State into which of these categories you would put (i) light waves, (ii) sound waves, and indicate how you would demonstrate experimentally that your statements are correct.

(b) In terms of the motion of the particles of a medium in which there are sound waves, describe *one* similarity and *two* differences between a progressive sound wave and a stationary sound wave.

(c) By means of suitable apparatus stationary waves of sound are set up in air. Describe how you would attempt to measure the wavelength of these waves.

(NEAB: *all other Boards except SEB*)

Points

(a) See 3.4. State the difference in terms of the direction of vibration related to the direction of travel of the wave. For light waves, describe a simple polarisation experiment to show that the wave vibrations are in one direction only. Give a simple diagram. For sound waves, describe a simple experiment in which sound waves are directed at a diaphragm. With a magnet attached to the diaphragm (must be a small magnet) and a coil near the magnet, vibrations of the diaphragm will induce a voltage across the coil terminals.

(b) The motion of the particles in each case is described in 3.4.

(c) Refer to your textbook for the 'resonance tube' experiment.

19 (a) State the necessary conditions for the establishment of a stationary wave. Explain how these conditions are fulfilled when a wire stretched between two fixed points is plucked in the middle.

(b) Describe how the amplitude and phase of the vibrations vary with position along the length of the wire when it is vibrating in (i) its fundamental mode, (ii) its first overtone.

(c) Describe an experiment using a sonometer to investigate how the fundamental frequency of transverse vibration of a fixed length of wire varies with diameter. You may assume that you are provided with a set of wires of the same material, but of different diameters, and any other essential apparatus. Show how you would establish the relationship by a graphical method and state the relationship you would expect to obtain.

(d) Two steel violin strings of the same length which are subjected to the same tension have fundamental frequencies of 440 Hz and 660 Hz, respectively. Calculate the ratio of the diameter of the two strings.

(**NEAB**: *all other Boards except SEB*)

Points

(a) The conditions for stationary waves are discussed in 3.4. When the wire is plucked in the middle, progressive waves travel in either direction away from the centre and reflect off the ends. You should describe how this leads to standing waves.

(b) Give a diagram showing how the wire would appear at maximum displacement for each mode to aid your explanation of how the amplitude (i.e. maximum displacement) varies with position. For the phase, remember that all points between adjacent nodes are in phase, whilst points on either side of a given node (as far away as the next node) are out of phase by 180°.

(c) For a suitable sonometer experiment, refer to your textbook.

(d) Use the relationship between f_0 and d which you have already derived in (c).

Answers: **1** A; **2** C; **3** C; **4** D; **5** A; **6** D; **7** E; **8** A; **9** D; **10** B; **11** B; **12** C; **13** D; **15**(c) 321m s^{-1}; **17**(b)(i) 2.86 cm (ii) 2.86×10^{-6} m; **19**(d) 3/2

CHAPTER 4

OPTICS

Units in this chapter

Chapter objectives

After working through the topics appropriate to your syllabus in this chapter, you should be able to:
- use the wave theory of light to explain interference
- explain the meaning of coherence
- describe how to measure the wavelength of light from Young's double-slit experiment
- explain the principle of a diffraction grating
- describe how to use a diffraction grating and a spectrometer to observe optical line spectra and measure wavelengths
- explain the interference patterns formed by different types of thin films
- use the law of refraction to work out light paths through blocks and prisms
- construct ray diagrams to explain image formation by mirrors, lenses and optical instruments, as appropriate
- use appropriate formulae to solve simple lens or mirror problems

4.1 WAVE NATURE OF LIGHT

The idea that light consists of waves was first put forward by **Huygens**. He assumed that a point source of light emits spherical wavefronts, and that every point on a wavefront acts as a secondary emitter. Using his theory, the propagation of light can be explained, as in Fig. 4.1. We need to assume that the new wavefront in the diagram, formed by the envelope of secondary 'wavelets', is formed ahead (and only ahead) of the old wavefront.

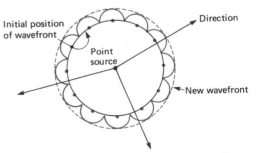

Fig. 4.1 Huygens' wavelets producing a spherical wavefront

Interference of light: Young's fringes

Direct evidence for the wave nature of light was first obtained by Young, more than a century after Huygens' time. What Young did was to observe the superposition of two sets of light waves derived from the same source, and he arranged this by allowing light from a point source to pass through two very close slits, as shown in Fig. 4.2.

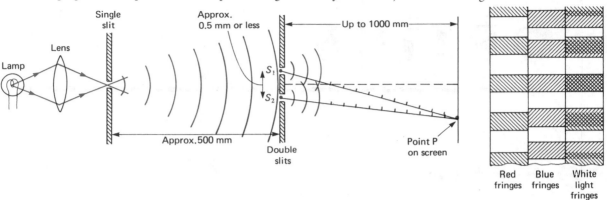

Fig. 4.2 Interference from Young's double-slit experiment

The pattern on the screen consists of alternate bright and dark fringes, parallel to the slits. The pattern can only be explained by using Huygens' wave theory; each slit acts as a source of waves, so that where two sets of waves overlap, an interference pattern can be produced.

A bright fringe is formed on the screen where the waves from one slit **reinforce** the waves from the other slit. In other words, a wave crest from one slit arrives at the bright fringe at the same time as a wave crest from the other slit. Referring to the diagram, the condition for a bright fringe to be formed at a given point P on the screen is:

E4.1 $S_1P - S_2P = m\lambda$

where $m = 0$ or 1 or 2 or 3, etc., λ = wavelength of the light used (m).

For a dark fringe to be formed, the light waves from one slit must **cancel** the waves from the other slit. In other words, a wave crest from one slit arrives at the dark fringe at the same time as a wave trough from the other slit.

E4.2 $S_1P - S_2P = (m + \frac{1}{2})\lambda$ (for a dark fringe)

By making suitable measurements, the wavelength of light can be calculated from the following equation (you should consult your textbook for a proof of this equation):

E4.3 $\lambda = \dfrac{Yd}{X}$

where Y = fringe spacing (i.e. distance between centres of two adjacent fringes) (m), λ = wavelength (m), d = slit spacing (centre to centre) (m), X = slits–screen distance (m).

A simple test to check that the pattern is caused by interference between light from the two slits is to block off one of the slits. The fringes will then disappear.

If a white light source is used, the central fringes will be white and black, but the outer fringes will be tinged with colours. The reason for this is that white light is composed of all the colours of the spectrum from red ($\lambda = 700$ nm) to blue ($\lambda = 400$ nm). Each wavelength component of the white light gives its own interference pattern on the screen, with spacing in proportion to the wavelength. Thus, each noncentral bright fringe will be tinged with red on its outer side and blue on its inner side.

Coherence   Two point sources of light are said to be coherent if they each emit light waves of the same frequency and with a **constant phase difference** between the two sets of waves. For an interference pattern to be observable in the double-slit experiment, the sources must be coherent. When two sets of waves overlap, cancellation and reinforcement take place. However, if the phase difference between the two sources changes, then the points of interference move. Thus, an overlap of light from two separate light bulbs will not produce an observable interference pattern because the two bulbs emit light waves with a randomly changing phase difference. However, in the double-slit experiment, waves are emitted by the two slits only when a wavefront arrives from the initial light source. Therefore, the two slits act as coherent sources, and an observable interference pattern can be produced. See also 3.5.

Because atoms emit light in short 'bursts' only, lasting of the order of nanoseconds, (i.e. 10^{-9} s), light from a given source will consist of 'wave trains' of length up to 0.3 m or so (= speed of light × time taken to emit). In Young's double-slit experiment, the two sets of light waves, arriving at the screen, will only be coherent (i.e. have a constant phase relationship) if the path difference $S_1P - S_2P$ is less than the length of a typical wave train. If the path difference exceeds the length of a typical wave train, then the coherence will be lost because the end of the wave train from one slit will arrive before the start of the simultaneously emitted wave train from the other slit. From a sodium light source, a typical wave train length is 1 cm, whereas for a laser source it is typically 0.40 m; hence, the laser produces an interference pattern with many more fringes visible.

4.2 EFFECTS DUE TO FILMS AND GAPS

Thin-film interference: near-normal incidence

When monochromatic light is incident upon a thin film, interference takes place between the light reflected from one boundary and that reflected from the other boundary. Fig. 4.3 shows

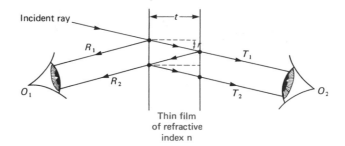

Fig. 4.3 Interference from a parallel thin film

a parallel monochromatic beam of light at near-normal incidence upon a thin film of thickness t. The two reflected beams are produced as a result of partial reflection and transmission at each boundary. If the two reflected beams are in phase, then they reinforce one another so that observer O_1 sees the film as a single colour corresponding to the incident light. The path difference, $2t \cos r$, must not exceed the length of a typical wave train in the film; hence thick films do not produce such effects.

The condition for **reinforcement** of the two reflected beams is:

E4.4　$2nt = (m + \frac{1}{2}) \lambda_0$

where $m = 0$ or 1 or 2, etc., n = refractive index of the film, t = thickness of film (m), λ_0 = wavelength *in air* (m), assuming angle r is small so that $\cos r \simeq 1$.

Given this condition, the two reflected beams will be in phase because:

❶ **external reflection** always causes a phase reversal of 180° (see 3.5), so the reflection of R_1 causes the extra half-wavelength;

❷ R_2 travels an extra distance $2t$ compared with R_1;

❸ the wavelength **in the film** is λ_0/n.

In this situation the two **transmitted beams** will be exactly out of phase (neither T_1 nor T_2 undergo phase reversal due to reflection) so that an observer at O_2 will see the film as entirely dark.

If the film thickness is such that the following condition holds:

E4.5　$2nt = m\lambda_0$

then the film will appear **dark by reflection** and bright by transmission. The reason for this is that the two reflected beams will be exactly out of phase because the path difference $2t$ is a whole number of wavelengths in the film, *and* R_1 undergoes phase reversal upon reflection. The two transmitted beams will be exactly in phase because $2t$ = whole number of 'film' wavelengths *but* neither T_1 or T_2 undergo phase reversal.

Thin-film interference: near-normal incidence for films with nonparallel sides

The simplest example is the **wedge**. The conditions for interference given by equations E4.4 and E4.5 still apply, but the film thickness t varies from one position to another. An observer will therefore see dark and bright fringes running along positions of equal thickness. Each fringe will therefore follow a contour of equal film thickness; for a wedge, the fringes will be alternate bright and dark lines parallel to its edge.

A more interesting example is provided by the film formed between a convex lens and a flat glass plate when the lens and plate are in contact. Whether the film is an air film or a liquid film, the same general pattern, known as **Newton's rings**, is produced. The pattern consists of alternate, concentric bright and dark rings, as in Fig. 4.4, centred upon the point of contact.

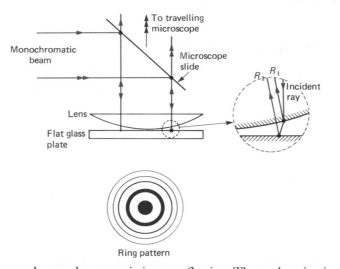

Fig. 4.4 Interference from lens and flat surface – Newton's rings

The ring pattern may be seen by transmission or reflection. The explanation is essentially the same as for the wedge and the parallel-sided film. In this case, the contours of equal thickness are rings centred upon the point of contact. The same conditions for bright and dark fringes as before hold (i.e. E4.4 and E4.5 apply); thus for a **dark ring by reflection** $2nt = m\lambda_0$. However, the diameter of the mth dark ring can be expressed by the following equation derived by geometrical considerations: $D_m^2/4 = 2Rt$, which gives:

E4.6 $$D_m^2 = \frac{4mR\lambda_0}{n}$$

where D_m = diameter of mth dark ring (m), R = radius of curvature of lens face (m), t = film thickness at ring position (m), m = 0 or 1 or 2, etc., λ_0 = wavelength (m) in air, n = refractive index of medium between lens and plate. E4.6 is known as the **Newton's rings equation**.

At the centre of the rings ($m = 0$), a dark spot will be seen because the reflected beams will be out of phase by 180°, owing to phase reversal upon external reflection. If the lens is slowly lifted off the plate, the central spot will first change from dark to bright (at spacing = $\frac{1}{4}\lambda_0/n$), then back to dark (at spacing = $\frac{1}{2}\lambda_0/n$), etc. When using Newton's rings to calculate the wavelength of light, a graph of $y = D_m^2$ against $x = m$ is usually plotted. The gradient of the graph is then measured, and since $4R\lambda_0/n$ equals the gradient, then λ_0 can be calculated provided values for R and n are known. The line may not pass through the origin since the lens may not be in actual contact with the plate; hence, it is better to use the graph method rather than to insert values directly into E4.6.

Single-slit diffraction

When a plane wavefront is incident upon a narrow slit, each point of the transmitted wavefront acts as a secondary emitter of waves. The net effect is to produce a diffraction pattern as shown in Fig. 4.5, with a central bright band, and bright and dark fringes parallel on either side. The formation of the fringes and central band is due to interference of the secondary wavelets

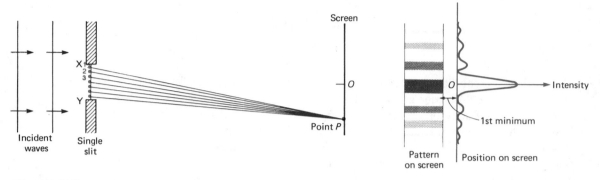

Fig. 4.5 Diffraction from a single slit

emitted by each point on the plane wavefront at the gap. The condition for a **minimum** in the diffraction pattern is:

E4.7 $\quad d \sin\theta_m = m\lambda$

where d = slit width (m), m = 1 or 2 or 3, etc., θ_m = angular position of the mth minimum from the centre, λ = wavelength (m).

The reason for this condition can be appreciated by considering point P on the screen at the first dark fringe. Each point on the wavefront XY acts as a secondary emitter, so that waves from point 1 arrive at P out of phase with waves from point 2, which arrive out of phase by the same amount compared with the waves from point 3, etc. Provided P is distant, the contributions from each point on XY are of the same amplitude; for P to be a minimum, the resultant of all these contributions will be zero. This will be so if corresponding points in each half give contributions out of phase by exactly 180°. The path difference from two such points, $\frac{1}{2}d \sin\theta_1$, must therefore be equal to one half-wavelength. The single-slit diffraction pattern causes intensity 'modulation' of the diffracted orders of a diffraction grating. See Fig. 4.9.

Diffraction gratings

With the aid of a spectrometer, diffraction gratings are used to study light spectra. A diffraction grating consists of many parallel, close slits which are ruled either on glass (transmission grating) or on metal (reflection grating). The effect of a diffraction grating upon a parallel beam of monochromatic light at normal incidence is shown in Fig. 4.6.

By measuring the angle of diffraction, θ_m, the wavelength of the light can be calculated from the diffraction grating equation:

E4.8 $\quad m\lambda = d \sin\theta_m$

where d = distance from the centre of one slit to the centre of the next adjacent slit (m), m = 0 or 1 or 2, etc., λ = wavelength (m), θ_m = angle of diffraction.

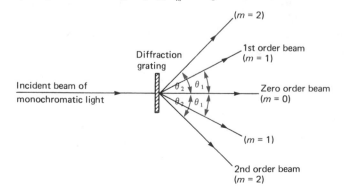

Fig. 4.6 Interference from a diffraction grating – monochromatic light

123

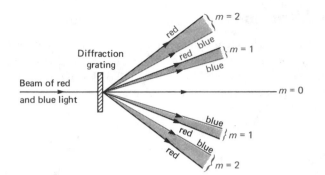

Fig. 4.7 Interference from a diffraction grating – red and blue light

When the incident beam contains several different values of wavelength (i.e. not monochromatic), the diffraction grating will split the beam so that the longer the wavelength, the greater will be the angle of diffraction, as in Fig. 4.7. If white light is used, then each order will contain the full spectrum with blue light diffracted least and red most.

The operation of a diffraction grating relies upon three phenomena. Firstly, there is the diffraction of a plane wavefront incident upon a narrow slit. Secondly, a series of narrow slits upon which plane waves are incident will produce sets of diffracted wavefronts, one set from each slit. Thirdly, the sets of wavefronts interfere constructively in certain directions only, as given by E4.8 and as illustrated by Fig. 4.8.

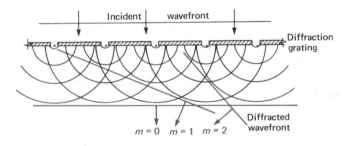

Fig. 4.8 Huygens' theory for a diffraction grating

The number of orders that a given grating can produce from monochromatic light is limited. Since $\sin \theta_m \leq 1$, then $m\lambda \leq d$, so m cannot exceed d/λ. In other words, since $\sin \theta_m$ cannot exceed 1, then the maximum order number must be the ratio d/λ rounded down to the nearest whole number. In addition, missing orders may also be caused by the single-slit diffraction pattern produced by the individual grating slits; depending upon the ratio slit width/slit space, certain orders will be fainter than others, as illustrated by Fig. 4.9.

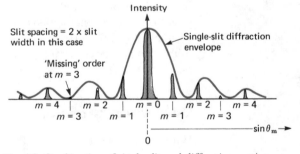

Fig. 4.9 Combination of single-slit and diffraction-grating patterns

4.3 RAY OPTICS

A light ray indicates the direction in which a wavefront travels. The rays from a point source spread out radially because the wavefronts are expanding spheres. Provided diffraction of wavefronts by gaps and obstacles is negligible, then the rays will be straight lines. Ray optics is based upon the assumption that diffraction can be neglected. Many aspects of optics can be more easily dealt with by considering the rays rather than the waves.

Reflection

The simple law of reflection from a **plane mirror** can easily be explained by considering light as a waveform as in Fig. 4.10. Incident wavefronts, assumed to be planar, travel forward by producing secondary wavelets. Thus points X and Y on the same wavefront send out wavelets

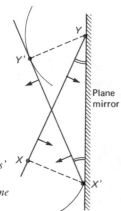

Fig. 4.10 Huygens' theory for a reflection at a plane mirror

at the same time. However, by the time that X's wavelet reaches the mirror at X', Y's wavelet will have travelled an equal distance to form part of a plane, reflected wavefront $X'Y'$. Therefore, the reflected wavefront $X'Y'$ makes the same angle to the mirror as the incident wavefront XY.

An image seen in a plane mirror is a **virtual image**, because rays reflected off the mirror to the viewer only **appear** to come from the image (a **real image** is one from which rays actually come **direct** to the viewer). The image formed by a plane mirror is always the same size as the object, and is as far behind the mirror as the object is in front of the mirror.

A **convex mirror** always forms a virtual image from a real object, although the image is diminished in size compared with the object. A **concave mirror** can either form a virtual image or a real image, depending upon how far the object is from the mirror.

To calculate image positions, the following equation is usually used:

E4.9 $\dfrac{1}{u} + \dfrac{1}{v} = \dfrac{1}{f}$

where u = object–mirror distance (m), v = image–mirror distance (m), f = focal length of mirror (m).

Also, the **linear magnification** is defined as the ratio

$$\frac{\text{image height}}{\text{object height}} = \frac{v}{u}$$

A sign convention must be employed to use the equation correctly. The most popular convention is: **real is positive, virtual is negative**. For real images and objects, numerical values of u and v are positive, and negative values of u and v indicate virtual images or objects. Also, concave mirrors are given positive values of f, and convex mirrors are given negative values of f. Remember that the focal length of a curved mirror is the distance from the pole (i.e. mirror centre) to the point where rays parallel to the axis (i.e. parallel to the normal through the pole):

❶ converge, after reflection, if the mirror is concave;

❷ appear to diverge from, after reflection, if the mirror is convex.

Scale ray diagrams are an alternative way of locating image positions, and some examples of these diagrams are shown in Fig. 4.11 and Fig. 4.12.

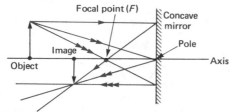

Fig. 4.11 Real diminished inverted image from a concave mirror

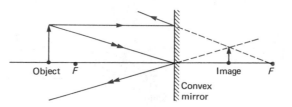

Fig. 4.12 Virtual diminished upright image from a convex mirror

Refraction

Huygens' wave theory can be used to explain refraction. In Fig. 4.13 wavefronts are incident on the boundary as shown. Points X and Y on the same wavefront emit secondary wavelets at the same time, but because the wavelet from X is in the more dense optical medium, it travels more slowly than that from Y. Thus the wavelet from Y reaches Y' in the same time as that from X reaches X'. The refracted wavefront $X'Y'$ travels therefore in a direction closer to the normal than the incident wavefront. From the diagram, the following equation can be proved:

E4.10 $\quad \dfrac{\sin i}{\sin r} = \dfrac{\lambda_0}{\lambda_m} = \dfrac{c_0}{c_m} =$ refractive index n

where i = angle of incidence, r = angle of refraction, λ_0 = wavelength in air (m), c_0 = wave speed in air (m s^{-1}), λ_m = wavelength in the medium (m), c_m = wave speed in medium (m s^{-1}).

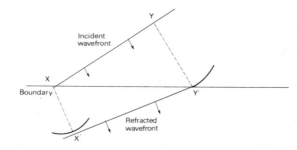

Fig. 4.13 Refraction of waves

This equation is, therefore, in agreement with **Snell's law**, an experimental law, which states that the ratio sin i/sin r is a constant. This constant is known as the **refractive index** (**n**), and by considering E4.10, it follows that the refractive index is equal to:

$$\frac{\text{wavelength in air } (\lambda_0)}{\text{wavelength in medium } (\lambda_m)} = \frac{\text{wave speed in air } (c_0)}{\text{wave speed in medium } (c_m)}$$

If the light ray is travelling from an optically more dense medium to an optically less dense medium, then **total internal reflection** will take place if the incident angle (i) exceeds the critical angle (c), where sin $c = 1/n$ from Snell's law.

The passage of a light ray through a prism involves use of Snell's law if the path is to be calculated, given the values of the refractive index and the incident angle. Fig. 4.14 shows the light path for a monochromatic beam; the angle of refraction at the first boundary is given by sin a_1/sin $g_1 = n$. For the secondary boundary, take care because the path is from glass to air, so the incident ray is in the medium (e.g. glass); the angle of refraction at the second boundary is therefore given by sin g_2/sin a_2 = $1/n$. It can be shown that the angle of deviation of the light beam from its original path is a minimum when the light passes symmetrically through the prism.

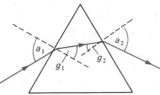

Fig. 4.14 Refraction through a prism

When white light passes through a prism, the beam is split into the colours of the spectrum, as shown in Fig. 4.15. This effect is known as **dispersion**. It occurs because the refractive

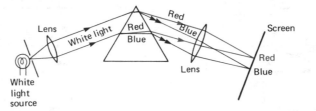

Fig. 4.15 Dispersion through a prism

index (of the prism) is greater for blue light than for red light. Hence, red light is deviated by the least amount.

Optical fibres

An optical fibre uses the principle of total internal reflection to guide light along the fibre. Fig. 4.16 shows the path of a light ray along an optical fibre. Each fibre is coated in a cladding which prevents leakage of light from the sides of a fibre. The cladding has a lower refractive index than the core so total internal reflection occurs at the interface between the core and the cladding.

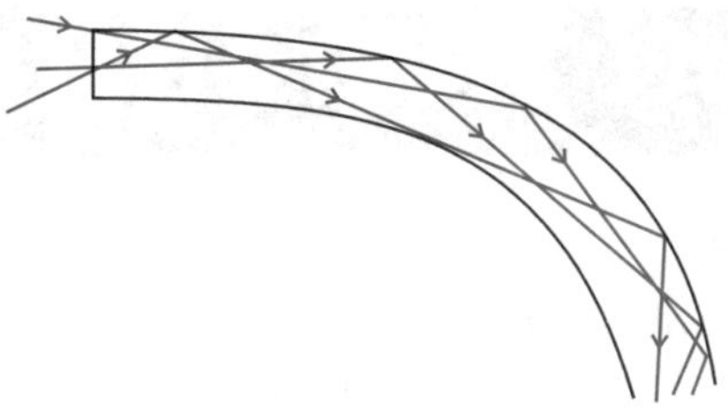

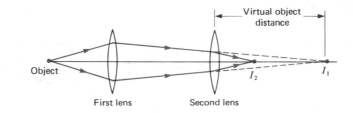

Fig. 4.16 Total internal reflection in an optical fibre

Optical fibres are used in medicine and in communications, where infrared 'light' pulses carry digital signals. The core of an optical fibre used for communications needs to be very thin, otherwise each pulse is lengthened as rays that repeatedly reflect off the sides take longer than rays that pass through direct. Also, laser light is used since it has a much smaller spread of wavelengths than does ordinary light. This reduces pulse spreading due to dispersion.

Lenses

Image formation by thin lenses can either be treated by ray diagrams, or by the use of E4.9. If the equation is used, then the sign convention should be employed; also, the focal length value for a convex lens is taken as positive, and for a concave lens as negative. Remember that the **focal length** (f) is the distance from the lens centre to the point at which rays parallel to the axis:

❶ converge, after refraction, if the lens is convex;

❷ appear to diverge from, after refraction, if the lens is concave.

Single-lens problems can usually be dealt with using the lens formula E4.9, although a sketch ray diagram is often valuable as well. In two-lens problems, either make an accurate ray diagram or do a 'rough' ray diagram *before* use of E4.9 so that you can check on approximate positions, etc. A two-lens problem may sometimes involve dealing with a 'virtual object', as in Fig. 4.17.

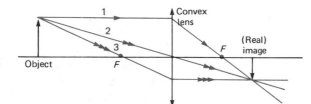

Fig. 4.17 Virtual object in a two-lens system

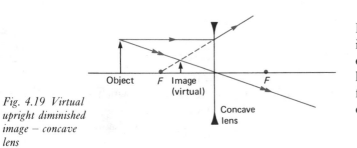

Fig. 4.18 The three key rays of geometric optics

The 3 key rays are;
1   parallel to the axis, to the lens, then through F
2   straight through the centre of the lens
3   through F to the lens, then parallel to the axis

Fig. 4.19 Virtual upright diminished image – concave lens

The first lens would form a real image at I_1 if the second lens was absent; refraction by the second lens takes place, so giving its image at I_2. The final image position can be calculated by treating I_1 as a virtual object for the second lens. For example, if the second lens is positioned such that its distance from I_1 is 10 cm, then the object distance for the second lens (u_2) is −10 cm.

The use of scale ray diagrams is shown in Fig. 4.18 and Fig. 4.19. The three 'key' rays in each diagram are shown clearly. Given the object distance and the focal length of the lens, the three key rays should be drawn in from the tip of the object so as to give the tip of the image.

4.4 OPTICAL INSTRUMENTS

The simple microscope

A convex lens may be used as a simple microscope (i.e. a magnifying glass) by placing the object to be viewed between the focal point and the lens. The viewer, looking from the other side of the lens, will see an enlarged virtual image, as shown in Fig. 4.20. If the object is brought closer to the lens, the apparent size of the image is increased because the viewer judges apparent size in terms of the angle subtended at the eye. However, if the object is brought too close to the lens, the image position will be closer than the viewer's near point of vision, so the image will appear blurred.

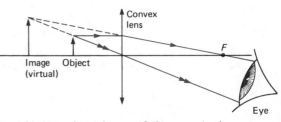

Fig. 4.20 Virtual upright magnified image – simple microscope

The refracting telescope

A simple refracting telescope can be constructed from two suitable convex lenses. The less powerful lens, the objective, forms a real image I_1 of the distant object being viewed. The eyepiece lens is then positioned so that, in **normal adjustment**, I_1 lies in its focal plane; then the viewer sees the final image I_2 at infinity. In this situation, the magnifying power (mp) is given by:

E4.11 $mp = \dfrac{\text{angle subtended by final image at infinity}}{\text{angle subtended by distant object}} = \dfrac{f_0}{f_e}$

where f_0 = focal length of the objective (m), f_e = focal length of the eyepiece (m).

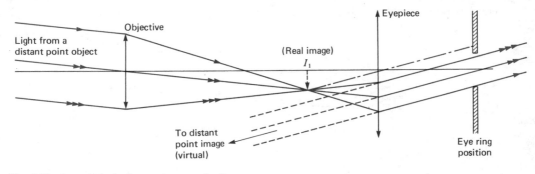

Fig. 4.21 Astronomical telescope in normal adjustment

The best position for the eye is at the **eye ring**, as shown in Fig. 4.21, since all the light that enters the objective from the object must pass out through the eye ring. It can be shown that the ratio f_0/f_e is equal to the ratio (object diameter)/(eye ring diameter). Usually, the telescope is designed to make the eye ring diameter no larger than the width of the pupil.

When used to view a point object, the brightness is considerably greater with the aid of the telescope than without; the larger diameter objective collects much more light than the unaided eye. Large objective diameters also enable the viewer to see greater detail of an extended object, or to resolve more easily two very close objects (e.g. binary stars). The **resolving power** of a telescope is theoretically given by the following equation:

E4.12 $rp = \dfrac{1.22\lambda}{D_0}$

where λ is the wavelength of the light (m), D_0 is the objective diameter (m).

This equation gives the angular separation, in radians, of two point objects at infinity which can just be resolved (i.e. the resolving power). If they are any closer, then the viewer sees one 'merged' image only. The basis of the formula is the diffraction of light as it passes through

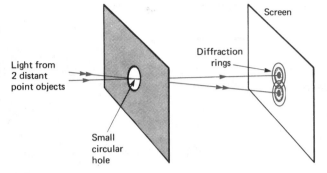

Fig. 4.22 Resolving distant sources

a gap, as explained in 4.2. For a circular gap (e.g. pupil of the eye or telescope objective), E4.12 applies; if the gap is the pupil, then D_0 is the pupil diameter. Because the objective of a telescope is usually several times wider than an average pupil, more detail can be seen with a telescope than without. Each point object gives a diffraction pattern of rings, as shown in Fig. 4.22. The rings from one point object will overlap the rings from the other point object if the two objects are too close; only one image will then be seen.

A **reflecting telescope** uses a concave mirror as its objective. Since increased objective diameter increases the brightness of point images, and also increases the ability to resolve detail, etc., large-diameter objectives are desirable. It is much easier to make large-diameter concave mirrors than large-diameter lenses. The Newtonian reflecting telescope, as shown in Fig. 4.23, is fitted with a small plane mirror to reflect light from the concave mirror into the eyepiece.

A **radio telescope** consists of a large parabolic reflecting dish with an aerial placed at its focal point. A coaxial cable connects the aerial to a high-gain amplifier which is connected to a computer to process the incoming signals. Electrical noise generated by the amplifier must be kept to a minimum, otherwise the signal from the aerial would be masked by the amplifier itself.

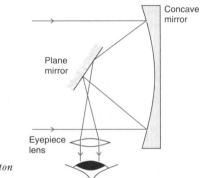

Fig. 4.23 The Newton reflecting telescope

The spectrometer

The use of a spectrometer with a diffraction grating is shown in Fig. 4.24. The essential parts of a spectrometer are the collimator tube, the turntable and the telescope. Before use, the telescope must be set to receive parallel light (by focusing on a distant object), and the collimator must be set to produce parallel light. This is done by viewing the slit through the correctly adjusted telescope via the collimator. The collimator is adjusted to bring the slit into focus. The turntable must be level, and the slit is then narrowed to a fine line.

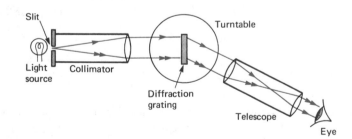

Fig. 4.24 The spectrometer

Chapter roundup

Make sure you have a sound grasp of interference, diffraction, refraction and the use of lenses before moving on. These topics are essential for further studies involving optical fibre communications, optical instruments and sight corrections.

Question bank

1   When light enters glass from air:
   A its frequency decreases and its velocity decreases.
   B its wavelength decreases and its velocity decreases.
   C its frequency increases and its velocity increases.
   D its wavelength increases and its velocity increases.

(–: all Boards)

Points

It is important to understand that the frequency of light on refraction cannot change; the same number of waves per second must pass a given point in the glass as pass a corresponding point in the air. This knowledge reduces the number of possible options and a final choice can be made using the knowledge that refractive index is equal to the ratio of velocities; see E4.10 if necessary.

2   In a double-slit experiment, light of wavelength 600 nm gave a fringe separation of 0.4 mm on a suitably placed screen. Use of a different light source, also monochromatic, gave ten fringes across a distance of 3.3 mm on the same screen at the same position. The wavelength of the second source, in nm, was:
   A 50   B 495   C 660   D 727   E 7300

(–:all Boards)

Points

Use E4.3; since slit spacing and slit–screen distance are constant, the wavelength ratio of the two sources is the same as the ratio of fringe separations. Remember that the fringe separation of the second pattern is *not* 3.3 mm.

3   Which one of the following effects provides direct experimental evidence that light is a transverse, rather than longitudinal, wave motion?
   A Light is diffracted by a narrow slit.
   B Two coherent light waves can be made to interfere.
   C The intensity of light from a point source falls off inversely as the square of the distance from the source.
   D Light is refracted by a glass prism.
   E Light can be polarised by reflection at a water surface.

(Cambridge: all other Boards except SEB, NICCEA)

Points

See 3.5 if necessary.

4   Monochromatic light of wavelength 600 nm is used in a spectrometer to illuminate a diffraction grating set normally to the collimator. The grating has 3×10^5 lines per metre. The telescope is used to scan the field to one side of the straight-through position. Not counting the 'straight-through' image, the maximum number of diffracted images of the slit visible to the observer will be:
   A 2   B 5   C 8   D 10   E 11

(Cambridge: all other Boards except SEB)

Points

See 4.2 if necessary.

5 Monochromatic light of wavelength λ falls normally on a diffraction grating, as shown in the diagram. The diffracted light is observed on a screen 400 mm away from the grating, the screen being parallel to the plane of the grating.

The second-order diffraction maximum occurs at P, 300 mm along the screen from the central maximum at Q. What is the spacing between the rulings of the grating (the grating element)?

    A $5\lambda/6$ B $4\lambda/3$ C $5\lambda/3$

D $8\lambda/3$ E $10\lambda/3$

    (NICCEA: *all other Boards*)

(diagram: laser → grating, 400 mm to screen, 300 mm from Q up to P)

Points

Work out the angle of diffraction from the 300 and 400 mm distances. Then use the diffraction grating equation E4.8 to work out the grating spacing in terms of the wavelength.

6 A convex lens of focal length 0.20 m is placed at a distance of 0.3 m from an illuminated object. A screen is moved until a clear image of the object is seen on the screen. Then, without moving object or image, the lens is moved towards the screen to a new position where a smaller clear image is seen on the screen. The distance through which the lens has been moved, in m, is:

    A 0.9   B 0.6   C 0.5   D 0.4   E 0.3

    (–: *London, AEB, O and C, Oxford, Cambridge, NEAB, NICCEA*)

Points

For the first position, use the lens formula ($1/u + 1/v$, etc.) to calculate the image distance. The second position corresponds to object and image distances interchanged; make a simple rough sketch to show the two positions, and then you ought to be able to calculate the distance between the two lens positions.

7 A narrow monochromatic beam of light passes through a triangular glass prism, being refracted at Q and R. Which of the following is correct?

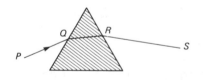

1 The angle between ray *PQ* and ray *QR* is known as the angle of deviation of the prism.
2 If the angle of incidence of ray *PQ* increases, the angle that ray *RS* makes with the normal at *R* decreases.
3 When the angle of deviation is a minimum, ray *QR* makes equal angles with the two refracting surfaces.

Answer:
A if 1, 2, 3 correct.   D if 1 only.
B if 1, 2 only.   E if 3 only.
C if 2, 3 only.

    (London: *and AEB, Oxford, NICCEA**)

Points

Deviation is the angle between the incident and *emergent* rays. As the angle of incidence of ray *PQ* increases, the angle of refraction at *Q* must also increase. By geometry, if follows that refraction at *Q* must also increase. By geometry, it follows that the angle *QR* makes with the normal must decrease, so ... For minimum deviation, see 4.3.

8   A lens forms a real image of an object. The distance from the object to the lens is

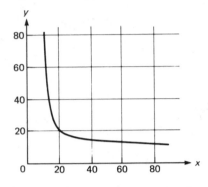

x cm and that from the lens to the image is *y* cm. The graph shows the variation of *y* with *x*. It can be deduced that the lens is:
A converging and of focal length 10 cm.
B converging and of focal length 20 cm.
C converging and of focal length 40 cm.
D diverging and of focal length 20 cm.
E diverging and of focal length 10 cm.
(**Cambridge**: *London, AEB, O and C, Oxford, NICCEA, NEAB*)

Points

Use E4.9

9   An illuminated object and screen are placed 90 cm apart. To produce a clear image on the screen twice the size of the object, the required lens must be:
A diverging, $f = -60$ cm.   B diverging, $f = -10$ cm.
C converging, $f = +20$ cm.   D converging, $f = +30$ cm.
E converging, $f = +60$ cm.
(–: *London, AEB, Oxford, Cambridge, NICCEA*, NEAB*)

Points

Use the lens formula, having first worked out *u* and *v* from the given facts that $u + v = 90$ cm and $v/u = +2$ (from the magnification).

10   The diagram represents two thin lenses L_1 and L_2 placed coaxially 30 cm apart. A beam of light parallel to the axis is incident on L_1. The final image formed by refraction through both lenses is:
A real and between L_1 and L_2.
B real and on the right of L_2.
C virtual and on the left of L_1.
D virtual and on the right of L_2.
E at infinity.
(**London**: *AEB, NICCEA, Oxford, Cambridge*)

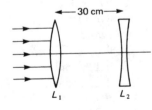

Points

Draw the continuation of the rays to where they would meet in the absence of L_2. Bearing in mind that L_2 will diverge the rays, you may be able to deduce the answer by elimination. If not, consider that L_1 will form a *virtual* object 10 cm beyond L_2 and apply $1/u + 1/v = 1/f$ to L_2 (watch those signs).

11 An astronomical telescope has an eye lens of focal length 20 mm. In normal adjustment, when the final image of a distant object is at infinity, the separation of the lenses is 500 mm. The angular magnification of the telescope under these conditions is:
A 22   B 23   C 24   D 25   E 26
(**London**: *NEAB, Oxford, SEB*)

Points

Refer to Fig. 4.19. Find the focal length of the objective and then use E4.12.

12 (a) The wave theory of light provides an explanation of many aspects of reflection, refraction and interference. (i) The three parallel lines in the sketch represent sections of a wave front of a parallel beam of light at times $t = 0$, $t = t_1$ and $t = 2t_1$. Using the principle of secondary wavelets, show how to construct one further wave front, for $t = 3t_1$. (A written explanation is not required.) (ii) The sketch shows a plane light wave encountering a glass block of refractive index 1.5. The four parallel lines represent the position of a single wave front at a series of equal time intervals. Draw six further wave fronts at the same equal time intervals showing how the light wave continues beyond AB. (Your drawing should be to scale.)

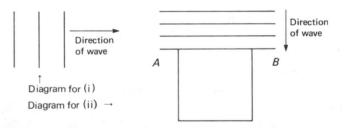

↑
Diagram for (i)

Diagram for (ii) →

(b) Three parallel narrow slits P, P_1 and P_2 are illuminated by the source S of monochromatic light of wavelength λ as shown in the diagram (which is not to scale). The distances of P_1 and P_2 from O are equal. A series of bright and dark fringes are formed on the screen at T, with a **bright** fringe at O', on the axis of the system, and a **dark** fringe at R. (i) The beams emerging from P_1 and P_2 are 'coherent'. What does this mean? (ii) What relationship must exist between the lengths P_1R and P_2R? (iii) When the wavelength of the source is 500 nm, the centre of the 120th **dark** fringe, counting from O', lies at R. Upon replacement of the source by one of unknown wavelength, R is found to be the location of the 90th **bright** fringe (counting from O' as zero). Find the wavelength of the unknown source.

(**NICCEA**: *all other Boards except O and C Nuffield, SEB*)

Points

(a) (i) Use Huygens' construction. See 4.1 and Fig. 4.1. (ii) Note that a *scale* drawing is required. Measure the distance between two positions of the incident wave front. Calculate the separation of the wave front positions in the glass from the comments after E4.10. Show the new positions of the wave front at each edge of the block as well as inside it.

(b) (i) See interference in 4.2. (ii) See E4.2. (iii) See your textbook for the derivation of E4.3. This will indicate that for the nth dark fringe $O'R = (n + \frac{1}{2})\lambda X/d$, and for the nth bright fringe $O'R = n\lambda X/d$. Substitute the data given and equate the two expressisons for $O'R$.

13 A refracting telescope has an objective of focal length 1.0 m and an eyepiece of focal length 2.0 cm. A real image of the sun, 10 cm in diameter, is formed on a screen 24 cm from the eyepiece. What angle does the sun subtend at the objective?

(**London**: *NEAB, Oxford, SEB*)

Points

If the final image if real, it must be beyond the eyepiece (i.e. on the opposite side to the objective). Apply $1/u + 1/v = 1/f$ to the eyepiece to find where the intermediate image must be (this acts as an object for the eyepiece to give the final image). Knowing that the linear magnification $= v/u$, you can now find the height of this intermediate image. As the sun is distant, this image I_1 will be in the focal plane of the objective, as shown in the diagram.

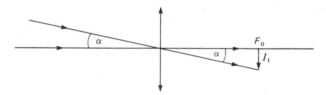

Using the approximation that $\tan \alpha = \alpha$, α can now be found.

14 A parallel beam of monochromatic light of wavelength 580 nm is incident normally on a diffraction grating having a large number of regular slits, each of width 0.70×10^{-6} m, as shown in the diagram. After passing through the grating the light will have, as a result of interference, intensity maxima in certain directions. Calculations predict that for the first, second and third order interference maxima, the values of the angle θ between the direction of the incident light and the directions of these maxima should be approximately 16°, 34° and 56°, respectively.

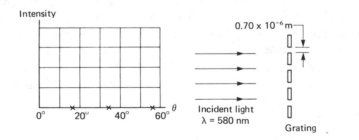

(a) At what value of θ would the light diffracted by a *single* slit of width 0.70×10^{-6} m have its first intensity *minimum*? Show the steps in your calculation.

(b) Produce a sketch graph showing how the intensity of light varies when plotted against θ. Make the range of θ vary from 0° to 60° and mark the 16°, 34° and 54° points on the axis. Plot the intensity on the vertical axis.

(**O and C Nuffield**: *NEAB* and all other Boards except SEB and WJEC*)

Points

(a) The questions refers to the single-slit diffraction pattern; use E4.7 with $m = 1$. The angle involved is measured in the same way as the angle in the diffraction grating formula.

(b) Two effects have to be allowed for in sketching the graph: diffraction from the individual slits and the interference pattern produced by the grating. The single-slit diffraction pattern should be drawn to form an 'envelope' to the interference pattern produced by the grating. The minima of the single-slit pattern are given by E4.7 and the maxima of the interference pattern are given in the question.

15 The phenomenon of Fraunhofer diffraction may be demonstrated by illuminating a wide slit by a parallel beam of monochromatic light and focusing the light that passes through the slit on to a white screen. A diffraction pattern may then be observed on the screen.

(a) Sketch the intensity variation in the diffraction pattern as a function of distance across it.

(b) What would happen to the intensity variation if the width of the slit were halved?

(**Cambridge***: all other boards except WJEC)

Points

(a) See 4.2 for a discussion of single-slit diffraction. For the intensity distribution, see Fig. 4.5.

(b) E4.7 can be used here. For fixed wavelength, consider the effect on the position of the first minimum caused by reducing the slit width. Give your answer by sketching the new intensity distribution, and comment on it and the initial distribution.

16 Both Young's double-slit interference experiment and the diffraction grating may be used to obtain a value for the wavelength of light. For each experiment:

(a) give a diagram of the experimental arrangement,

(b) state what measurements need to be made, and

(c) show how a value of the wavelength of light used is obtained from the measurements.

In Young's double-slit experiment a sodium lamp emitting light of wavelength 589 nm illuminates two slits with centres 0.60 mm apart. Interference fringes are observed on a screen 150 mm from the slits. Calculate the spacing of the maxima in the interference pattern.

Describe qualitatively the changes in the pattern that would be observed if each of the following changes were separately made to the Young's double slit experiment. (i) One slit only is covered with a thin sheet of mica which delays the light through the slit by one half of a period. (ii) The slits are made narrower without changing the separation of the centres of the slits. (iii) One slit is closed.

(**O and C**: all other Boards except SEB)

Points

(a) The arrangement for Young's double-slit experiment is shown in Fig. 4.2. For the diffraction-grating experiment, see the spectrometer in Fig. 4.24.

(b) For Young's double-slit experiment, E4.3 will tell you which measurements need to be made. For the fringe spacing, you would need to measure across several fringes to obtain an accurate value of spacing between two adjacent fringes. For the diffraction-grating experiment, E4.8 will tell you which measurements need to be made.

(c) See E4.3 and E4.8.
For the calculation of the spacing between maxima in the double-slit experiment, use E4.3.

(i) Before insertion of the mica over one slit, each bright fringe on the screen is at a point where arrival of a 'wave crest' from one slit coincides with arrival of a wave crest from the other slit. At one of these points, the effect of the mica is to make wave crests from one slit arrive at the same time as wave troughs from the other slit. You must describe how this will cause the observed pattern to change. (ii) See Fig. 4.9. The key point here is about single-slit diffraction. If single-slit diffraction could be ignored, the bright fringes would all be of the same intensity; however, the effect of single-slit diffraction is to cause 'missing orders' (as discussed for the diffraction grating in 4.2).

17 This question is about diffraction.

The passage below presents three sets of ideas about diffraction. For each of the sections (i) to (iii) you are asked to write a more complete explanation of the ideas. Your explanation may include:

fuller explanations of the theory;

quantitative calculations to illustrate the ideas;

discussion of possible experiments.

Passage

(i) Light from a point source appears to cast sharp shadows and this leads to the familiar idea that it travels in straight lines. However, this is not exactly true: the shadows are not perfectly sharp, although special experiments are needed to show the effect because it is so small. This unfamiliar property is called diffraction and is explained by a wave model of light.

(ii) The consequence is that the eye, or a camera, or even the best possible telescope, doesn't produce a perfect image. Instead it gives an image which is slightly blurred. When we try to make a telescope magnify more to show finer details of the stars this blurring effect can become an obstacle.

(iii) But diffraction can also be put to good effect. Diffraction gratings are made to enhance the effect and make use of it to give a powerful method of investigating spectra.

(O and C Nuffield: *all other Boards except SEB*)

Points

This question is in three parts (i), (ii) and (iii) and each part refers to a passage and should contain three sections: fuller explanation, quantitative calculations, possible experiments. However, these three sections need not be written separately and will often end up 'interwoven' as you answer each part of the question. However, it is a good idea to produce a rough plan based on the three sections and such a rough plan (deliberately slightly incomplete!) provides the rest of this commentary.

(i)   Fuller explanation – diagram and discussion of plane waves passing obstacle and being bent at the edges.

Quantitative calculations – diffraction only really noticeable when obstacle of comparable size to wavelength, but wavelength of light makes it difficult to get suitable objects.

Possible experiments – ripple tanks for wave properties, lasers (intense parallel beam) to project shadow over large enough distance to make diffraction noticeable (diagram).

(ii)   Fuller explanation – 'obstacles' behave like 'holes', optical instruments have 'hole' (objective lens) at front and hence must diffract. Two stars very close together can blur into one (resolution diagrams).

Quantitative calculations – use resolving power formula (E4.12) applied to eye or telescope (diagram to illustrate the angle involved).

Possible experiments – eye looking at lines on paper, telescope with variable aperture focused on adjacent point sources.

(iii) Fuller explanation – each slit diffracts light from a parallel beam so it spreads over a wide angle interfering with beams from other slits. Constructive interference only at certain specific angles (diagram).

Quantitative calculations – show that it is possible to get light at different wavelengths to interfere constructively at measurably different angles, use formula (E4.8) to show this.

Possible experiments – describe how to use gratings to measure wavelengths, remember need for parallel beam, importance of source of light, mention reflection gratings used on X-rays.

18 Describe the steps in outline you would take in setting up a diffraction grating in combination with a spectrometer to examine the spectrum of visible light. Explain the function of each part of the system and show on a clear diagram the passage of light rays from the source, through the system, to the observer's eye.

When the spectrum of light containing violet and red components only is examined with a diffraction grating, it is found that the fourth line from the centre (not counting the zero order line) is a mixture of red and violet. Explain this. If the grating has 500 lines per mm, and the diffraction angle for the composite line is 43.6°, find the wavelengths of the violet and red components.

What will be the fifth line in the spectrum and at what diffraction angle will it occur?

(**WJEC**: *all other Boards except SEB*)

Points

See 4.4 and Fig. 4.24 for the spectrometer used with a diffraction grating. You should explain the function of (i) the collimator, (ii) the turntable and levelling screws, (iii) the diffraction grating, (iv) the telescope. You should include the following 'setting up' steps: (i) adjusting the telescope to receive parallel light, (ii) adjusting the collimator to give parallel light, (iii) levelling the turntable. Refer to your textbook if necessary.

To explain the overlap of red and violet in the fourth line, remember that, for a given order, the longer wavelength is diffracted most, so you need to find out if red light has a longer wavelength than violet light. Then decide if the red part of the fourth line is second or third order, and if the violet part is second or third order; to overlap, red and violet must belong to different orders. See Fig. 4.7 if necessary. Once you have determined which order each of the two colours belongs to, then use E4.8 to determine the wavelength of each colour.

You should now know what order the red part of the fourth line belongs to, so calculate the angle of diffraction for the next order red line. Then do likewise for the violet (i.e. calculate the next order diffraction angle for violet), and then you should be in a position to state if the fifth line is red or violet.

19 (a) A narrow parallel-sided beam of white light is dispersed by a diffraction grating which is placed perpendicular to the beam, i.e. parallel to the incident wavefronts. Explain this effect in detail.

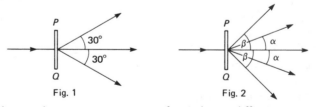

Fig. 1   Fig. 2

(b) Fig.1 shows the action, in air, of a plane diffraction grating *PQ* on a monochromatic beam of light which falls normally on the grating. Fig. 2 shows the whole arrangement immersed in water, of refractive index 1.33. The beam is now diffracted as shown, making angles α (<30°) and β (>30°) with the normal. Explain these changes produced by the water.

Given that the grating has 8.0×10^5 lines per metre, calculate (i) the wavelength of the light in air, (ii) the wavelength of the light in water, (iii) the angle α, (iv) the angle β.

(**London**: *all other Boards*)

Points

(a) The question asks you to *explain* the effect in *detail*, so you need to explain diffraction at each slit producing secondary wavelets, which reinforce in certain directions. With the aid of a diagram show $m\lambda = d \sin \theta_m$, from which it can be deduced that different wavelengths of the white light beam will reinforce at different angles (i.e. white light will be dispersed). Describe what will be seen – see Fig. 4.7.

(b) As the refractive index of water is 1.33, the speed of light in the water will be reduced by a factor of 1.33. As $c = f\lambda$, what will happen to λ? (and hence θ, from the diffraction formula?). This explains the magnitude of α and also why β is possible (see 4.2 for the number of orders possible). (i) Use E4.8 where $d = 1/(\text{no. of lines per metre})$. (ii) See 4.3 if necessary. (iii) Use the wavelength value from (ii) in E4.8, assuming $m = 1$. (iv) What is the value of m for the beam at angle β? Use E4.8 again.

20 (a) (i) Draw a diagram showing the path of rays (from a nonaxial point on a distant object) through a simple refracting telescope to the eye. (ii) Explain what is meant by the 'magnifying power' of the telescope, and show how it is related to the focal lengths f_o of the objective and f_e of the eyepiece.

(b) A refracting telescope has an objective focal length 900 mm and an eyepiece of focal length 12mm. (i) What is the distance apart of the lenses when the telescope is in normal adjustment and what is its magnifying power? (ii) If the telescope is used to project a sharp image of the Sun onto a screen that is at a fixed distance of 150 mm from the eyepiece, what adjustment to the telescope will be needed? (iii) Given that the Sun's diameter subtends an angle of 9×10^{-3} rad, calculate the diameter of the image on the screen.

(Oxford: *NEAB, SEB*)

Points

(a) (i) See Fig. 4.21 if necessary. (ii) See E4.11. To show how the mp is related to the focal lengths, you need to state and prove E4.11. You will find a proof in your textbook.

(b) (i) In normal adjustment, the eyepiece is adjusted so that the intermediate image (formed in the focal plane of the objective) lies in the focal plane of the eyepiece. It should be clear from your diagram how the distance between the lenses is related to the focal lengths. (ii) In this situation the eyepiece lens is further from the objective lens than in normal adjustment. The intermediate image formed by the objective still lies in the objective's focal plane, but because the eyepiece lens has been moved away, the intermediate image now lies outside the focal point of the eyepiece. Since the intermediate image acts as an object for the eyepiece lens, calculate the object distance (for the eyepiece lens) given the image distance (+150 mm) and the focal length of the eyepiece. Then you can state how far the eyepiece lens must be moved from its 'normal adjustment' position. (iii) Proceed in two stages: firstly, calculate the height of the intermediate image; secondly, given object and image distances for the eyepiece lens, calculate the linear magnification of the eyepiece (see E4.9). Then you can calculate the height of the final image (= lin. mag. × height of intermediate image). To calculate the height of the intermediate image, use of the small angle approximation, $\tan \alpha = \alpha$, and reference to the diagram may be helpful.

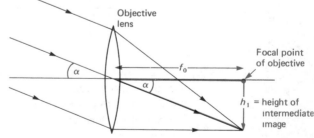

Objective lens

Focal point of objective

f_o

α

α

h_1 = height of intermediate image

Answers: 1 B; **2** B; **3** E; **4** B; **5** E; **6** E; **7** C; **8** A; **9** C; **10** B; **11** C; **12**(b)(iii) 664 nm; **13** 0.01 rad; **14**(a) 56°; **16** 0.15 mm; **18** 460 nm, 690 nm, 67° (violet); **19**(b)(i) 625 nm (ii) 470 nm (iii) α = 22.1° (iv) β = 48.8°; **20**(b)(i) 912 mm, × 75 (ii) increase separation by 1.04 mm (iii) 93.2 mm

CHAPTER 5

MATERIALS

Units in this chapter

Chapter objectives

After working through the topics appropriate to your syllabus in this chapter, you should be able to:

- define stress, strain and Young's modulus
- describe how to measure Young's modulus for a wire
- sketch and explain stress–strain curves for different materials
- understand the nature and effect of different types of bonds
- describe the difference between laminar and turbulent flow
- explain and use the continuity equation
- understand the difference between viscous and nonviscous flow
- solve simple flow problems using Bernoulli's equation
- work out the flow rate of a viscous fluid through a uniform pipe

5.1 STRENGTH OF SOLIDS

Stress is defined as force per unit area applied to a surface. The unit of stress is $N\,m^{-2}$ (or the pascal, Pa). Note that the unit of pressure is the same.

Strain is defined as change of length per unit length. Since strain is a ratio, it has no unit.

When stress is applied to a solid, the solid will change shape, even if only by a small amount. Consequently, the applied stress will cause a strain within the solid. Applied stress is treated by considering its normal and tangential components separately (i.e. perpendicular and parallel to the surface). Normal stress can either be **tensile**, if the solid is stretched, or it can be **compressive**. Tangential stress, if excessive, will cause the solid to shear. Knowledge of stress–strain characteristics is vital to ensure correct use of materials, either alone or in composite form.

Stress–strain curves

Various methods are available for obtaining information. The equipment must be able to apply regular increments of stress or strain to the sample, and it must enable accurate measurement to be made of both stress and strain. For metals, the sample is often in the form of a thin wire. In this form, large stresses may be applied by means of suspending the wire vertically and then hanging weights of up to 100 N from its lower end.

Consider the test wire shown in Fig. 5.1. The applied stress is given by the ratio (weight of load/initial cross-sectional area). The strain is calculated from the ratio (increase in length/initial length). Results are usually displayed graphically, with y = stress and x = strain. Fig. 5.2 shows how the strain in a wire varies with stress. With the type of test apparatus in Fig. 5.1, the wire would snap at B when the applied stress from the load exceeds the breaking stress.

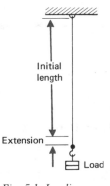

Fig. 5.1 Loading a wire

An alternative method of obtaining stress–strain information is to use a more specialised piece of apparatus known as a 'tensometer'. This equipment acts like a 'rack' because it extends the test material by measured amounts (i.e. the strain is increased in steps), allowing the tension to be measured at each step. Consequently, the material can be taken beyond the strain corresponding to the 'breaking stress'; the stress will fall as the material loses its strength beyond B.

The essential features of stress–strain curves for metals are shared by most metals, and are as follows (see Fig. 5.2):

❶ OP is a straight line, thus **stress is proportional to strain** (i.e. load is in proportion to extension) in agreement with Hooke's law.

❷ P is the **limit of proportionality**, beyond which the ratio stress/strain is no longer constant.

❸ E is the **elastic limit**. If the elastic limit is not exceeded when the material is stretched, then the material will regain its initial shape; otherwise, it will suffer permanent distortion.

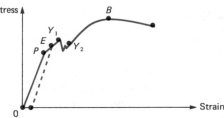

Fig. 5.2 Stress–strain curve for a typical metal

❹ Y_1 to Y_2 is the range over which **plastic flow** begins. From E to the lower yield point Y_1, removal of stress will reduce the strain, although not completely. If the wire was unloaded between E and Y_1, stress and strain would follow the dotted line of Fig. 5.2. Beyond Y_2, the strain would be permanent. For some metals, Y_1 and Y_2 are very close (i.e. a single yield point).

❺ B is the point of maximum stress, referred to as the **breaking stress** (or sometimes as the 'ultimate tensile strength', UTS).

For **nonmetals**, stress–strain curves vary widely from one material to another. Fig. 5.3 shows curves for rubber and polythene. The curves show that both materials readily depart from Hooke's law, but rubber regains its initial shape and so behaves elastically; polythene becomes permanently distorted from its initial shape, so it behaves in a 'plastic' manner. The term **hysteresis** is used to describe the difference between the 'loading' and 'unloading' curves. This means that when the stress is reduced from point A to point B on the diagram, the strain remains greater than if there was no hysteresis effect (i.e. strain 'lags' behind stress).

Knowledge of stress–strain behaviour is essential in giving scientific meaning to common terms used to describe the strength of materials. For example, a **ductile** material is one which can easily be lengthened and worked into 'shape' without fracture. This is because the stress–strain relationship has an extensive

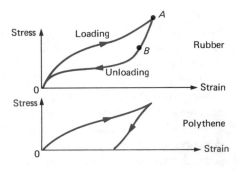

Fig. 5.3 Elastic loading (with hysteresis) and inelastic loading

plastic section, as shown by Fig. 5.4. Lead is a good example of a ductile material.

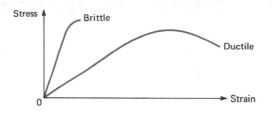

Fig. 5.4 Stress–strain curves for brittle and ductile materials

A **brittle** material, such as glass, will easily snap after reaching the elastic limit. Plastic behaviour of brittle materials is almost nonexistent, as the stress–strain curve of Fig. 5.4 shows.

5.2 ELASTIC BEHAVIOUR

Elasticity is the physical property that enables a solid to regain its initial shape after removal of an applied stress.

Young's modulus of elasticity (E) is defined as $\dfrac{\text{tensile stress}}{\text{tensile strain}}$

provided the limit of proportionality is not exceeded. The unit of E is the same as that of stress (N m^{-2} or Pa). For a loaded vertical wire, this definition gives:

E5.1 $\quad E = \dfrac{Tl}{Ae}$

where E = Young's modulus (N m^{-2}), T = tension due to load (N), l = initial length (m), A = cross-sectional area (m^2) = $\pi d^2 / 4$ (d = diameter), e = extension from initial length (m).

The value of E is obtained from the stress–strain curve by determining the gradient OP in Fig. 5.2. Also, from E5.1, the **spring constant** (k) may be defined by $T = ke$, where k is given by AE / l for a wire.

The **work done** to stretch a wire can be determined from a graph of y = tension, x = extension, as in Fig. 5.5. Since work done = force × distance, the area under the line gives the work done.

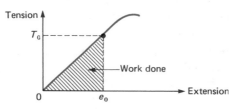

Fig. 5.5 Work done during Hooke's law loading

Provided that the limit of proportionality is not exceeded, the work done to stretch the wire to extension e_0 is therefore given by $\frac{1}{2} T_0 e_0$ where $T_0 (= EAe_0 / l)$ is the corresponding tension.

E5.2 $\quad$ Work done $= \dfrac{1}{2}\left(\dfrac{EA}{l}\right) e_0^2 = \dfrac{1}{2} k e_0^2$

The area under a stress–strain curve gives the work done per unit volume; so the area of the hysteresis 'loop' for rubber in Fig. 5.3 gives the energy absorbed (per unit volume) by the material in one cycle of stretching and relaxing the material. **Resilient** materials are those which can be stretched and relaxed without dangerous overheating; they must therefore have as small a hysteresis area as possible.

5.3 STRUCTURE OF SOLIDS

Solids can be classified in terms of their structure as either **crystalline** or **amorphous**. The atoms of a crystalline solid are arranged in a regular pattern, known as a crystal lattice. Some materials are obviously crystalline, but others, known as polycrystalline materials, are composed of many tiny crystals (e.g. metals). Amorphous solids, such as glass, totally lack any regularity of atomic arrangement in the sense of a lattice. Materials such as rubber are composed of molecules in the form of long chains of atoms, and the chains are tangled together in a haphazard way. Long 'chain' molecules are called **polymers**. Polymeric materials can be either crystalline (where the molecules fold against each other in an ordered pattern) or amorphous (where the molecules are tangled together).

Composite materials such as fibreglass are a physical mixture of two separate materials, chosen so that the weaknesses of each separate material are reduced by the presence of the other material.

Interatomic forces

The forces which bond atoms together are essentially electrostatic in origin, although there are several different varieties. The more important are:

① **Ionic** bonds, formed in crystals such as sodium chloride. When formed, each sodium atom gives up an electron to a chlorine atom; the resulting structure is a lattice of alternate sodium ions (Na^+) and chlorine ions (Cl^-), as in Fig. 5.6.

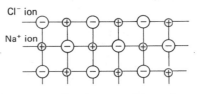

Fig. 5.6 *An ionic solid – sodium chloride*

② **Covalent** bonds, formed in many organic compounds and in semiconductors such as silicon. Each bond involves the sharing of two electrons, one from each atom, so that each atom's electron shells are full. For example, the outer shell of a silicon atom takes eight electrons when full; an isolated atom of silicon has only four electrons in its outer shell. In the solid state, each silicon atom forms four covalent bonds, one bond with each of its four neighbours, so satisfying the 'full shell' requirement (see Fig. 5.7).

③ **Metallic** bonds, in which metal atoms give up their outermost electrons, so as to form a lattice of positive metal ions in a 'sea' of electrons.

④ **Van der Waals** bonds exist between neutral atoms, and are thought to be due to the pull on the electrons of one atom by the nucleus of the other atom.

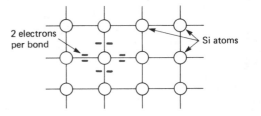

Fig. 5.7 *A covalent solid – silicon*

Atoms and elasticity

The various types of bonds share several common features, including long-range attraction, short-range repulsion and an equilibrium separation, as illustrated by Fig. 2.5(a). At small displacements from equilibrium, the force is in proportion to the displacement (i.e. the curve is linear near equilibrium), but this is not so at greater displacements. Thus, Hooke's law and the limit of proportionality can be explained.

In a crystalline material at its elastic limit, crystal planes begin to slip over one another, so that the solid will not return to its initial shape when freed from stress.

In an amorphous polymer, the long-chain polymer molecules are tangled together. Near the melting point, the material loses its elasticity because its molecules slide over one another under stress. At lower temperatures, the material is rubbery because the cross-links between adjacent molecules are strong enough to prevent the molecules sliding past each other when the material is under stress. Below the **glass transition temperature** T_g, an amorphous polymer becomes brittle like glass because the molecules lock together in a rigid structure.

Atomic spacing

The Avogadro constant (N_A) is defined as **the number of atoms in exactly 12 grams of carbon-12**. Note that the mole is the amount of substance that contains as many elementary units as there are carbon atoms in 12 grams of carbon-12 (the elementary unit might be an atom, a molecule, an ion, etc.). Thus N_A is given per mole. X-ray methods have been used to determine the Avogadro constant precisely.

Atomic spacing can be estimated if values are given for density (ρ), relative atomic mass (M) and the Avogadro constant (N_A). Since the mass of a single atom of a solid element is given by M/N_A, then the volume occupied by one atom is $M/\rho N_A$. Hence the appropriate spacing between atoms of a solid element is given by $D^3 = M/\rho N_A$ where D is the approximate distance between the centres of two adjacent atoms. Typically, values of the order of 0.1 nm are obtained.

5.4 FLUIDS

Fluids at rest

The **pressure** in a static fluid acts equally in all directions and increases with depth. At a certain depth beneath the surface of a liquid, the pressure due to the liquid is given by depth × density × g. This equation is used to work out atmospheric pressure from the reading of a mercury barometer. For example, if the height of the mercury column is 0.760 m, the equation gives 0.760 m × 13 600 kg m^{-3} (density of mercury) × 9.8 N kg^{-1} = 101 kPa for atmospheric pressure.

Any object immersed in a fluid experiences an **upthrust** due to upward pressure on its base. The upthrust is equal to the weight of fluid displaced. This is known as **Archimedes' principle**.

Fluid flow

A description of fluid flow, whether for a liquid or gas, is usually made by considering the motion of a small element of the fluid. The size of the fluid element must be small enough to enable it to follow the flow accurately, but if the fluid element is too small, then molecular motion becomes significant.

A **flowline** is a line along which a fluid element moves. A **streamline** is a stable flow line, along which every element follows the same path. Thus, a fluid element that passes through a given point will describe a flowline. If every subsequent fluid element which passes through that same point takes the same path as the first fluid element, then the flowline is stable, and is therefore a streamline.

A flow pattern can exist with both stable and unstable flow in different parts of the fluid, as in Fig. 5.8. Where the flow is stable, it is described as **laminar** (or streamlined) flow; unstable flow is referred to as **turbulent** flow. Thus streamlines cannot exist in areas of turbulence; in streamlined flow, fluid elements cannot cross streamlines.

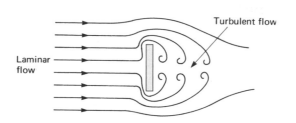

Fig. 5.8 Laminar and turbulent flow

Continuity

Flow of fluid along a stream-tube (i.e. a tube bounded by streamlines) will be fastest where the tube is narrowest. The mathematical statement of this fact is known as the **continuity equation**:

E5.3 $\rho A v$ = constant (= mass flow per unit time)

where ρ = fluid density (kg m^{-3}), A = area of cross-section of the tube (m^2), v = fluid speed (m s^{-1}).

For two points X and Y along a stream-tube, the mass per second flowing past X is equal to the mass per second flowing past Y. Since mass flow per second can be proved to be given by $\rho A v$ (see your textbook), then $\rho_x A_x v_x = \rho_y A_y v_y$. If the fluid is incompressible, then its density is constant (i.e. $\rho_x = \rho_y$) so the product vA is then constant (i.e. $v_x A_x = v_y A_y$).

5.5 NONVISCOUS FLOW

Bernoulli's equation

In the absence of viscous effects, laminar flow of a fluid can be described by Bernoulli's equation:

E5.4 $P + \rho g H + \frac{1}{2} \rho v^2$ = constant

where P = pressure (N m^{-2}), ρ = density (kg m^{-3}), H = height (m) measured from a fixed level, v = fluid speed (m s^{-1}).

For two points X and Y along a stream-tube, as in Fig. 5.9, Bernoulli's equation gives:

$$P_x + \rho_x g H_x + \frac{1}{2} \rho_x v_x^2 = P_y + \rho_y g H_y + \frac{1}{2} \rho_y v_y^2$$

Each of the three terms can be seen in relation to total energy of a fluid element, as follows:

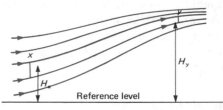

Fig. 5.9 Laminar flow

❶ $\frac{1}{2} \rho v^2$ is the **ke per unit volume**.

❷ $\rho g H$ is the **gravitational pe per unit volume**.

❸ P is the **work done per unit volume from work done by pressure**.

Thus, E5.4 is an expression of the fact that the ke + gravitational pe + energy due to pressure of a fluid element is constant. It follows that frictional energy loss due to viscous effects must be negligible for the equation to apply.

An example of the application of E5.4 is provided by laminar flow along a horizontal pipe which has a narrow section and a wide section, as in Fig. 5.10. For points X and Y as in the

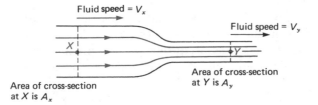

Fig. 5.10 Flow in a narrowing pipe

diagram, $H_x = H_y$; assuming the fluid is incompressible (i.e. $\rho_x = \rho_y$), then E5.3 gives $v_x A_x = v_y A_y$. Thus, $P_x - P_y = \frac{1}{2} \rho_y v_y^2 (1 - (A_y^2 / A_x^2))$. Since the fluid moves fastest in the narrow section, it follows that the pressure in the narrow section is less than in the wide section.

Pitot tubes

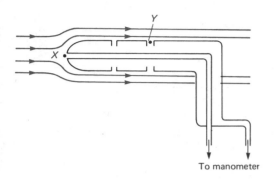

Fig. 5.11 Pitot-static tube

These tubes (also known as pitot-static tubes) are used to measure the flow speed of laminar flow which may be in a pipe or in the 'open' fluid. The pitot tube itself consists of a central narrow tube inside a second coaxial tube, as in Fig. 5.11. The tube is turned into the flow so that the opening X of the inner tube is a **stagnation point** (i.e. fluid speed is zero at X). Point Y, at one of the holes into the outer tube, is at a point where the fluid speed is the same as the upstream speed (U). A manometer is used to measure the pressure difference between points X and Y. Use of Bernoulli's equation then gives $P_x - P_y = \frac{1}{2}\rho U^2$ (assuming $H_x = H_y$), so that the fluid speed U can be calculated if the fluid density ρ is known.

5.6 VISCOUS FLOW

Fluid friction exists between layers of a fluid moving at different speeds. Consider laminar flow in a uniform channel, as in Fig. 5.12. The fluid in the centre moves fastest, whereas at

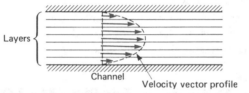

Fig. 5.12 Viscous flow in a pipe

the side the fluid will be stationary. Therefore, a **velocity gradient** will exist between the centre and the sides. In the absence of viscosity, the velocity at the centre would be the same as at the sides; the effect of viscosity is to cause the slow-moving surface layers to drag on fast-moving inner layers which in turn drag on the fastest moving central layers. The viscosity of a fluid is a measure of its resistance to flow.

The **coefficient of viscosity** (η) is defined as the tangential stress (T), in a fluid, required to produce unit velocity gradient across the fluid:

E5.5 $\quad T = \eta \dfrac{dv}{dx}$

where T = tangential stress (N m^{-2}), η = coefficient of viscosity (N s m^{-2}), $\dfrac{dv}{dx}$ = velocity gradient (s^{-1}).

Fig. 5.13 illustrates the meaning of this definition.

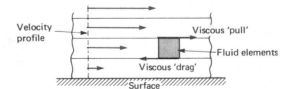

Fig. 5.13 Liquid–surface interface

Poiseuille's law

For viscous flow through a horizontal uniform pipe, Poiseuille derived the following equation:

E5.6 $\quad Q = \left(\dfrac{\pi r^4}{8L\eta}\right)P$

where η = coefficient of viscosity (N s m^{-2}), P = pressure difference across the ends of the pipe (N m^{-2}), r = internal radius of the pipe (m), L = length of the pipe (m), Q = volume per unit time (m^3 s^{-1}).

Stokes' law

When an object passes through a viscous fluid, the object experiences a frictional drag. For a uniform sphere moving at constant velocity through a viscous fluid, Stokes derived the following equation for the drag force:

E5.7 $F = 6\pi\eta r v$

where F = viscous force (N), η = coefficient of viscosity (N s m^{-2}), r = sphere radius (m), v = speed of sphere relative to the undisturbed fluid (m s^{-1}).

Molecular explanation of viscosity

In any fluid, molecules move in random directions with a wide range of speeds. Consider the interchange of molecules between fluid layers moving at different speeds, as in Fig. 5.14. The number of molecules that transfer each second from one layer to an adjacent layer will be unchanged from one part of the fluid to any other part. Thus, the number of molecules per second that transfer from layer A to layer B will be the same as the number that transfer in the reverse direction in the same time. However, since layer B moves faster than layer A, B gains slow-moving molecules from A while A gains fast-moving molecules from B. Since exchange of fast- for slow-moving molecules involves change of momentum, the

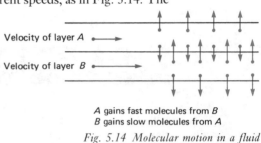

Velocity of layer A →

Velocity of layer B →

A gains fast molecules from B
B gains slow molecules from A

Fig. 5.14 Molecular motion in a fluid

loss of momentum from layer B to layer A means that A drags on B whereas B pulls A forward. To maintain the flow at a steady rate, a pressure gradient is required along the direction of flow, but the work done by the pressure force is dissipated as heat (i.e. increased molecular motion). Hence, viscosity corresponds to friction forces.

5.7 SURFACES

A surface is characteristic of either a solid or a liquid, and is the interface between the solid (or liquid) and the surrounding fluid. For a liquid–liquid surface, such as an oil–water surface, the two liquids must not mix with one another. For a solid, the surface is well defined, with little change in the atomic arrangement at the surface compared with the interior. However, solid surfaces easily become contaminated with 'foreign' atoms. Also, the state of the surface is an important factor in the strength of materials such as glass; in these, small cracks easily spread causing fracture.

For a liquid, the molecules move in a random manner although in contact with one another much more frequently than in a gas. There is a difference between the molecular arrangement at the surface and that at the interior; surface molecules have more space between them than interior molecules, and this increased separation gives rise to a state of tension between surface molecules (see Fig. 5.15).

Gas

Surface of liquid

Interior of liquid

Fig. 5.15
Molecules at
a liquid–gas
interface

Surface tension effects are evident in simple experiments such as floating a needle on a clean water surface. To explain such effects, it is necessary to assume that the surface is in a state of tension, like a stretched drumskin.

The **coefficient of surface tension** (γ) is defined as the surface tension force per unit length acting along the surface, perpendicular to a line in the surface. The unit of γ is N m^{-1}. Using this definition and Fig. 5.16:

E5.8 $F = \gamma L$

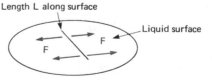

Length L along surface

Liquid surface

Fig. 5.16
Coefficient of
surface tension

where L = the length along the surface (m), F = surface tension force (N) acting on length L.

Capillary action

The rise of water up the inside of a narrow vertical capillary tube, with its lower end in a beaker of water, is because of surface tension forces acting on the water surface where it is in contact with the glass, as in Fig. 5.17.

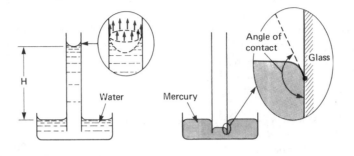

Fig. 5.17 Meniscuses of mercury and water

E5.9 $\gamma = \frac{1}{2} H r \rho g$

where H = height as shown (m), r = internal radius (m), ρ = density of the liquid (kg m^{-3}).

The equation is arrived at by equating the weight of the liquid in the capillary thread ($\rho H \pi r^2 g$) to the surface tension force acting upon the perimeter of the meniscus at the top of the thread ($\gamma . 2\pi r$).

Angle of contact

This is the angle (in the liquid) that a liquid surface makes with a solid surface at the point of contact. For water, the angle of contact is zero when in contact with glass. Thus the surface tension forces in the capillary action example, where water is in contact with glass, act upwards parallel to the tube axis and hence the water has to rise. For mercury on clean glass, the angle of contact exceeds 90°, so mercury in a capillary tube with its lower end in a beaker of mercury will be pushed downwards, as in Fig. 5.17.

Bubbles

The internal pressure of a bubble must be greater than the external pressure, so that the pressure difference will be sufficient to enable the surface tension forces (which try to collapse the bubble) to be withstood. By considering the forces on half the bubble, as in Fig. 5.18, the excess pressure is given by:

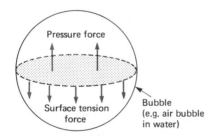

Forces shown acting on top half

Fig. 5.18 Forces in a bubble

E5.10 $P = \dfrac{2\gamma}{r}$

where P = excess pressure (N m^{-2}),
γ = coefficient of surface tension (N m^{-1}),
r = bubble radius (m).

The equation is derived by equating the surface tension force on the perimeter of the hemisphere to the pressure force on the flat area of the hemisphere (i.e. $\gamma . 2\pi r = P\pi r^2$). For a soap bubble or any other bubble with *two* surfaces, the excess pressure is given by $4\gamma/r$.

Surface energy

When a surface is extended, work must be done against the surface tension forces. Consider a surface, as in Fig. 5.19, which is extended by moving one of its edges. The surface tension force on the edge is γL, where L is the edge length.

If the edge is moved by distance d, perpendicular to its length, so as to extend the surface, then the work done ($=$ force $\times$ distance) must be equal to $\gamma L d$. Therefore, the work done to extend the surface by unit area is γ. If there is no change of temperature, then the increase of surface energy per unit area, σ, is equal to the work done per unit area, γ. In other words, the work done to extend a surface under isothermal conditions (i.e. constant temperature) can be considered as surface energy.

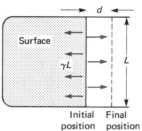

Fig. 5.19 *Stretching a liquid film*

Molecular explanation of surface tension

The variation of pe with separation for a pair of molecules is shown in Fig. 5.20. In the interior of a liquid, molecules will be at equilibrium separation, so there will not be an overall force on any given 'interior' molecule; their pe will be a minimum. At the surface, molecules will be at a slightly greater separation so weak attractive forces will link surface molecules, thus preventing escape unless a surface molecule has sufficient ke.

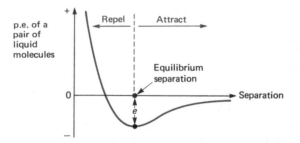

Fig. 5.20 *Potential plotted against separation for two molecules*

When a surface is extended, molecules will be brought from the interior to the surface. Each such molecule will lose half its number of neighbours, so that if n is its number of neighbours (on average) in the interior, then at the surface it has neighbours on one side only, giving it (on average) $\frac{1}{2}n$ neighbours. Removal of the molecule from one of its neighbours requires energy e to be supplied, so the removed molecule and the neighbour each require energy $\frac{1}{2}e$. Therefore, if the removed molecule loses half its number of neighbours, then it requires energy $\frac{1}{2}n \cdot \frac{1}{2}e$. Finally, if the number of molecules per unit area of surface is A, then the energy required to increase the surface by unit area must be $\frac{1}{4}neA$. Since the surface energy per unit area is equal to γ, then it follows that:

E5.11 $\gamma = \frac{1}{4}neA$

where A = number of molecules per unit area (m^{-2}),
n = number of neighbours of each interior molecule,
e = equilibrium pe between two molecules (J),
γ = coefficient of surface tension (N m^{-1}).

Chapter roundup

Your studies in this chapter involve the nature of materials as well as their physical properties. The nature of materials is developed further in Chapter 7 in the context of electrical conductors and semiconductors. Also, make sure you understand the concept of the mole and the nature of the interaction between atoms before moving onto Chapters 6 and 8.

Question Bank

1 The diagram shows stress–strain curves for three different materials taken to 'fracture'. Which curve best shows the behaviour of a copper wire, and which one best shows the behaviour of a 'glass fibre'?

Stress

I

II

III

0 — Strain

	A	B	C	D	E
Copper wire	I	I	II	III	II
Glass fibre	II	III	III	I	I

(–: *all Boards except SEB*)

Points

Remember that copper is ductile, and glass is brittle. See 5.1 if necessary.

2 The following data were obtained when a wire was stretched within the elastic region:
Force applied to wire 100 N.
Area of cross section of wire 10^{-6} m^2.
Extension of wire 2×10^{-3} m.
Original length of wire 2 m.
Which of the following deductions can be correctly made from this data?
1 The value of Young's modulus is 10^{11} N m^{-2}.
2 The strain is 10^{-3}.
3 The energy stored in the wire when the load is applied is 10 J.
Answer:
A if 1, 2, 3 correct.   D if 1 only.
B if 1, 2 correct.   E if 3 only.
C if 2, 3 correct.

(**London**: *and all Boards except SEB*)

Points

For 1, use E5.1. For 2, see 5.1 if necessary. For 3, see Fig. 5.5 and the subsequent discussion if necessary.

3 A helical spring extends 40 mm when stretched by a force of 10 N, and for tensions up to this value the extension is proportional to the stretching force. Two such springs are joined end-to-end and the double-length spring is stretched 40 mm beyond its natural length. The total strain energy, in J, stored in the double spring is:
A 0.05   B 0.10   C 0.20   D 0.40   E 0.80

(**NICCEA**: *all other Boards except SEB H*)

Points

Since stress/strain remains constant for a given material, within the limits of proportionality (see 5.1), it follows that a double-length spring will give double the extension for the same applied force. The energy stored in the stretched spring will equal the work done in stretching it. Remember that the average force will be half the applied force.

4 A certain wire which obeys Hooke's law is extended by a force *F*. If the wire is replaced by a similar piece of wire which is twice as long as before, and which is subjected to the same force *F*, which of the following statements is/are true?

1 The stress in the longer wire is the same as that in the shorter.
2 The strain of the longer wire is the same as that of the shorter.
3 The work done in extending the longer wire by the force *F* is the same as for the shorter.

A 1 only.   B 2 only.
C 1 and 2 only.   D 1 and 3 only.
E 2 and 3 only.

(**NICCEA**: *all other Boards except SEB*)

Points

For 1, assume that the two wires are of the same diameter. See 5.1 for the definition of stress if necessary.
For 2, remember that strain = stress/Young's modulus.
For 3, use E5.2 but take care since the extensions are not equal.

5 The diagram below shows how the extension of five wires, each of different material, changes with tension. The diagram also shows the initial length and diameter of each wire. Which wire exhibits least strain for a given stress (i.e. for the same stress applied to each wire)?

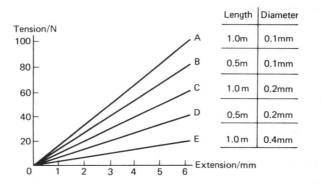

	Length	Diameter
A	1.0m	0.1mm
B	0.5m	0.1mm
C	1.0m	0.2mm
D	0.5m	0.2mm
E	1.0m	0.4mm

(*–: all Boards except SEB*)

Points

You must pick the wire with the greatest value of Young's modulus of elasticity. See 5.2 for the definition of Young's modulus if necessary.

6 Which of the wires A–E of the previous question has most energy stored in it when a load of 20 N is applied to each wire in turn?

(*–: all Boards except Oxford, SEB*)

Points

Remember that energy stored in this situation can be calculated from the area under the line of the tension–extension graph. Take care when you compare areas because you must consider the area under each line up to the point where tension = 20 N.

7 The force *F* required to extend a length of rubber by a length *x* was found to vary as in the diagram. The energy stored in the rubber for an extension of 5 m was:

A less than 100 J.
B 100 J.
C between 100 J and 200 J.
D 200 J.
E 500 J.

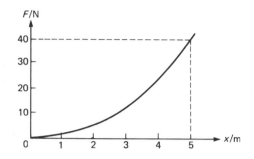

(**Cambridge**: *and all other Boards except Oxford, SEB*)

Points

You must consider the area under the curve up to $x = 5$ m. It might be helpful to mark the point on the curve at $x = 5$ m, and then draw a line from that point to the origin. Compare the area under the curve with the area under the straight line which you have drawn in; you ought to be able to easily calculate the area under the straight line.

8 A uniform wire, fixed at its upper end, hangs vertically and supports a weight at its lower end. If its radius is r, its length L and the Young's modulus for the material of the wire is E, the extension is:
1 directly proportional to E.
2 inversely proportional to r.
3 directly proportional to L.
Answer:
A if 1, 2, 3 correct   D if 1 only
B if 1, 2 correct   E if 3 only
C if 2, 3 correct

(**London**: *NICCEA and all other Boards except SEB*)

Points

Rearrange E5.1 with the extension e as the subject. Remember area $= \pi r^2$.

9 A sealed U-tube contains nitrogen in one arm and helium (at pressure p) in the other arm. The gases are separated by mercury of density ρ with dimensions as shown below.

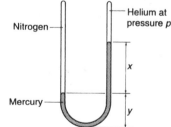

Which one of the following gives the pressure of the nitrogen?
A p
B $x\rho g$
C $p - x\rho g$
D $p + x\rho g$
E $p + (x + y)\rho g$

(**Oxford**: *and all other Boards*)

Points

See 5.4 if necessary.

10 The average male has a mass of 70 kg and the minimum cross-sectional area of the bone in each leg is approximately $5.0 \times 10^{-4} m^2$. The compressive breaking stress of bone is approximately $1.0 \times 10^7 N\ m^{-2}$.

   (a) Assuming that a man is standing with his weight equally shared by each leg, showing the steps in your working, calculate (i) the maximum stress in his leg bones and (ii) the ratio (maximum stress in bones)/(breaking stress).

   (b) Now suppose that all the linear dimensions of the human body are increased by a factor of nine. (i) What is the new value of the ratio found in (a) (ii)? Show and explain how you reach your answer. (ii) Explain what would happen to this 'giant' if he attempted to chase a normal human being.

   (O and C Nuffield: *NICCEA and all other Boards except SEB*)

Points

(a) (i) Stress is defined as force/area, so maximum stress from a given force will occur when it is applied to the minimum area. Remember that the force on the bones comes from 'weight' (and not 'mass') and that the weight must be divided equally between each leg. (ii) Divide the answer to (a) (i) by the value quoted for the breaking stress. As this is a ratio, remember that it will have no units.

(b) (i) The breaking stress is a property of the material and will be unaffected by the 'size' of the bones. However, to determine what has happened to the 'maximum stress' you need to consider the effect on the 'force' and 'area' as a result of the nine times linear growth. The area has increased 81 times but what about the force? The force depends on the mass of the man which is proportional to his volume. (ii) The way in which this question is phrased suggests that the running giant will break his bones whereas the man will not. The answers to (a) (ii) and (b) (i) must be used to justify this. A sensible guess as to the extra stress imposed as a result of running rather than standing must be made.

11 (a) Define the Avogadro constant.

   (b) The relative atomic mass of aluminium is 27.0 and the density is 2.70×10^3 kg m^{-3}. If the Avogadro constant is 6.02×10^{23} mol^{-1}, calculate the number of atoms in one cubic metre of aluminium. *Estimate* the distance between centres of neighbouring aluminium atoms, stating any assumptions you make.

   (NEAB: *and all other Boards*)

Points

   (a) The definition is in terms of carbon-12. Refer to your textbook if necessary.
   (b) See 5.3.

12 This question is about the way a karate expert breaks blocks of wood or concrete by bringing down the side of his fist vertically at high speed as shown in the diagram.

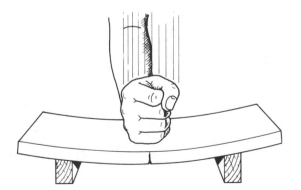

(a) Estimate the kinetic energy a fist and forearm could have on impact, explaining how you arrive at your estimate.

(b) Show how the following results can be derived for a uniformly stretched block of material: (i) energy stored = $\frac{1}{2}$(final force × extension); (ii) energy stored per unit volume = $\frac{1}{2}$(stress × strain); (iii) energy stored per unit volume at breaking point = $\frac{1}{2}S^2/E$ where E is Young's modulus and S the stress at breaking.

(c) Use the results from (b) to calculate the energy required to break the blocks of wood and concrete specified in the data given at the end of the question.

(d) The impact is a collision in which a moving mass (the fist and forearm) hits a stationary mass (the block), and the two then move forward together. Using the data given at the end of the question and your estimate in (a), estimate how much of the initial kinetic energy is converted into other forms of energy. Explain how this result might be used in relating the results of parts (a) and (b) above.

(e) The energies actually needed to break blocks turn out in practice to be smaller than the arguments in (c) and (d) above would predict. Suggest *two* possible reasons for this and describe in outline how your suggestions could be checked by experiment.

(f) When it strikes a concrete block of the size quoted below, a karate fist slows down in about 3 ms. Estimate the mean force exerted and then the average power involved.

(g) How would you expect the processes of impact and breaking for a wood block to differ from these processes for a concrete block?

Useful Data	Wood	Concrete
Block size	30 cm × 15 cm × 2 cm	40 cm × 20 cm × 4 cm
Density	350 kg m^{-3}	2150 kg m^{-3}
Young's modulus	1.4×10^8 N m^{-2}	2.8×10^9 N m^{-2}
Breaking stress	3.6×10^6 N m^{-2}	4.5×10^6 N m^{-2}

Estimate any quantities that you need and that are not given.

(**O and C Nuffield**: *all other Boards except SEB*)

Points

(a) Use $\frac{1}{2}mv^2$ where m is the mass of the fist and forearm and v is the speed at which they are travelling on impact.

(b) (i) Use the explanation of E5.2 in the text; or use work done = average force × distance, where average force is half the final force for Hooke's law stretching. (ii) This is easy to prove as long as the definition of stress and strain (5.1) can be recalled. (iii) This follows from rewriting the formula proved in (ii) as $\frac{1}{2}$(stress)^2 × (strain/stress).

(c) Calculate the volume of each block and use the formula of (b) (iii).

(d) This is a difficult question to follow. It points out that after the collision the hand and block continue to move downward together. If you can estimate the speed at which they move you can calculate the kinetic energy now in the system (using the table of information to help you find the mass of the block). The kinetic energy loss will have been used to break the block; part (a) helps you in the kinetic energy calculations, part (b) supplies an independent method of assessing the energy needed to break the block.

(e) The predictions of part (d) are only correct if all the missing kinetic energy is used to break the block, but is it? However, part (b) is based on the properties of the material and can only be wrong if the material were to break below the theoretical value; three reasons could be put forward for this. One could argue that the calculation is for a 'perfectly' formed slab, but all specimens in practice may be flawed. The other two arguments are better. Firstly, the calculations of part (b) are based on static tests but the experiment is dynamic (consider the effect of a sudden jerk on a string compared with a steady static pull). Secondly, the experiment involves bending the material, but the calculation uses Young's modulus which can only be applied to tensile and compressive forces. Whatever two choices you make you will need to make all the measurements experimentally both in a static and dynamic case. Load the slab with weights until it breaks, taking appropriate measurements to check the theory of (b). Drop a weight (equal to fist and forearm) from the right height to ensure that it hits the block with an appropriate speed; make sure the weight 'sticks' to the block. Stroboscopic photography will aid measurement.

(f) It is easiest to find the average force by halving the breaking force calculated from the breaking stress. The power is found easily from dividing the energy delivered to the block by the time taken to deliver it (3 ms).

(g) Consider the different behaviour of wood and concrete under bending stresses, the different energies needed to break them and the different nature of the fractures involved.

13 Fig. 1 shows the results of an experiment to investigate the variation of stress with strain for steel wire.

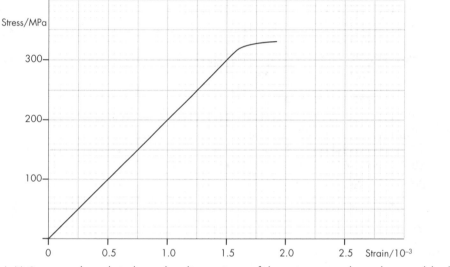

Fig. 1

(a) (i) State and explain how the dimensions of the wire are selected to enable the uncertainty in measuring the extension to be as low as possible. (ii) State the quantities you would need to *measure* to determine values of stress and strain. Name the instruments you would use to determine these quantities.

(b) Use the graph to determine a value for the Young modulus of steel.

(c) (i) In one experiment steel wire, with the stress–strain characteristics shown in Fig. 1, had a cross-sectional area of 0.25×10^{-6} m^2 and an initial length of 3.0 m. Determine the strain energy when it is extended to the limit of proportionality. (ii) Explain why it is sensible to wear safety spectacles when performing this experiment.

(d) Fig. 2 shows the force–extension graph for a *brittle* material to the point at which it breaks.

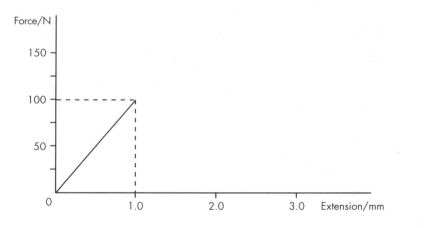

Fig. 2

Copy the graph. Draw other graphs on the same set of axes for a sample of brittle wire of the same length, which:
(i) has twice the Young modulus but the same diameter and breaking strain (label this A);
(ii) is made of the same material but has half the diameter (label this B).

(**AEB** Jun 93: *all Boards*)

Points

(a) Consider E5.1 to decide how the extension depends on the wire's dimensions and to decide what measurements to make.
(b) and (c) See p.141 if necessary.
(d) (i) The breaking strain and length are unchanged so the extent at breaking point is the same. How does the force depend on E? (ii) The breaking stress and E are the same hence the breaking strain is the same. How does the force depend on the diameter?

14 (a) Draw a sketch graph showing how the force between two atoms separated by a distance x varies with x over the range up to about $x = 5$ atomic diameters. Using the same axes, draw a second graph to show how the potential energy of the system varies with x. Use your graphs to explain: (i) why most solids obey Hooke's law for small extensions; (ii) why a solid expands when its temperature is raised.

(b) The density of copper is $8.92 \times 10^3 \, kg \, m^{-3}$ and one mole of copper atoms has a mass of $6.35 \times 10^{-2} kg$. Calculate: (i) the number of atoms in a cubic metre of copper; (ii) the separation (a) of nearest-neighbour atoms, given that the effective volume occupied by an atom is $\left(1/\sqrt{2}\right)a^3$. The specific latent heat of sublimation of copper at 25 °C is $5.34 \times 10^6 \, J \, kg^{-1}$. (iii) Express this in joules per atom and explain how the result is related to the energy of the bonds between the copper atoms. (Take the value of the Avogadro constant N_A to be $6.02 \times 10^{23} \, mol^{-1}$.)

(**Oxford**: *and London, NEAB, Cambridge, WJEC and NICCEA*)

Points

(a) See 2.4 and Fig. 2.5(a) and Fig. 2.5(b).
(b) (i) Remember that 1 mole of copper contains N_A atoms. As a first step, calculate the volume of 1 mole of copper, given its density and the mass of 1 mole. (ii) Use your answer to (b) (i) to calculate the volume occupied by a single atom of copper, then use the given formula to calculate separation a. (iii) To express the specific latent heat in $J \, atom^{-1}$, first of all express the specific latent heat in $J \, mol^{-1}$ instead of $J \, kg^{-1}$ by using the fact that each mole of copper has a mass of $6.35 \times 10^{-2} kg$. Then use the given value of N_A to convert to $J \, atom^{-1}$. See 6.2. Note that the final value for specific latent heat in $J \, atom^{-1}$ is *not* the bond energy, but it is related to the bond energy. E6.5 shows how to calculate the bond energy, but since you are not given the number of neighbours (n) of each copper atom, then you are not expected to calculate the bond energy. Instead, show how the bond energy (e) and the number of neighbours (n) are related to the latent heat value in $J \, atom^{-1}$. Remember that the latent heat value in $J \, atom^{-1}$ is the energy that must be supplied to a single atom of copper to remove it from the 'lattice' of copper atoms and take it away completely. Remember that energy $\frac{1}{2}e$ is given to the removed atom when each bond between it and a neighbour is broken. One final point; sublimation is change of state from a solid directly to a vapour.

15

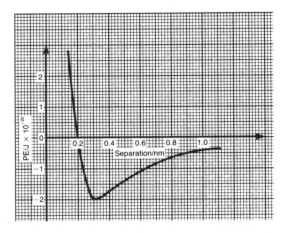

(a) The graph shows the variation of potential energy with separation for two atoms of a particular element. Explain how the following quantities can be obtained from the graph and give values for the element concerned:
(i) the equilibrium separation of the atoms;
(ii) the energy required to dissociate the atoms.
(b) Using the data on the graph, construct a graph of the variation of the force between the atoms with separation. Put numerical values on your graph where possible.
(c) Use your graph or the energy graph to discuss the elastic properties of the element in its solid form.
(d) Explain each of the following observations in terms of the microstructure of the solid concerned.
(i) A sheet of glass can be cut by scratching its surface and bending along the scratch.
(ii) Concrete is strong in compression but not in tension.
(iii) An annealed copper wire becomes stiffer when it is bent back and forth.
(O and C*: *NEAB*, London*, O and C Nuffield, Oxford, Cambridge, WJEC)*

Points

(a), (b) and (c) See 2.4.
(d) (i) The scratch creates tiny cracks on the surface. When the sheet is bent, stress concentrates at the cracks so the cracks grow and join together. See your textbook for further details.
(ii) In tension, stress builds up at internal boundaries, and cracks and fissures develop and grow inside the concrete. In compression, stress is transmitted through internal boundaries. See your textbook for further information.
(iii) See your textbook.

16 This question is about explaining the flow of water from the tank and about designing an investigation of this flow. Approximately one-third of the marks are reserved for the investigation plan in part (e).

A student is investigating the flow of water through a hole at *B* in the vessel shown. He arranged a device to keep the head (*h*) of water constant.
(a) Explain each of the following steps (i), (ii) and (iii) in a theory which the student has worked out:
(i) The mass per second (*m*) is related to the volume per second (*q*) and the density (*p*) as follows: $m = pq$
(ii) The value of *q* should be $q = Av$, where *A* is the area of cross section of the stream of water and *v* is its velocity of flow at *B*.
(iii) The value of *v* should be given by $v = \sqrt{2gh}$.

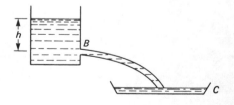

(b) In order to check his theory, the student sets h at 9.0 cm and collects the water for 100 s; the mass collected is 6.91 kg. He knows that the diameter of the circular opening at B is 1.0 cm. Show that, if he assumes that the area A is the same as that of B, his measurements do not agree with his theory.

(c) The student decides that either v or A do not have the expected values. He decides to check on the velocity by making measurements of the drop y in the level of the jet when it is a distance x from the hole and using the equation $v = x \sqrt{g / 2y}$. He finds from such measurements that v is within 5% of this theoretical value. Explain the theory of his method to check the velocity.

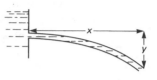

(d) Say what conclusions you could draw from the results so far and discuss the proposal that there must be a difference between the assumed and actual values of A. Consider the size of the difference that might be required and whether this could provide a reasonable explanation.

(e) Outline a plan for further experimental investigation designed to check your ideas in (d) and to extend the understanding of them as much as possible. Credit will be given for thinking of a variety of systematic tests, and for stating quite precisely what you would measure and what you would do with the measurements to relate them to your ideas about the situation. You should also consider how you would make the measurements and whether you could obtain adequate accuracy.

(O and C Nuffield: *and all other Boards*)

Points

There is normally one long-answer question (for Nuffield) each year that tests your skill at handling algebra. This question is an example, and it follows the usual trend of describing an experiment or system that you will not have encountered during normal teaching. Some candidates foolishly ignore such questions because they wrongly imagine that they have not covered the subject material. If you are good at manipulating algebra and equations, and found it easy to follow the proofs you have been shown during the course, then you should be able to score well on this type of question.

(a) (i) Mass = density $\times$ volume, so mass per second = … (ii) Water with velocity v in one second will produce a stream v long with cross-sectional area A from a nozzle of area A; this is a cylinder of length v, cross-sectional area A and volume … (iii) gh normally suggests potential energy in a uniform gravitational field (mgh). Equate the pe lost by the water to the kinetic energy it gains and …

(b) You should expect the practically obtained value of v from (a) (ii) to be less than the theoretical value calculated from (a) (iii); this may help you if you make a mistake in your calculations.

(c) This is a standard projectile problem; the water should have a constant horizontal velocity v and be falling vertically with a constant acceleration g. This enables the formula to be derived; see 1.2.

(d) Part (c) makes it clear that the velocity v is correct (within 5%) so the value of A used in the equation is wrong. Calculate what value of A should be used in (a) (ii) to give the correct value of q and compare with the measured value for A. Is it possible that error in measuring the diameter of the nozzle would account for the discrepancy? Perhaps 'friction' between the water and the tube at the nozzle effectively reduces the value of A which has been measured with good accuracy.

(e) With one-third of the marks available for this part, clearly some depth is needed. Depending on how you have answered part (d) the emphasis of the answer to this part may vary. Key points in any set of experiments should involve varying both the nozzle diameter and the water height, as well as choosing suitable instruments to measure the diameter. You may choose to try and measure the actual diameter of the water stream as it leaves the nozzle. Explain the analysis of the results as well as discussing the practical details and whether you can obtain sufficient accuracy.

17 (a) Explain the terms 'lines of flow' and 'streamlines' when applied to fluid flow and deduce the relationship between them in laminar flow.

(b) State Bernoulli's equation, define the physical quantities which appear in it and the conditions required for its validity.

(c) The depth of water in a tank of large cross-sectional area is maintained at 20 cm and water emerges in a continuous stream out of a hole 5 mm in diameter in the base. Calculate: (i) the speed of efflux of water from the hole; (ii) the rate of mass flow of water from the hole.

Density of water = 1.00×10^3 kg m^{-3}.

(**NEAB*** *and AEB, Cambridge**)

Points

(a) See 5.4.

(b) See E5.4 and subsequent comments.

(c) Sketch the arrangement and consider two points as follows: point X on the water surface of the tank and point Y at the hole. Since both points are open to the atmosphere, then the pressure at X = pressure at Y = atmospheric pressure. Measure levels from a horizontal line through the hole. Apply E5.4 to the situation, with terms for X on one side of the equation and terms for Y on the other side. To calculate the fluid speed at Y (i.e. the speed of efflux), you need to make an assumption about the fluid speed at X. Base this assumption on the fact that the level at X would drop very slowly for a large tank.

To calculate the mass flow per second, use E5.3.

Answers: 1 E; **2** B; **3** B; **4** C; **5** A; **6** E; **7** A; **8** E; **9** D; **10**(a)(i) 7×10^5 N m^{-2} (ii) 0.07 (b)(i) 0.63; **11** 6.02×10^{28} m^{-3}, 2.55×10^{-10} m; **12** (a) about 300 J (c) 42 J for wood, 12 J for concrete (f) about 7 kN and 4 kW; **13**(b) 2.0×10^{11} N m^{-2} (c)(i) 0.34 J; **14**(b)(i) 8.46×10^{28} m^{-3} (ii) 2.03×10^{-10} m (iii) 5.63×10^{-19} J atom^{-1}; **15**(a)(i) 0.3 nm (ii) 2.0×10^{-20} J; **16**(b) from (a)(ii) $v = 0.9$ m s^{-1} and from (a)(iii) $v = 1.3$ m s^{-1} (d) measurements suggest diameter should be 0.8 cm and not 1.0 cm; **17**(c)(i) 2 m s^{-1} (ii) 0.039 kg s^{-1}

CHAPTER 6

HEAT AND GASES

Units in this chapter

Chapter objectives

After working through the topics appropriate to your syllabus in this chapter, you should be able to:
- define the absolute and Celsius scales of temperature
- explain how different types of thermometers are used
- define and measure specific heat capacity and solve simple problems
- define and measure specific latent heat and solve simple problems
- define thermal conductivity and compare it to electrical conductivity
- solve simple thermal conduction problems including thermal conductors in series
- outline the nature and properties of thermal radiation
- state and use the ideal gas equation
- prove and use the kinetic theory equation
- relate the mean kinetic energy of the molecules of an ideal gas to the absolute temperature
- solve simple problems on the thermodynamics of ideal gases

6.1 TEMPERATURE

Temperature is a measure of the degree of hotness of an object, and can be thought of as a measure of the mean ke of the random movements of its molecules. **Heat** is energy transferred on account of temperature difference. Since heat is a form of energy, it is measured in joules. Like work, heat should always be considered in the context of transfer of energy, so it is discussed in terms like 'heat gained', etc. However, heat has no meaning for individual molecules, unlike work. In terms of energy, an object can possess either mechanical energy (i.e. ke and pe due to bulk movement of the whole mass) or internal energy (i.e. ke and pe due to molecular behaviour). An object can do work by using up some of its mechanical energy or some of its internal energy or both. Likewise, its energy can be transformed into heat released by the object.

Two objects in thermal contact will be at the same temperature if there is no resultant flow of heat between them. If there is a resultant flow of heat from one object to the other, then the object that is the 'net' source of heat must be at a higher temperature than the object that accepts the heat energy (i.e. the 'net' heat sink).

Temperature scales

Every thermometer is based upon a physical property which changes continuously with change of temperature. Such a property is known as the **thermometric property**, and some examples include length (for a liquid-in-glass thermometer) and electrical resistance (for a resistance thermometer). Each thermometer must be calibrated so that readings of the thermometric property give numerical values of temperatures on a chosen scale. The calibration requires measurement of the thermometric property at two standard degrees of hotness, known as **fixed points**.

The **centigrade scale** of temperature is based upon the melting point of pure ice as the lower fixed point (defined as $0\,°C$) and steam temperature at 1 standard atmosphere of pressure as the upper fixed point (defined as $100\,°C$). For a thermometer with thermometric property X, the following equation gives its temperature θ:

E6.1 $\theta = \dfrac{\left(X_\theta - X_0\right)}{\left(X_{100} - X_0\right)} \times 100$

where X_θ = value of the thermometric property X at $\theta\,°C$, X_0 = value of X at ice point, X_{100} = value of X at steam point.

By definition, all thermometers calibrated on the same scale give the same readings at the fixed points. Between the fixed points, readings will usually differ from one thermometer to another.

The **Celsius scale** is the centigrade scale achieved by calibrating a constant-volume gas thermometer (see Fig. 6.1) at the ice and steam points. Because its readings are based on the properties of an ideal gas, it serves as the standard thermometer against which other types of thermometers are compared.

The **thermodynamic scale** (sometimes called the **absolute scale**) is defined in terms of two fixed points which are:
❶ **absolute zero** (0 kelvin (K) or $-273.15\,°C$)
❷ the **triple point** of water ($273.16\,K$ or $0.01\,°C$)

A constant-volume gas thermometer may be used to measure thermodynamic temperatures; the thermometric property is the product of pressure and volume of an ideal gas. In practice, pV at low pressure is used and thermodynamic temperature is calculated from the equation:

E6.2 $T = \dfrac{(pV)}{(pV)_{Tr}} \times 273.16$

where $(pV)_{Tr}$ = the value of (pV) at the triple point, T = temperature (K)

The use of this equation means that temperature according to the Celsius scale = $T - 273.15$.

Types of thermometers

The table in Fig. 6.2 lists the more important features of five types of thermometers commonly used in scientific work. The constant-volume gas thermometer can be used to measure ideal gas temperatures directly, but above 90 K other thermometers are more suitable in practice. The international practical scale of temperatures (IPTS) gives the type of thermometer which should be used for accurate scientific work between defined fixed points.

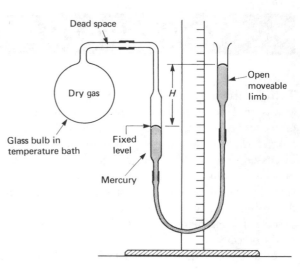

Fig. 6.1 Constant-volume gas thermometer

Fig. 6.2 *Types of thermometers*

Type of thermometer and its range	Thermometric property	Advantages	Disadvantages	Particular uses
Mercury-in-glass −39 °C to 450 °C (234 K to 723 K)	Length of column of mercury in capillary tube	(i) Quick and easy to use (direct reading) (ii) Easily portable	(i) Fragile (ii) Small size limits precision (iii) Limited range	(i) Everyday laboratory use where high accuracy is not required (ii) Can be calibrated against constant-volume gas thermometer for more accurate work
Constant-volume gas thermometer −270 °C to 1500 °C (3 K to 1750 K)	Pressure of a fixed mass of gas at constant volume. See Fig. 6.1	(i) Very accurate (ii) Very sensitive (iii) Wide range (iv) Easily reproducible	(i) Very large volume of bulb (ii) Slow to use and inconvenient	(i) Standard against which others are calibrated (ii) He, H_2 or N_2 used depending on range (iii) Can be corrected to the ideal gas scale (iv) Used as standard on IPTS below −183 °C (90 K)
Platinum resistance −180 °C to 1150 °C (90 K to 1400 K)	Electrical resistance of a platinum coil. See 7.5 for the Wheatstone bridge method of measuring resistance	(i) Accurate (ii) Wide range	Not suitable for varying temperatures (i.e. is slow to respond to changes)	(i) Best thermometer for small steady temperature differences (ii) Used as standard on IPTS between −183 °C and 630 °C (90 K and 903 K)
Thermocouple −250 °C to 1150 °C (25 K to 1400 K)	Emf produced between junctions of dissimilar metals at different temperatures. See 7.4 for measurement of 'thermal' emfs	(i) Fast response because of low heat capacity (ii) Wide range (iii) Can take remote readings using long leads	Accuracy is lost if emf is measured using a moving-coil voltmeter (as may be necessary for rapid changes when potentiometer is unsuitable)	(i) Best thermometer for varying temperatures (ii) Can be made direct reading by calibrating galvanometer (iii) Used as standard on IPTS between 630 °C and 1063 °C (903 K and 1336 K)
Radiation pyrometer Above 1000 °C (1250 K)	Colour of radiation emitted by a hot body	Does not come into contact with temperature being measured	(i) Cumbersome (ii) Not direct reading (needs a trained observer)	(i) Only thermometer possible for very high temperatures (ii) Used as standard on IPTS above 1063 °C (1336 K)

6.2 HEAT CAPACITIES

The **specific heat capacity** (c) of a material is defined as the heat necessary to raise the temperature of unit mass by one degree, without change of state. The unit of c is usually J kg^{-1} K^{-1}, although mass in moles is frequently encountered so giving the unit of molar heat capacity as J mol^{-1} K^{-1}.

E6.3 $\quad \Delta Q = Mc\Delta\theta$

where ΔQ = heat energy supplied or removed (J), M = mass (kg), c = specific heat capacity (J kg^{-1} K^{-1}), $\Delta\theta$ = temperature change (K or °C).

Note that the **heat capacity** of an object is the heat energy which will raise its temperature by one degree. The symbol C is used for heat capacity, and its unit is J K^{-1}. If the object is made from a single material of specific heat capacity c, then $C = Mc$, where M is the mass of the object.

Specific latent heat (l) of vaporisation (or fusion or sublimation) is defined as the heat necessary to change the state from liquid to vapour (or from solid to liquid, or from solid to vapour directly) of unit mass of material, without change of temperature. The unit of l is usually $J\ kg^{-1}$, although $J\ mol^{-1}$ is sometimes used.

E6.4 $\Delta Q = Ml$

where ΔQ = heat energy (J), M = mass (kg) that changes its state, l = specific latent heat $(J\ kg^{-1})$.

Molecular interpretation of latent heat

Heat supplied to an object will either change its temperature or change its state. Increased temperature mainly corresponds to increased ke of the molecules, whereas change of state involves energy associated with breaking or forming bonds between molecules. For example, when water freezes, latent heat is given off because bonds are formed; to melt ice, latent heat must be supplied because bonds must be broken. A rough estimate of the energy required to break a single bond can be obtained from E6.5, which relates bond energy to latent heat:

E6.5 $l = \frac{1}{2}Nne$

where l = specific latent heat $(J\ kg^{-1})$, N = number of molecules per unit mass (kg^{-1}), n = number of neighbours per molecule, e = bond energy (J).

The bond energy is equal to the minimum pe of a pair of molecules, as shown in Fig. 5.20. Note that if l is given in $J\ kg^{-1}$, then N must be expressed as the number of molecules per kg; if M is the relative molecular mass and N_A is Avogadro's number, then the number of molecules per unit mass will be N_A/M.

Experimental methods

Most modern methods for determining specific heat capacities and specific latent heats of materials involve electrical heating. A typical circuit for supplying electrical power to a heater is shown in Fig. 6.3. A rheostat is used to maintain a constant value of current, recorded on the ammeter. The pd across the heater is recorded from the voltmeter, and the time for which electrical power is supplied must be measured. The electrical energy supplied is given by current $(I)\times$ pd $(V)\times$ time (t), where t must be in seconds.

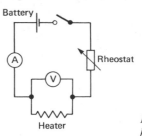

Fig. 6.3 Electrical heating

Measurement of c of a solid by Nernst's method

The heat capacity of the container is not required, since the solid is suspended in an evacuated container with silvered walls, as in Fig. 6.4. The heating coil is also used as a resistance thermometer.

E6.6 $IVt = Mc\Delta\theta$

where I = current (A), V = pd (V), t = time (s), M = mass of solid (kg), c = specific heat capacity of solid $(J\ kg^{-1}\ K^{-1})$, $\Delta\theta$ = rise in temperature of solid (K).

Measurement of c of a liquid by electrical calorimetry

A measured quantity of liquid is heated electrically in an insulated calorimeter of known heat capacity (C). The temperature rise (θ), current, pd and heating time must all be measured. To calculate c, use:

E6.7 $IVt = Mc\Delta\theta + C\Delta\theta$

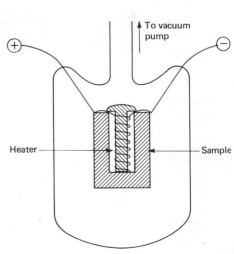

Fig. 6.4 Nernst's apparatus

where C = heat capacity of calorimeter (J K^{-1}), M = mass of liquid (kg), c = specific heat capacity of liquid (J kg^{-1} K^{-1}), $\Delta\theta$ = temperature rise of liquid (K), I = current (A), V = pd (V), t = time (s).

The method can be extended to determine c of a solid by putting a known mass of the 'unknown' solid in the calorimeter, and then adding a known mass of liquid (in which the solid is insoluble) of known specific heat capacity.

Measurement of c of a liquid by continuous flow calorimetry

The liquid is passed at a steady rate through an insulated tube containing an electrical heater, as in Fig. 6.5. For a measured rate of flow, electrical power is supplied at a constant rate so

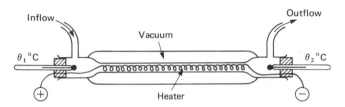

Fig. 6.5 Continuous-flow calorimeter

that steady temperatures θ_1 and θ_2 are recorded at the inlet and outlet. Thus the electrical energy supplied per second (IV) is given by the equation:

$$(IV) = \left[\left(\frac{m}{t}\right)c(\theta_2 - \theta_1)\right] + H$$

where m/t is the mass flow per second and H is the heat loss per second. The rate of flow is then altered (and remeasured) and the power supply is adjusted so that the inlet and outlet temperatures remain at the *same* values as before. Thus the new rate of supply of electrical energy (IV)' is given by:

$$(IV)' = \left[\left(\frac{m}{t}\right)'c(\theta_2 - \theta_1)\right] + H$$

where $(m/t)'$ is the new rate of flow of mass. Note that H is the same as before because the temperatures are the same as before. By subtraction, the following equation can be used to calculate c:

E6.8 $\quad (IV)' - (IV) = \left[\left(\frac{m}{t}\right)' - \left(\frac{m}{t}\right)\right]c(\theta_2 - \theta_1)$

where V = pd (V), I = current (A), θ_2 = outlet temperature, θ_1 = inlet temperature, c = specific heat capacity of liquid, (m/t) = mass flow per second (kg s^{-1}).

Note that the heat capacity of the container is not required, and that heat loss is accounted for by taking the difference between two rates, as in E6.8.

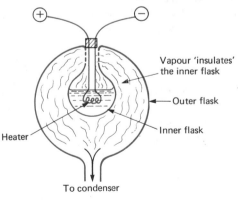

Fig. 6.6 Apparatus to measure latent heat of vaporisation

Measurement of l of a liquid

The liquid is heated electrically in a jacketed container, as shown in Fig. 6.6. The vapour in the outer jacket cuts down heat loss from the inner vessel, and condensed vapour is collected as shown. The rate of condensation is measured for two different power supply rates. For each rate, the electrical energy supplied per second $(IV) = (m/t)l + H$, where (m/t) is the rate of condensation of mass of vapour and H is the heat loss per second. Because H is the same at each rate (remember that H depends only upon the liquid temperature, which is the same at both rates because the liquid is at boiling point in each case), then by subtraction, the following equation gives l:

E6.9 $(IV)' - (IV) = \left[\left(\dfrac{m}{t}\right)' - \left(\dfrac{m}{t}\right)\right] l$

where I = current (A), V = pd (V), (m/t) = mass condensed per second (kg s⁻¹), l = specific latent heat (J kg⁻¹).

6.3 THERMAL COEFFICIENTS

For a physical property Z, the thermal coefficient of Z is defined by:

E6.10 $\alpha = \dfrac{(Z_\theta - Z_0)}{Z_0\theta}$

where α = thermal coefficient (°C⁻¹), Z_0 = value of Z at 0 °C, Z_θ = value of Z at θ °C.
   Some of the more common examples of thermal coefficients include the following:

Linear expansivity

$\alpha = (L_\theta - L_0)/(L_0\theta)$ = expansion/(length at 0 °C × temperature rise from 0 °C).

Cubic expansivity

$\gamma = (V_\theta - V_0)/(V_0\theta)$ where V is the volume, etc. The two coefficients are linked by the relationship $\gamma = 3\alpha$.

Resistance

$\alpha_R = (R_\theta - R_0)/(R_0\theta)$ where R is electrical resistance. For metals, resistance increases with increased temperature so α_R is positive. For semiconductors, resistance falls with increased temperature so α_R is negative.

Gas pressure

$\alpha_P = (P_\theta - P_0)/(P_0\theta)$ where P is gas pressure. For an ideal gas, α_P is equal to 1/273.

6.4 THERMAL CONDUCTION

There are three methods of heat transfer:
❶ **Convection**, which involves bulk motion of a fluid (liquid or gas), and is usually caused by hot fluid (being less dense) rising and displacing cold fluid.
❷ **Radiation**, which involves emission and absorption of electromagnetic radiation.
❸ **Conduction**, which take place in solids and in fluids, regardless of any bulk motion of the fluid. Conduction, like convection, requires a temperature difference to be established within the material which conducts.

   Consider heat conduction along a solid bar of uniform cross-section, as in Fig. 6.7, which has one end maintained at temperature θ_1 and the other at θ_2. If the sides of the bar are well insulated, heat flows along straight lines as shown, and the temperature falls uniformly along the axis, as on the accompanying graph.

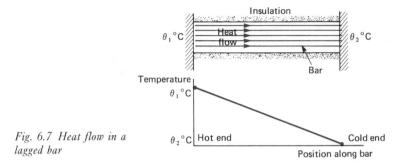

Fig. 6.7 Heat flow in a lagged bar

Temperature gradient is defined as the change of temperature per unit change of distance. The unit is K m^{-1}. For the insulated bar, the temperature gradient is constant and equals $(\theta_1 - \theta_2)/L$, where L is the bar length.

If the bar was not insulated along its sides, then the direction of heat flow would not be constant. The temperature gradient would be different at different positions along the bar, as shown in Fig. 6.8.

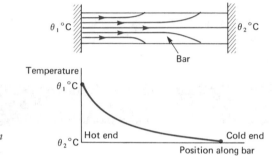

Fig. 6.8 Heat flow in an unlagged bar

For the insulated bar, the heat flow per second (dQ/dt) is proportional to (i) the temperature gradient and (ii) the area of cross-section (A). Introduction of a constant of proportionality (k) gives:

E6.11 $\quad \dfrac{dQ}{dt} = kA \dfrac{(\theta_1 - \theta_2)}{L}$

where (dQ/dt) = heat flow per second (J s^{-1}), A = area of cross-section (m^2), $\theta_1 - \theta_2$ = temperature difference (K or °C), L = length (m).

The constant k is known as the **coefficient of thermal conductivity** of the material of the bar, and it is defined as the heat flow per second per (metre)^2 of cross-section through which the heat flows normally, per unit temperature gradient along the flow direction. The unit of k is W m^{-1} K^{-1}.

Thermal conductors in series

Consider two heat conductors placed so that all the heat which flows through one then passes through the other, as in Fig. 6.9. The interface temperature (x) can be calculated by using E6.11 for each conductor separately, and then equating the heat flow per unit time of one body to the corresponding quantity for the other body, i.e.:

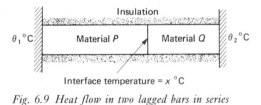

Fig. 6.9 Heat flow in two lagged bars in series

$$k_P A_P \frac{(\theta_1 - x)}{L_P} = k_Q A_Q \frac{(x - \theta_2)}{L_Q}$$

This situation is not unlike electrical conductors in series, in that a flow of heat requires a temperature difference, whereas a flow of electrical current requires a potential difference. Thus (dQ/dt) is equivalent to current I, and $(\theta_1 - \theta_2)$ is equivalent to pd V; hence the quantity L/kA is equivalent to electrical resistance R. The quantity L/kA is therefore known as the **thermal resistance** of the material.

U-values

The U-value of a wall (or window or floor or roof) in a building is defined as the heat flow per second per m² produced by a temperature difference of 1 K. Hence, heat per second passing through can be calculated from U-value × surface area × temperature difference. Note the unit of U-value is W m⁻² K⁻¹.

Comparison of the nature of conduction in good and poor conductors

For solids that are poor conductors, heat transfer is due to lattice vibrations, so that when a solid gains heat at one point, the atoms near that point vibrate more. Thus the neighbouring atoms vibrate more, which then makes atoms further away vibrate more, etc., spreading the heat energy gradually through the solid.

Good conductors, such as metals, conduct both heat and electricity well. The reason for this is the presence of free (conduction) electrons. When one part of a metal is heated, the atoms in that part vibrate more, so free electrons in that part gain extra ke by collision with atoms. Since the electrons are free to move through the solid, they move away and transfer their ke to other electrons (in other parts) when they collide. Transfer by lattice vibrations still takes place, but energy transfer by movement of free electrons is much greater.

6.5 THERMAL RADIATION

Thermal radiation is electromagnetic radiation, and is emitted by all objects at temperatures greater than absolute zero. The higher the temperature of an object, the greater is the amount of radiation energy emitted per second. When a hot object is placed in a cold room, the object radiates to the room walls at a greater rate than it receives from the walls. Thus, the object loses energy and cools until it is at the same temperature as the walls.

Absorption of radiation by a given surface depends upon:

❶ the nature of the surface, in terms of smoothness and colour;

❷ the quantity and quality (i.e. wavelength) of the incident radiation.

Emission of radiation by a given surface depends upon:

❶ the surface temperature;

❷ the nature of the surface.

When radiation is incident upon a surface, it is either absorbed, transmitted or reflected. Surfaces which are good absorbers are also good emitters of thermal radiation.

Black-body radiation

A surface that absorbs all wavelengths of inci-dent radiation will appear black. Such a surface can emit all wavelengths of radiation, and is known as a **black-body emitter**. The quality and quantity (sometimes called the 'spectral distribution') of the radiation from a black-body emitter changes with temperature, as indicated by Fig. 6.10.

The total energy radiated away per second from a surface is given by **Stefan's law**:

E6.12 $W = \sigma e A T^4$

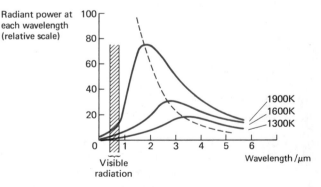

Fig. 6.10 Black-body radiation curves

where W = radiated power (J s⁻¹ or W), A is the surface area (m²), T = **absolute** temperature of the surface (K), e = surface emissivity, σ = Stefan's constant (W m⁻² K⁻⁴).

The **emissivity** is the ratio of black-body power to actual power for the same area and temperature. Therefore, $e = 1$ for a black body; for non-black bodies e can vary from 0 to 1.

The net power radiated away from a black body (at temperature T) in cooler surroundings (at temperature T_0) can be written as $(\sigma A T^4) - (\sigma A T_0^4)$, where the last term is the power radiated from the surroundings and absorbed by the body. Note that both terms contain the surface area A of the body; the area of the surrounding surfaces does not enter the equation.

It is worth noting that the black-body radiation curves cannot be explained using classical wave theory; the idea of electromagnetic radiation in terms of photons leads to an explanation of the shape of the curves (and of Stefan's law).

6.6 IDEAL GASES

The word 'ideal' used in the study of gases has a special meaning in that it is used to describe gases under conditions such that the experimental gas laws are valid. The gas laws are summarised by the **ideal gas equation**:

E6.13 $PV = nRT$

where P = gas pressure (N m^{-2}), V = gas volume (m^3), n = number of moles, T = temperature (K), R = molar gas constant (J mol^{-1} K^{-1}).

The gas constant R, expressed in terms of moles (i.e. J mol^{-1} K^{-1}), has the same value for all ideal gases, and this important point is based on the experimental result that 1 mole of any gas under 'ideal' conditions at stp (i.e. 0 °C and 1 standard atmosphere of pressure) occupies 0.0224 m^3. Thus, substitution of $V = 0.0224$ m^3, $T = 273$ K, $n = 1$ mole and $P = 1.02 \times 10^5$ N m^{-2} (= 1 standard atmosphere) gives the value of R as 8.31 J mol^{-1} K^{-1}.

From ideal gas equation, you can see how the individual gas laws 'fit in':

❶ **Boyle's law:** n and T are fixed, hence PV = constant (i.e. $P_1 V_1 = P_2 V_2$).

❷ **Charles' law:** n and P are fixed, hence V/T = constant (i.e. $V_1 / T_1 = V_2 / T_2$).

❸ **Combined gas law:** n is fixed, hence PV/T = constant (i.e. $P_1 V_1 / T_1 = P_2 V_2 / T_2$).

Brownian motion

When smoke particles in air are observed with the aid of a microscope, the path of each individual particle is seen to continually change direction in an unpredictable way. The cause of these unpredictable movements is the random bombardment of the smoke particles by much smaller (hence invisible) fast-moving gas molecules. Any one smoke particle will be subjected to uneven bombardments over its surface, since the gas molecules move about at random.

The kinetic theory of gases

The theory is based upon several assumptions:

❶ A gas consists of **point molecules**; in other words, the volume of the molecules themselves compared with the container volume is negligible.

❷ Gas molecules are in a state of **continual random motion**. This is based upon Brownian motion observations.

❸ Molecules collide **elastically** (i.e. no loss of total ke) with one another and with the container walls. If the collisions were inelastic, the gas would need to be supplied with energy to maintain its motion.

❹ Except when in collision, molecules exert **negligible forces** upon one another; in other words, intermolecular attraction is so small that it does not affect the pressure of impacts on the walls.

❺ The **duration** of a collision is much less than the time between collisions.

With the above assumptions, the laws of mechanics and statistics give the **kinetic theory equation** for gas pressure:

E6.14 $PV = \frac{1}{3} Nm\overline{c^2}$

where P = gas pressure (N m^{-2}), V = gas volume (m^3), N = number of molecules, m = mass of a single molecules, $\overline{c^2}$ = mean square speed of the gas molecules (m^2 s^{-2}).

You should consult your textbook for the proof of this equation and should try to see at which stage each assumption is used. Most proofs consider N molecules in a rectangular box, and treat the proof in two main stages:

❶ Use of the laws of mechanics to prove that the pressure on any one face is given by $PV = Nm\overline{u^2}$, where $\overline{u^2}$ is the mean value of (velocity components normal to that face)^2. Assumptions 1, 3, 4 and 5 are used here. Note that $\overline{u^2}$ is a shorthand way of writing $(u_1^2 + u_2^2 + u_3^2 +\ldots+u_N^2)/N$. Your textbook proof should show you how to deal with this stage in detail.

❷ Use of the laws of statistics to prove that $\frac{1}{3}\overline{c^2}$ is equal to $\overline{u^2}$, so giving the equation as stated. The use of statistics is based upon the assumption that the molecules are in random motion. Again, look at your textbook for details.

Note that the **root mean square** (rms) **speed** of the gas molecules is the square root of the mean value of the (molecular speeds)^2; in other words,

$$\text{the rms speed} = \sqrt{\left(\frac{c_1^2 + c_2^2 + c_3^2 +\ldots+c_N^2}{N}\right)}$$

where $c_1, c_2, c_3, \ldots, c_N$ are the individual values of molecular speeds.

Kinetic energy and temperature

By combining the ideal gas equation (E6.13) with the kinetic theory equation (E6.14), the mean ke of a gas molecule, $\frac{1}{2}m\overline{c^2}$, can be shown to be equal to $3RT/2N_A$, where N_A is Avogadro's number (and is not to be confused with the number of moles, n). Thus the total ke of 1 mole of an ideas gas is $\frac{3}{2}RT$.

6.7 THERMODYNAMICS OF IDEAL GASES

When heat is supplied to an isolated system, the amount of heat energy equals the increase of internal energy plus the mechanical work done by the system. This general rule is known as the **first law of thermodynamics**. Note that a gas can only do work when its volume changes: the **work done by a gas = pressure × change of volume**, since pressure is force/area and volume change is area × distance moved (so that work done = force × distance = pressure × volume change).

E6.15 $\Delta Q = \Delta U + P\Delta V$

where ΔQ = heat supplied (J), ΔU = change of internal energy (J), P = gas pressure (N m^{-2}), ΔV = change of gas volume (m^3).

The symbol Δ is used to mean a change, and if the change is a +ve value, then this is always taken to mean an increase. If the change is a −ve value, then this is taken to mean a decrease. For example, if ΔV is given or calculated as −4.0 m^3, then the volume has decreased by 4 m^3. Again, if the gas is supplied with 15 J of heat energy and the gas volume increases by 10^{-4} m^3 at a constant pressure of 10^5 N m^{-2}, then it follows by use of E6.15 that the internal energy increases by 5 J.

Constant volume

No work is done since ΔV is zero, so heat supplied goes entirely to increasing the internal energy of the gas. If c_v is the molar specific heat capacity at constant volume, then a temperature raise of ΔT for n moles of gas will require heat supply $\Delta Q = nc_v\Delta T$. Therefore, the increase of internal energy will be $nc_v\Delta T$.

Constant pressure

Since $PV = nRT$, then $P\Delta V = nR\Delta T$, since the pressure is constant. Thus, the work done $= nR\Delta T$. Now the increase of internal energy for temperature rise ΔT is $nc_v\Delta T$. It follows that to increase the temperature at constant pressure, the heat supplied must be equal to $(nR\Delta T)$ $+ (nc_v\Delta T)$. If c_p represents the molar specific heat capacity at a constant pressure, then the heat supplied is $nc_p\Delta T$ so that $nc_p\Delta T = nc_v\Delta T + nR\Delta T$, giving:

E6.16 $c_p - c_v = R$

where $c_p =$ molar specific heat capacity at constant pressure, $R =$ molar gas constant, $c_v =$ molar specific heat capacity at constant volume.

In other words c_p is greater than c_v because heat supplied at constant pressure is used to increase the internal energy *and* to enable the gas to do work when it expands. **Isobaric changes** occur at constant pressure.

Adiabatic changes

An adiabatic change is one in which no heat enters or leaves the system (i.e. $\Delta Q = 0$). For example, the sudden expansion of an ideal gas from a high-pressure container into low pressure would not involve heat transfer on account of the rapidity of the expansion. The work done by the gas in expanding comes from a fall in the internal energy of the gas.

Consider an adiabatic change of an ideal gas which involves a small change of temperature ΔT. The work done is $P\Delta V$ and the change of internal energy ΔU is $nc_v\Delta T$. Because the change is adiabatic, $\Delta Q = 0$ so use of E6.15 gives $P\Delta V = -nc_v\Delta T$. By combining this equation with the ideal gas equation $PV = nRT$ the following equation for adiabatic changes can be derived:

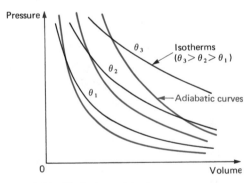

Fig. 6.11 *Adiabatics and isotherms*

E6.17 $PV^\gamma =$ constant, where $\gamma = c_p/c_v$

Adiabatic changes for an ideal gas can be compared with **isothermal changes** ($\Delta T = 0$) by appropriate sets of curves on a P–V diagram, as in Fig. 6.11. Note that the adiabatic curves are always steeper then isotherms, since the slope of an isotherm is $-P/V$ whereas the slope of an adiabatic curve is $-\gamma P/V$, and γ is always greater than 1.

Heat engines

A heat engine such as the petrol engine works in a continuous cycle of operations in which a working substance is heated, used to do work, then cooled before being reheated. In a petrol engine, internal combustion of petrol vapour heats the working substance which is air. The indicator diagram in Fig. 6.12 shows how the pressure changes during one complete cycle of a petrol engine. This is an idealisation since it is based on a fixed mass of air, which is not the case in practice since oxygen in the air and petrol vapour burn in the cylinder.

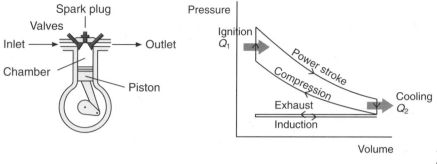

Fig. 6.12 Indicator diagram for a four-stroke petrol engine

Any heat engine absorbs heat Q_1 from a source at high temperature T_1 and rejects a lesser amount of heat Q_2 to a sink at lower temperature T_2. The work done by the engine $W = Q_1 - Q_2$. This is represented by the area of the loop of the indicator diagram.

The thermal efficiency of the engine $= W/Q_1$ ($\times 100$ to convert to %)
$$= (Q_1 - Q_2)/Q_1$$

The most efficient engine is a theoretical engine known as a 'reversible engine' which can reverse the energy changes without any energy being wasted. As explained in 6.10, this corresponds to no change of entropy (i.e. disorder). This requires Q_1/T_1 to be equal to Q_2/T_2.

Since $Q_2/Q_1 = T_2/T_1$ for a reversible engine, the maximum possible efficiency of an engine operating between temperatures T_1 and T_2 is therefore $(T_1 - T_2)/T_1$.

6.8 REAL GASES

The term 'real gas' is used to describe gases under conditions where the ideal gas equation does *not* hold.

Andrews' experiments

The behaviour of gases under a wide range of pressures and temperatures was thoroughly investigated by Andrews. The apparatus used, as in Fig. 6.13, was designed to withstand

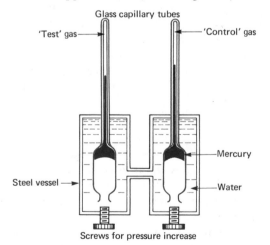

Fig. 6.13 Apparatus for testing gases

pressure up to 400 atmospheres. The volume of gas under test (e.g. CO_2) can be measured directly from the length it occupies. The pressure is calculated by first measuring the volume of the 'control' gas (e.g. dry air), and then using Boyle's law to calculate its pressure. The test gas tube is contained in a water bath so that sets of readings are made at different temperatures.

The results are usually presented as a set of PV isotherms (i.e. curves of $y = P$, $x = V$ at constant temperature for each curve) as in Fig. 6.14. The important features of the graph are:

❶ The **critical isotherm**, above which the gas cannot be liquefied, no matter how much pressure is applied.

❷ The **critical point**, which is the point of inflexion on the critical isotherm (i.e. the point on the critical isotherm at which the slope is zero).

❸ Each isotherm below the critical isotherm exhibits common features. For example, consider the isotherm $ABCD$. (i) Section AB is unsaturated vapour. (ii) Section BC is saturated vapour in equilibrium with liquid. At B, the material is 100% vapour; at C, it is 100% liquid. (iii) Section CD corresponds to liquid.

The **critical temperature** is defined as the temperature above which a gas cannot be liquefied by increase of pressure alone. Note that the word 'vapour' is reserved for the gas when its temperature is less than the critical temperature. Each gas has its own critical temperature; for example, the critical temperature of CO_2 is 31 °C, as shown in Fig. 6.14.

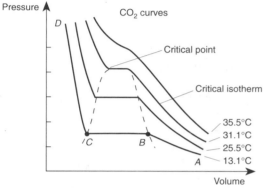

Fig. 6.14 PV curves for carbon dioxide

Explanation of real gas behaviour

An ideal gas is one that obeys the ideal gas equation. To illustrate the *difference* between ideal gas and real gas behaviour, Andrews' results can be plotted as isotherms of $y = PV$ against $x = P$, as in Fig. 6.15.

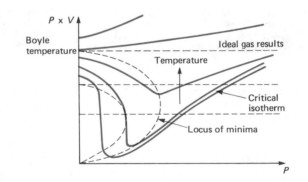

Fig. 6.15 PV against P curves

Ideal gas behaviour would give results as indicated by the dotted lines parallel to the x-axis (since $PV = \text{constant} \times T$). Actual gas behaviour only conforms to ideal gas behaviour for a limited temperature range (different for different gases) at low pressure. The temperature for which the curve of PV against P is initially flat is known as the **Boyle temperature** (T_B).

❶ $T > T_B$: PV increases as P increases.

❷ $T < T_B$: PV falls then rises as P increases.

The explanation of the experimental results lies in reconsidering two of the assumptions of kinetic theory, as follows:

❶ Point molecules – at high density, the actual volume of the molecules themselves is a significant fraction of the measured gas volume.

❷ Intermolecular attraction – at low temperatures, the molecules move more slowly, so the weak forces of attraction (van der Waals forces) between molecules have longer to act. Hence the force of impact on the container sides is reduced.

Above T_B, as P increases, the density increases and the molecules become closer to one another. Consequently, the measured volume increasingly overestimates the available volume. Hence PV rises.

Below T_B, as P increases, the molecules again come closer together. However, intermolecular attractions become significant, making the measured pressure less than the 'kinetic theory' pressure. Hence PV falls until further increase of pressure causes the volume factor to become more important; then PV stops falling and starts to rise.

The **Van der Waals equation** is an attempt to represent real gas behaviour:

E6.18 $\left(P + \dfrac{a}{V^2}\right)(V - b) = RT$

where b depends on the molecular volume, and a/V^2 represents the pressure reduction caused by intermolecular forces between molecules hitting the container walls and 'bulk' molecules (i.e. those far from any surfaces). The constant b is called the **co-volume** and is about 4 times the actual volume of the molecules.

Saturated vapours

A **saturated vapour** is one that is in equilibrium with liquid. In molecular terms, equilibrium means that the number of molecules per second that transfer from the liquid to the vapour state (evaporation) is equal to the number per second transferring from vapour to liquid (condensation).

Saturated vapour pressure (svp) is: (i) independent of volume (as shown by Fig. 6.14), (ii) increased by increase of temperature and (iii) equal to external pressure when the liquid boils.

In molecular terms, reduction of the volume of a saturated vapour makes the vapour molecules temporarily more concentrated. Therefore, the number per second that return to the liquid state increases. Since the rate of transfer in the opposite direction is unchanged, the concentration of molecules in the vapour state falls until it returns to its value before the volume reduction. So svp returns to its previous value (i.e. value before volume reduction).

6.9 LOW-TEMPERATURE PHYSICS

Absolute zero is the lowest possible temperature. Physicists have achieved temperatures lower than 0.001K. Several different physical processes may be used to achieve low temperatures:

❶ Cooling by evaporation

A volatile liquid such as ether evaporates rapidly when air is blown over its surface. The more energetic molecules leave the liquid and are carried away by the air stream. The less energetic molecules remain in the liquid, which is therefore cooler.

In a refrigerator, a volatile liquid at low pressure absorbs energy and vaporises in one part of the refrigerator; it then releases energy when it condenses under pressure in a different part.

Oxygen and nitrogen were first liquefied using refrigerators 'in series', in which a first refrigerator was used to cool the condenser of a second. The latter contained a different refrigerant liquid with a lower boiling point and was used to cool the condenser of a third refrigerator, which contained a liquid with an even lower boiling point.

❷ Cooling by adiabatic expansion

If a gas is compressed, it heats up owing to work done to compress it. If it is allowed to regain its original temperature, then released suddenly so it expands rapidly into the atmosphere, it cools because it does work pushing back the atmosphere. Both oxygen and nitrogen have been liquefied by this procedure.

❸ The Joule Kelvin effect

If a gas is forced through a narrow outlet from high pressure to low pressure, intermolecular forces cause its temperature to change. This effect, known as the **Joule Kelvin effect**, may be used to achieve temperatures less than 1 kelvin (Fig. 6.16).

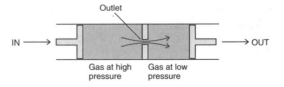

Fig. 6.16 The Joule Kelvin effect

The temperature always increases in a Joule Kelvin expansion unless the initial temperature is below the **inversion temperature** of the gas. This is the temperature above which no cooling effect is possible.

Continuous liquefaction of a gas such as nitrogen is possible by cooling the gas through forcing it through a narrow valve, then making the outgoing cool gas reduce the temperature of the incoming gas in a countercurrent heat exchanger. Fig. 6.17 shows the idea. The temperature is reduced so much that the gas liquefies. Helium and hydrogen need to be precooled before liquefaction can occur. This is because their inversion temperatures (202 K for hydrogen, 33 K for helium) are below room temperature.

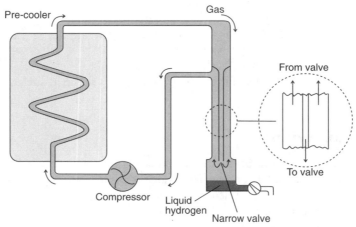

Fig. 6.17 Liquefying hydrogen

The electric resistance of certain metals and alloys falls sharply to zero as the material is cooled below a certain temperature, referred to as its critical temperature, T_c (Fig. 6.18). The material is a **superconductor** at or below its critical temperature.

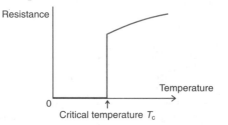

Fig. 6.18 Resistance v. temperature near T_c

The viscosity of liquid helium falls to zero abruptly at 2.2 K as it is cooled. Below this temperature, referred to as its **lambda point**, liquid helium is a superfluid since it has no viscosity. In its superfluid state, it is capable of emptying itself from a beaker by creeping up the side and over the rim.

6.10 ENTROPY AND EQUILIBRIUM

Energy transformations

Conservation of energy is insufficient to explain why certain processes happen whereas others do not. For example, the energy of an oscillating simple pendulum is gradually converted into internal energy of its surroundings. The reverse process (increase of pendulum energy by using some of the internal energy of its surroundings) is most unlikely to happen, even though conservation of energy could be satisfied.

Another example is provided by considering the direction of heat flow between a hot and a cold object. When unaided, heat always flows from hot to cold, never in the reverse direction. Heat flow in either direction could satisfy the requirement of conservation of energy; however, the chance of heat flow from cold to hot is negligibly small when unaided.

Chance

The law of chance (in the form that all possibilities are equally likely to occur) applies when dealing with large numbers of objects that behave in a random manner. Its role in the kinetic theory of gases can be extended beyond the development of the kinetic theory equation (E6.14) by considering an ideal gas of N molecules in a rectangular container as shown in Fig. 6.19.

Initially, all N molecules are confined to one half of the box. Once released the molecules can move to any position in the box, so each molecule can be found in either half with equal likelihood (i.e. a 1 in 2 chance of being found in the left half). The probability of *all* N molecules returning to the initial half at the same time is 1 in 2^N, and since N is typically of the order of Avogadro's number, such a probability is vanishingly small.

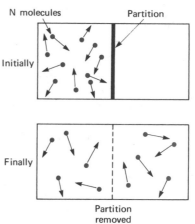

Fig. 6.19 Diffusion

Thermal equilibrium

Some understanding of the role of chance in thermal equilibrium can be achieved by using the **Einstein model** of a solid. Each atom is considered to have a set of equally spaced energy levels; each energy level can be occupied by a single quantum of energy, or by several quanta of energy. Further, because of random interactions between atoms, the quanta move about at random. The number of quanta of any one atom will be independent of the number of quanta of any neighbouring atom. Therefore, there will be many **ways** of distributing the quanta amongst the atoms so that, at any instant, the actual distribution could be any one of these ways (i.e. they are all equally probable). For example, the possible ways of distributing four quanta amongst three atoms are shown in Fig. 6.20.

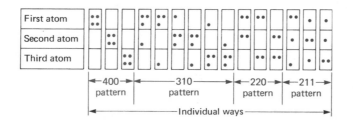

Fig. 6.20 Distributing four quanta to three atoms

In this example, one of the **distribution patterns** occurs in more ways than any other pattern. For a large fixed number of atoms (N) and of quanta (Q), the quanta move about at random so that the arrangement of quanta amongst atoms continually changes from one **way** to another. However, as in the simple example, one set of ways (corresponding to a given **distribution pattern**) occurs more frequently (in fact, far more frequently for large N, Q) than any other pattern; the equilibrium distribution will therefore conform to that pattern.

The equilibrium distribution can be represented by a graph of y = number of atoms with n quanta, $x = n$; the shape of the curve is always an **exponential decrease** as in Fig. 6.21. Probability theory gives the following equation for curve steepness:

E6.19 $$\frac{\text{No. of atoms with } n \text{ quanta}}{\text{No. of atoms with } (n+1) \text{ quanta}} = 1 + \frac{N}{Q}$$

where N = total number of atoms, Q = total number of quanta.

The equation shows that if the total number of atoms is fixed, then when more quanta are supplied (by heat flow into the solid), the exponential decrease becomes less steep. Thus, high temperature corresponds to a less

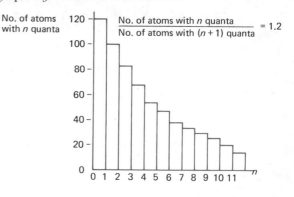

Fig. 6.21 Distribution of quanta within a solid

steep exponential curve. In other words, at high temperature, energy is shared out more evenly, although even sharing is unattainable since this would correspond to infinite temperature.

Achieving equilibrium

Consider two Einstein solids A and B in individual thermal equilibrium at different temperatures. Assume equal numbers of atoms for each, and that A is initially hotter than B. Hence A has initially more quanta than B, so that initially there are more ways of distributing quanta in A than in B.

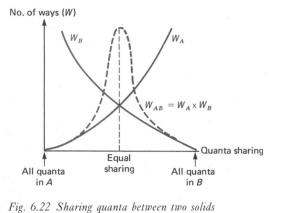

Fig. 6.22 *Sharing quanta between two solids*

Now suppose that A and B are brought into thermal contact with one another so that quanta can move in either direction (i.e. A to B or B to A), but the total number of quanta in A and B must stay the same. The variation of the number of ways of distributing quanta in A and B separately, and together, is shown in Fig. 6.22.

Equilibrium between A and B is attained when the number of ways, W, of distributing quanta between A and B together reaches a maximum, and then the distribution pattern for A (and hence temperature) will be the same as for B. In other words, the final distribution pattern is the most frequent of the most ways. Thus heat flows from hot to cold because that is the most likely direction.

Further insight can be achieved by using the result from probability theory that the ratio:

$$\frac{W \text{ for } (Q+1) \text{ quanta amongst } N \text{ atoms}}{W \text{ for } Q \text{ quanta amongst } N \text{ atoms}} = 1 + \frac{N}{Q}$$

so if $N_A = N_B = 400$, and Q_A and Q_B have initial values of 400 and 100, respectively, then the transfer of 1 quantum from A to B makes W_A change by $\frac{1}{2}$ (since $1 + \frac{N}{Q} = 2$ for A, so loss of one quantum means that W_A falls by a factor of 2); at the same time B gains a quantum from A so W_B changes by $\times 5$. Thus W_{AB} changes by $\times \frac{5}{2}$. Now consider the initial situation again, with the transfer of a single quantum from B to A this time; W_{AB} changes by $\times \frac{2}{5}$ by the same method of reasoning as above. Hence, the transfer from A to B of 1 quantum is $6\frac{1}{4}$ times (i.e. $\frac{5}{2} \times \frac{5}{2}$) more likely than transfer in the reverse direction.

Entropy

Equilibrium involves maximising the number of ways in which energy can be distributed through a system. In other words, **disorder** always tends to increase. This general result goes far beyond the simple model of a solid used here, and it is a determining factor in deciding the direction of energy flow involving heat. The disorder of a system is measured by its **entropy**, as follows:

E6.20 $\quad S = k \ln W$

where S = the entropy (J K^{-1}), W = the number of ways of arranging the system, ln is the natural log, k = the Boltzmann constant ($= R / N_A$)(J K^{-1}).

Entropy is defined in this way so that heat transfer Q to or from a system produces a change of entropy $\Delta S = Q/T$. Provided no work is done, the change of internal energy of the system, ΔU, is equal to the heat transferred, Q. Thus, for A and B in contact, with the temperature of A (T_A) initially greater than the temperature of B (T_B), transfer of internal energy ΔU from A to B causes the entropy of A to fall by $\Delta U / T_A$ and that of B to rise by $\Delta U / T_B$. Since $T_A > T_B$, the total entropy increases. The flow of energy from A to B continues until the total entropy stops increasing (i.e. until the number of ways of distributing the energy has reached its maximum value). The system will then be in thermal equilibrium. Note that the unit of entropy is J K^{-1}.

Chapter roundup

In this chapter, you should have developed a thorough grasp of the nature and properties of an ideal gas. The models and theories in this chapter are important examples of how ideas in Physics are developed mathematically and used to predict and explain measurable properties.

Question Bank

1 A coil of wire has a resistance of 2.00 Ω, 2.80 Ω and 3.00 Ω at temperatures of 0 °C, 100 °C and t, respectively. What is the value of t on the scale defined by this resistor?
A 20 °C   B 25 °C   C 80 °C   D 105 °C   E 125 °C
(**London**: *All other Boards except O and C Nuffield, SEB*)

Points

Substitute resistance values into E6.1 to give t.

2 One junction X of a thermocouple is placed in melting ice at 273 K and the other junction Y in steam at 373 K. The emf measured is 1.0 mV. Junction Y is then transferred to a bath whose temperature is 398 K. Assuming the variation of emf with temperature difference is linear, the emf, in mV, recorded will now be:

A $\dfrac{25}{100}$ B $\dfrac{398}{373}$ C $\dfrac{125}{100}$ D $\dfrac{398}{273}$

(**AEB** June 90: *all Boards*)

Points

Use E6.1.

3 An unlagged cubic tank containing hot water loses heat to its surroundings at a rate of 900 W. This loss is reduced to 60 W if all the faces of the tank are covered with a layer of lagging. What will be the rate of loss of heat if one face is left unlagged? (The temperatures of the water and surroundings are unaltered. You may assume that heat is lost only from the faces and that the rate of loss of heat from a face is unaffected by whether it is vertical or horizontal, top or bottom.)
A 210 W   B 200 W   C 190 W   D 160 W   E 150 W
(**Cambridge**: *and all other Boards*)

Points

Since the unlagged tank loses heat at a rate of 900 W, calculate the heat loss per second from each unlagged face. Since the lagged tank loses heat at a rate of 60 W, calculate the heat loss per second from each lagged face. Use your calculations to determine the heat loss per second from five lagged faces and one unlagged face.

4 A composite rod of uniform cross section has aluminium and copper sections of the same length in good thermal contact. The ends of the rod, which is well lagged, are maintained at 100 °C and at 0 °C, as shown in the diagram. The thermal conductivity of copper is twice that of aluminium.

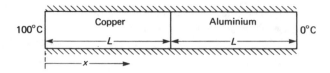

Which one of the following graphs represents the variation of temperature t with distance x along the rod in the steady state?

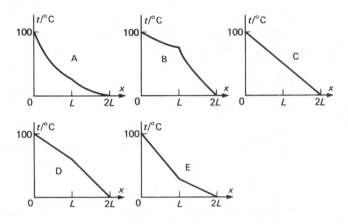

(**Cambridge**: *and all other Boards except SEB*)

Points

There are two key points to consider. (i) With no heat loss from the sides, is the temperature gradient constant or not? See 6.4 and Fig.6.7 and Fig 6.8 if necessary. (ii) Bearing in mind that copper is a better heat conductor than aluminium, is the junction temperature likely to be more than, equal to, or less than 50 °C?

5 A mass of an ideal gas undergoes a reversible isothermal compression. Its molecules will then have, compared with the initial state, the same:
1 root mean square velocity.
2 mean momentum.
3 mean kinetic energy.
Answer:
A if 1, 2, 3 correct.
B if 1, 2, correct.
C if 2, 3 correct.
D if 1 only.
E if 3 only.

(**London**: *all other Boards except O and C Nuffield*)

Points

What does isothermal mean? See 6.7 if necessary. For 1 and 3, see 6.6. For 2, remember that momentum is a vector (i.e. its *direction* is important) and that we are dealing with a *very large* number of molecules moving with *random* motion. What will the *average* momentum be?

6 The kinetic theory of gases gives the formula $PV = \frac{1}{3} Nm\overline{v^2}$, for the pressure P exerted by a gas enclosed in a volume V. The term Nm represents:
A the mass of a mole of the gas.
B the mass of the gas present in the volume V.
C the average mass of one molecule of the gas.

D the total number of molecules present in volume *V*.
E the total number of molecules in a mole of the gas.

(**SEB**: *and all other Boards*)

Points

See E6.14 if necessary.

7 Two molecules of a gas have speeds of 1 km s⁻¹ and 9 km s⁻¹, respectively. What is the root mean square speed of these two molecules?

A 2 km s⁻¹
B √3 km s⁻¹
C 4 km s⁻¹
D √41 km s⁻¹
E √82 km s⁻¹

(**SEB**: *and all other Boards*)

Points

See 6.6 if necessary for the formula for rms speed in terms of individual speeds.

8 An ideal gas is heated at constant volume until its pressure doubles. Which one of the following statements is correct?
A The mean speed of the molecules doubles.
B The number of molecules doubles.
C The mean square speed of the molecules doubles.
D The number of molecules per unit volume doubles.
E The rms speed of the molecules doubles.

(–: *all Boards except O and C Nuffield*)

Points

Assume the mass of the gas is unchanged. Use the kinetic theory equation E6.14 to relate the various factors above to the pressure change at constant volume.

9 Which of the following pressure (*P*)–volume (*V*) diagrams best describes the behaviour of a gas which is returned to its original state after first being expanded adiabatically, then heated at constant volume and finally compressed isothermally?

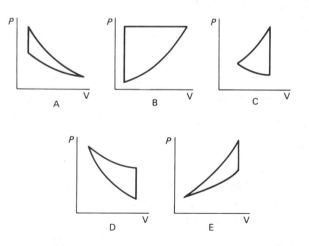

(**AEB**: *and London, WJEC, Oxford, Cambridge and O and C*)

Points

For this problem you need to be able to distinguish between the *P–V* curves for adiabatic and isothermal changes. Remember that for an ideal gas, the curve for an adiabatic change is steeper than that for an isothermal change. Take care to put the sequence of events in the correct order.

10 (a) A window is made from glass 6 mm thick and measures 2.0 m × 1.0 m. Use the

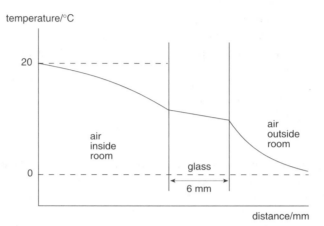

U-value in the data booklet (see commentary) to calculate the heat flow rate through the window when the temperature inside the room is 20 °C and the outside temperature is 0 °C.

(b) The temperature variation near to a single-glazed window similar to that described in (a) is shown on the graph.
(i) State where the temperature gradient is constant but not zero.
(ii) State where the temperature gradient has the greatest value.
(iii) The thermal conductivity of glass is 1.0 W m⁻¹ K⁻¹. Use this and your answer to (a) to calculate the temperature difference between the glass surfaces of the window.

(c) State, with a reason, what type of thermometer you would use to investigate the variation of air temperature with distance from the window, within about 100 mm from either face.

(**NEAB**: *and all other Boards except SEB*)

Points

(a) The U-value for 6 mm window glass is 5.6 W m⁻² K⁻¹. The heat flow rate is equal to the U-value × the window area × the temperature difference.
(b) (i) and (ii) The temperature gradient is the slope of the graph. For (i), where is the slope constant? For (ii), where is the slope greatest? (iii) See E6.11 if necessary.
(c) See 6.1.

11 (a) In an air-conditioned building the air is changed continuously at a rate that is equivalent to two complete changes per hour. In cold weather fresh air drawn in must be warmed up to a comfortable temperature. (i) Use the following data to calculate the rate at which heat must be supplied to warm the air for one room in the building:

Dimensions of the room = 5.0 m × 4.0 m × 3.0 m
Temperature of the fresh air drawn in = 5.0 °C
Temperature of the room = 20 °C
Average density of the air = 1.3 kg m⁻³
Specific heat capacity of air = 1.01 × 10³ J kg⁻¹ K⁻¹

(ii) State whether you would expect the actual rate of heating to be greater or less than the value you have calculated and give a reason to justify your statement.

(b) Give any two reasons why it is necessary to refer to the average density of the air in (a) (i).

(c) Outline an experiment you would carry out to measure the density of air.

(**SEB**: *and all other Boards*)

Points

(a) (i) First, calculate the mass of air in the room. Given that there are two complete changes per hour, calculate the mass of air to be heated up in 1 s. Then use E6.3 to calculate the heat supplied per second. (ii) You must consider heat losses due to conduction and radiation.

(b) Density of air depends upon pressure and temperature. There would almost certainly be a temperature difference between the top and the bottom of the room.

(c) See your textbook.

12 In this question you are expected to make sensible estimates for various quantities and then combine them in calculating the required answer. Give your answer to a justifiable number of significant figures and show the units of your estimated quantities and of your final answer.

What would be a typical mean rise in temperature of the brake-blocks when the brakes are applied to bring a bicycle to rest?

(a) List the quantities that you have estimated and need to enable you to calculate the answer.

(b) Carry out the calculation of the estimate, using the quantities that you have listed in (a).

(c) A practical test showed a higher temperature rise than you have calculated, give two possible reasons for this.

(**O and C Nuffield**: *and all other Boards*)

Points

In any estimation it is first necessary to think about the physics of the problem. Where does the energy that heats up the brake blocks come from? Is it possible to calculate the energy stored by a moving bicycle (and its rider)?

For part (a) you are going to need to estimate the mass of the bicycle and rider, the speed of the bicycle, the mass of the four brake blocks and the specific heat capacity of rubber.

For part (b) you must equate the kinetic energy lost to heat energy gained by the rubber. Don't forget to put in all the units and quote the final answer either to the nearest power of ten (order of magnitude) or to *one* significant figure if you consider that your answer is unlikely to be wrong by a factor of '10'. To estimate the one difficult quantity (the specific heat capacity of rubber) use about $\frac{1}{3}$ of the specific heat capacity of water (the specific heat capacity of water is known to be an unusually high value compared with other materials).

To answer part (c) examine your calculation to part (b) and decide which quantities affect the temperature change and the ways in which two of them could be altered to give a higher temperature rise. You should also ask yourself whether all the energy put into brakes is from stopping the bicycle, or is there another energy source? In our answer it has been assumed that the bicycle only has translational kinetic energy, but perhaps it also has rotational ke (see 1.8)? What if the bicycle was going downhill as it was brought to a halt?

13 (a) (i) A steady current *I* flows through a conductor of resistance *R* when the steady potential difference between its ends is *V*. Write down an equation for the current. (ii) The ends of a perfectly lagged bar of length L and cross-sectional area A are maintained at steady temperatures θ_1 and θ_2 where $\theta_1 > \theta_2$. If the thermal conductivity of the material of the bar is k, write down an expression for the rate of flow of heat along the bar in the steady state. (iii) A property of the bar called its 'thermal resistance', R_θ, can be defined in a similar way to the resistance of the conductor. By comparing your equations stated in (i) and (ii), deduce an equation for R_θ.

(b) The base of an aluminium kettle is 2.4 mm thick and has a cross-sectional area of 0.020 m². The inside surface is coated with a uniform layer of scale 0.50 mm thick. When water is boiling inside the kettle at a steady temperature of 100 °C, heat flows normally through the base at a rate of 2.0 kW. By considering the thermal resistance of the arrangement, or otherwise, calculate the temperature of the underside of the kettle base.

Thermal conductivities of aluminium and the scale are 240 W m⁻¹ K⁻¹ and 1.0 W m⁻¹ K⁻¹, respectively.

(**NEAB**: *and all other Boards except SEB*)

Points

(a) (i) See E7.7 if necessary. (ii) See E6.11 if necessary. (iii) See 6.4.
(b) Thermal resistance method: Calculate the thermal resistance of the aluminium, then of the scale. Since all the heat flow through the scale also passes through the aluminium, then the scale and the aluminium are 'in series'; hence, add their individual thermal resistances to give the total thermal resistance. Then, multiply the total thermal resistance by the 'heat current' (i.e. the rate of flow of heat) to give the temperature difference across the scale and aluminium. If the kettle is heated internally, the underside temperature will be lower than 100 °C, but if heated externally (via the base) the underside temperature will be higher than 100 °C. Since the form of heating is not specified, either answer is acceptable. See 6.4 for further details of this method.

Otherwise: Let the interface temperature be X. Assume internal heating. The temperature difference across the scale is $(100 - X)$, so given the rate of heat flow through the scale is 2 kW, use E6.11 to calculate X. Then the temperature across the aluminium can be written as $(X - Y)$ where Y is the underside temperature (to be determined). Use E6.11 applied to the aluminium to calculate Y.

14 Find the rms speed of the air molecules in a room under typical conditions (pressure 10^5 Pa, density of air 1.3 kg m⁻³). When ammonia is released, it takes several seconds for the smell to be noticed by a person in another part of the laboratory. Explain why there is no real contradiction between this experimental fact and the answer to the first part of the question.

(**WJEC**: *and all other Boards except O and C Nuffield*)

Points

First part: use E6.14. Assume R = 8.3 J mol⁻¹ K⁻¹.
Second part: you should explain the process of diffusion in molecular terms. Use your textbook to find out why gas molecules moving at high speeds (of the same order as the speed of sound in air) only spread out slowly when released into another gas. The key lies in collisions between gas molecules.

15 At a temperature of 100 °C and a pressure of 1.01×10^5 Pa, 1.00 kg of steam occupies 1.67 m³, but the same mass of water occupies only 1.04×10^{-3} m³. The specific latent heat of vaporisation of water at 100 °C is 2.26×10^6 J kg⁻¹. For a system consisting of 1.00 kg of water changing to steam at 100 °C and 1.01×10^5 Pa, find:

(a) the heat supplied to the system.
(b) the work done by the system.
(c) the increase in internal energy of the system.
(Cambridge: *and London, AEB, Oxford, NEAB, NISEAC)*

Points

(a) Use E6.4.
(b) For work done by a gas, see 6.7. Calculate the volume change from the volume of steam and the volume of water (for mass = 1.00 kg in both cases).
(c) Use E6.15.

16 Explain what is meant by a scale of temperature, discussing how a scale would be established.

The following table gives an example of a thermometer, the associated physical property and the measurement taken. Make a list giving the entries for the blank space (a)–(d) in the table.

Instrument	Physical property	Measurement taken
Constant-volume gas thermometer	Pressure change of gas	Difference in manometer levels supported by the pressure of a fixed volume of nitrogen
Liquid-in-glass thermometer	(a)	(b)
Thermocouple	(c)	(d)

Why is it *not* convenient to use a constant-volume gas thermometer in most practical situations? To what is it actually put to use?

The masses of hydrogen and oxygen atoms are 1.66×10^{-27} kg and 2.66×10^{-26} kg, respectively. What is the ratio of the 'average' speed of hydrogen and oxygen atoms at the same temperature? What is usually meant by the term 'average' speed in this case?

For the thermocouple give, with reasons, a situation for which it would be particularly suitable.

(London: *and all other Boards except O and C, Nuffield)*

Points

To explain what is meant by a 'scale of temperature', your answer should explain thermometric property, fixed points and the defining equation (E6.1).

For (a) to (d), see Fig. 6.2.

The advantages and disadvantages of the constant-volume gas thermometer are discussed in Fig. 6.2.

For the ratio of 'average' speeds, etc., assume that 'average' used here means rms speed. From E6.13 and E6.14, combine the two equations to eliminate PV so that you can establish the link between rms speed and molecular mass at constant temperature. Given the mass ratio, use the link to determine the ratio of rms speeds. The meaning of rms speed is given in 6.6.

For the advantages of a thermocouple, see Fig. 6.2.

17 (a) (i) Describe how you would determine the coefficient of thermal conductivity of copper. Your account should include details of how each measurement is taken.

(ii) State one way in which the behaviour of the apparatus used differs from that assumed in calculating the thermal conductivity. What is the cause of the difference and how does it affect the result?

(b) The cabin of a light aircraft can be regarded as an approximately rectangular box, sides 1.5 m by 1.5 m by 2 m. The windows are made of perspex, 10 mm thick, and have a total area of 3 m²; the remainder of the cabin wall is made of thin aluminium alloy lined with insulating material, 20 mm thick. The cabin heater is able to maintain a temperature of 20 °C in the cabin when the outside temperature is –10 °C.

Assuming that the temperature difference across the aluminium alloy may be neglected, calculate: (i) the power of the heater; (ii) the percentage of the total energy which is lost through the perspex.

Explain why, in practice, the power of the heater needed would be much less than that calculated.

(Take the thermal conductivity of perspex to be 0.2 W m^{-1} K^{-1} and that of the insulating material to be 3.5×10^{-2} Wm^{-1} K^{-1}.)

(**Oxford**: and AEB, London, O and C, NICCEA and Cambridge)

Points

(a) (i) Your textbook should have a suitable method. (ii) When you calculate thermal conductivity (k), you need to know the heat flow per second, the area of cross-section and the temperature gradient. For heat flow per second it is assumed that the measured value is the actual heat flow per second all the way along the sample; in other words, heat loss per second from the sides is assumed negligible. You should discuss whether or not heat loss causes the calculated value of k to be greater or smaller than the value you would obtain if there was no heat loss.

(b) (i) The heater power equals the heat flow per second through the perspex + the heat flow per second through the insulation. Use E6.11 to calculate the heat flow per second through each of the two materials in turn. (ii) Use the values calculated in (b) (i).

To explain why the heater power in practice would be much less than your calculated value, consider the effect of still air either side of the cabin walls. Outside the cabin, the air behind (and possibly underneath) the cabin would be relatively undisturbed and would provide another layer of insulation between the cabin wall and the cold air further away. Inside the cabin, even the presence of a very thin layer of still air against the walls will act as an effective layer of further insulation. Don't forget that people act as heaters!

18 State the condition necessary for (a) heat and (b) electrical charge to flow along a solid.

Write down equations which give the rate of flow in each case, and from these point out analogies between relevant thermal and electrical quantities.

Explain on the microscopic scale why metals are good conductors of both heat and electricity.

The external wall of a brick house has an area of 16 m² and a thickness of 0.3 m. The indoor and outdoor temperatures are 20 °C and 0 °C, respectively. Find the rate at which heat is lost through the wall. What is the rate of loss when the internal surface of the wall is covered with expanded polystyrene tiles of thickness 20 mm, and what is the temperature at the brick–tile interface?

(Thermal conductivity of brick = 0.5 W m^{-1} K^{-1}; thermal conductivity of expanded polystyrene = 0.03 W m^{-1} K^{-1}.)

(**WJEC**: and all other Boards except SEB)

Points

See 6.4 for analogies between heat flow and flow of electric charge. The equations required are E6.11 and E7.7.

Conduction of heat in metals is explained microscopically in 6.4.

To calculate the rate of loss of heat, use E6.11; the temperature difference is the same in °C as in K so there is little point in converting the temperatures.

With the polystyrene as above, start by drawing a cross-section of part of the wall, marking the temperature on the outside of the brick as 0 °C, on the outside of the polystyrene as 20 °C and at the interface as X °C.

To determine the heat loss per second, the straightforward method is to use E6.11 to obtain an expression for the heat loss per second through the brick in terms of X, and then repeat the procedure to obtain an expression for the heat loss per second through the polystyrene also in terms of X. Since the heat flow per second through the brick is the same as that through the polystyrene, then equate the two expressions to find X. Then use the value of X in either expression for heat flow to complete the problem.

An alternative method for the heat loss per second with polystyrene is to follow the thermal resistance method given in the points of question 13.

19 (a) The equation $P = \frac{1}{3}\rho\overline{c^2}$, relating the pressure P and density ρ of an ideal gas, may be derived by making certain assumptions about the behaviour of the molecules of the gas. (i) List four of the assumptions that are essential to this derivation. (ii) Explain what is meant by the symbol $\overline{c^2}$. (iii) Using the above equation and the ideal gas equation $PV = RT$, find an expression for the total translational kinetic energy of one mole of ideal gas molecules in terms of the temperature T.

An athlete of mass 70 kg can run 100 m in 10 s. Find the ratio of the athlete's average kinetic energy to the total transitional kinetic energy of one mole of ideal gas molecules at 300 K. (Take the molar gas constant as $R = 8$ J mol^{-1} K^{-1}.)

(b) (i) How do the properties of the molecules of a 'real' gas differ from those which are attributed to the molecules of an ideal gas? (ii) Draw labelled graphs of pressure against volume for a fixed mass of a 'real' gas at a number of temperatures, including the critical temperature. On these curves, label the critical isotherm, and an isotherm along which liquefaction could *not* be achieved. Indicate also on an appropriate isotherm the part of the curve for which the substance is entirely liquid.

(c) When fuel from a cylinder of liquefied butane (such as is used for camping stoves) is released through an opening which is not itself a pressure regulator, the temperature of the cylinder being kept constant, the pressure in the cylinder remains steady until the cylinder is almost empty, when the pressure falls continuously. Account for this in terms of a molecular model.

(NICCEA: *and all other Boards for (a))*

Points

(a) (i) See 6.6 if necessary. (ii) See 6.6 and E6.14 if necessary. (iii) See 6.6 (kinetic energy and temperature).

(b) (i) See 6.8 for an explanation of real gas behaviour. (ii) See 6.8 and Fig. 6.14 in particular.

(c) The liquefied butane in a cylinder is in equilibrium with its saturated vapour. A molecular explanation of the fact that svp is independent of volume is given in 6.8; however, take care since the given explanation is for a *reduction* of the volume of a saturated vapour, and allowing vapour to escape is equivalent to an *increase* of volume. The explanation needs only minor changes to meet the conditions of the question.

Answers: 1 E; 2 C; 3 B; 4 D; 5 A; 6 B; 7 D; 8 C; 9 D; 10(a) 224 W (b)(iii) 0.67 K; 11(a)(i) 657 W; 12 About 20 K; 13(b) 49 °C (internal heating assumed); 14 480 m s^{-1}; 15(a) 2.26 MJ (b) 0.17 MJ (c) 2.09 MJ; 16 4.00; 17(b)(i) 2.51 kW (ii) 72%; 18 530 W, 250 W, interface temp. = 9.5 °C; 19(a) 35/36

CHAPTER 7

ELECTRICITY AND ELECTRONICS

Units in this chapter

Chapter objectives

After working through the topics appropriate to your syllabus in this chapter, you should be able to:
- understand the nature of current, pd and resistance
- define charge, pd, resistance, emf and internal resistance and their units
- know and apply Kirchhoff's circuit laws
- understand the potential divider principle
- use the Wheatstone bridge to work out an unknown resistance
- define capacitance and its unit
- work out pd and charge on capacitors in dc circuits
- work out problems on the discharge of a capacitor through a fixed resistor or to a second capacitor
- define the rms value of an alternating current or pd
- explain the operation of rectifying circuits
- understand the effects of reactive and resistive components in ac circuits
- understand the nature of conduction in conductors and semiconductors
- explain the operation of a transistor as a current amplifier and a current-operated switch
- sketch op-amp circuits on open loop and with resistive feedback and work out the output pd for a given input pd
- describe the action of different logic gates and various digital devices such as multivibrators and counters

7.1 CURRENT AND CHARGE

An electric current is a flow of charge. In a metal, the charge is carried by 'conduction' electrons which are not attached to any given fixed ion of the metal. The unit of electric current is the

ampere, defined in 2.11. All other electrical units are defined in terms which relate back to the ampere.

1 **coulomb** (C) of electric charge is the charge which passes a given point when 1 A of steady current flows for 1 s. Thus a steady current of 3 A for a time of 5 minutes means that a charge of $3 \times 5 \times 60 = 900$ C has been passed in total. The link can be expressed in terms of the following equations:

E7.1 $Q = It$ for steady direct currents

E7.2 $I = \dfrac{dQ}{dt}$ for changing currents (dc and ac)

where Q = charge (C), I = current (A), t = time (s).

Note that the prefixes milli- (m = 10^{-3}) and micro- ($\mu = 10^{-6}$) are commonly used.

Conduction in metals

The reason for the presence of conduction electrons (sometimes called free electrons) in metals is that the outer shell electrons of metal atoms are easily removed. In the solid state, the metal ions form a lattice structure, as in Fig. 7.1, and the free electrons move about at high speeds.

When a potential difference is applied across the metal, the free electrons are made to 'drift' towards the +ve terminal, thus giving a flow of charge through the metal. In an insulator, the vast majority of the electrons remain firmly attached to atoms.

Consider a metal wire of uniform cross-sectional area (A) along which a steady current (I) passes. Conduction electrons carry the current by 'drifting' along the wire towards the +ve terminal, as in Fig. 7.2. Let u be the average drift speed of a

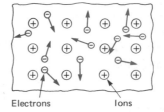

Electrons   Ions

Fig. 7.1 Free electrons in a metal

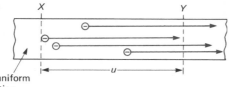

Wire of uniform cross-section

Fig. 7.2 Conduction by electrons

conduction electron. Referring to the diagram, in 1 s a conduction electron will move a distance u from point X to point Y. So all the free electrons between X and Y will pass Y in 1 s. The volume of the wire section XY is uA, and so between X and Y there are nuA free electrons, where n is the number of free electrons per unit volume. Hence, the charge flow per second (I) is $nuAe$, where e is the charge carried by an electron.

E7.3 $I = nuAe$

where I = current (A), n = number of free electrons per unit volume (m^3), A = area of cross-section (m^2), u = drift velocity (m s^{-1}), e = electron charge (C).

7.2 POTENTIAL DIFFERENCE

To use potential difference (pd) effectively in circuit theory, its basic nature must first be understood. Remember that electric potential (see Chapter 2) is essentially the potential energy of a unit +ve charge. If unit +ve charge is allowed to move from a point of high potential (V_1) to a point of low potential (V_2) then the energy given up by the charge is ($V_1 - V_2$) J.

1 **volt** (V) of potential difference exists between two points of a circuit when 1 J of work is done by 1 C of charge in moving from one point to the other. Note that 1 V is identical

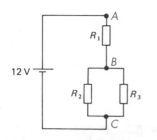

Fig. 7.3 Nature of potential difference

to 1 J C^{-1}. It is important to remember that pd can only exist between *two* points (e.g. across a resistor's terminals). To illustrate the nature of pd, consider the network shown in Fig. 7.3. The pd from A to C is 12 V, so that when 1 C of charge passes from A to C, either via R_2 or via R_3, it loses 12 J of energy. An individual charge passing from A to C uses one of two possible routes, which are (i) through R_1 and R_2, and (ii) through R_1 and R_3. Now suppose the pd across R_1 is 4 V; then, unit charge will use up 4 J on passing through R_1, and will give up its remaining 8 J in either R_2 or R_3 according to its route. Thus, the pd across R_2 = pd across R_3 = 8 J C^{-1}. This simple example illustrates two key points in connection with pd:

❶ The pd across parallel components (R_2 and R_3 above) is *always* the same.

❷ The pd across AC = pd across AB + pd across BC.

Power taken by an electrical component is given by the following equation:

E7.4 $W = IV$

where W = power taken (watts, W), I = current (A), V = pd across device terminals (V).

The equation follows from the fact that a pd of V V means that each coulomb of charge which passes through the device gives up to V J of energy. Since current I means that I C of charge pass through the device each second, then IV J of energy are delivered each second.

The **electromotive force (emf)** of a cell or battery of cells is the energy converted into electrical energy per coulomb of charge inside the cell. In the case of a solar cell the energy is converted from light energy, while in the case of a dynamo, the energy is converted from mechanical energy. Thus, an emf of 12 V means that each coulomb from the cell will deliver 12 J of energy. If the cell has **internal resistance** then some of the electrical energy will be converted to heat energy inside the cell when current is drawn from the cell. Consequently, the amount of energy available for the external components will be reduced. For a cell of emf E and internal resistance r, connected up to an external 'load', as shown in Fig. 7.4, the pd across the cell terminals when current I flows will not be E but $E - Ir$. The reason for this is the loss of energy of the charge inside the cell as it tries to flow out through the cell's internal resistance. The 'lost voltage' (Ir) represents this loss of electrical energy inside the cell. As an example, consider the circuit of Fig. 7.4 in which a 6 V cell (i.e. $E = 6$ V) with internal resistance 4 Ω is connected to an external 8 Ω resistor. Since the total circuit resistance is 12 Ω, the current from the cell will be 0.5 A. Hence, the pd across the external resistor will be 0.5 $\times$ 8 = 4.0 V, and the 'lost voltage', at 0.5 $\times$ 4 = 2.0 V, will make up the difference between the cell emf (6 V) and the pd (4 V) across its terminals. In energy terms, each coulomb of charge which passes through the cell will be given 6 J of energy but will use 2 J in moving through the interior of the cell; thus, only 4 J of energy will be delivered to the

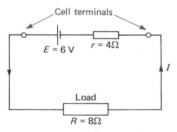

Fig. 7.4 Internal resistance

external resistor. Note that the more current which a cell with internal resistance delivers, the greater is the 'lost voltage', so the pd across the cell terminals falls below the emf value by an increasing amount when the current is increased. It follows that the accurate measurement of the emf of a cell requires that the pd across its terminals be determined when no current is taken from the cell (since there is no 'lost voltage' when no current is taken).

E7.5 and E7.6 summarise the ideas above:

E7.5 $V = E - (Ir)$

where V = pd across cell terminals (V), E = cell emf (V), r = internal resistance (Ω), I = current drawn (A).

If the external load has a total resistance R, and since the external load is connected across the cell terminals so $V = IR$, then E7.5 may be written as $IR = E - (Ir)$ or $E = (IR) + (Ir)$. By multiplying each term in this equation by I, the power distribution becomes clear:

E7.6 $IE = I^2R + I^2r$

where IE = power generated by the cell (W), I^2R = power dissipated by R (W), I^2r = power 'wasted' inside the cell (W).

Kirchhoff's second law states that the sum of the emfs round a closed loop is equal to the sum of the pds round the loop. Since a source of emf produces electrical energy and a pd is where electrical energy is used, Kirchoff's second law is essentially another way of stating that all the electrical energy created by the sources of emf in a closed loop equals the total electrical energy used.

7.3 LIMITS AND USES OF OHM'S LAW

Resistance is defined by the following equation:

E7.7 $R = V/I$

where R = resistance (Ω), V = pd (V), I = current (A).

Note that 1 kilohm (kΩ) = 10^3 ohms and 1 megohm (MΩ) = 10^6 ohms.

I–V Curves

The graphs of Fig. 7.5 show the current–voltage relationship for several different circuit components. They are not to the same scale. Since resistance $R = V/I$, it should be clear from

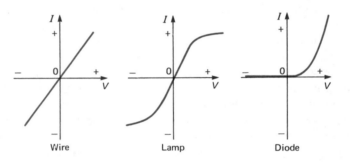

Wire   Lamp   Diode

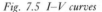

Fig. 7.5 I–V curves

the graphs that only the wire has a resistance which is independent of current; in other words, only the $I–V$ curve for the wire is a straight line. For the filament lamp, the graph shows that the ratio V/I increases as the current increases; in other words, the resistance of the filament lamp increases with increased current. The reason for this is that the filament lamp becomes hotter at increased current, and the resistance of metals increases with increased temperature (i.e. +ve temperature coefficient of resistance, see 6.3). The silicon diode has a resistance that is very large in the 'reverse' direction; in the forward direction, the resistance is large until the pd exceeds approximately 0.5 V after which the resistance falls.

Ohm's law For a metallic conductor, the current is proportional to the applied pd for constant physical conditions. It follows that the resistance of an 'ohmic' conductor is constant, and does not change when the current changes (e.g. the wire in Fig. 7.5). Note that the equation $R = V/I$ is not an expression of Ohm's law; it simply defines R and does not express the fact that V/I (= R) is independent of I for an 'ohmic' conductor.

Conductivity is defined as 1/resistivity, and **resistivity** is defined by the following equation:

E7.8 $\rho = RA/L$

where R = resistance (Ω), ρ = resistivity (Ω m), L = length of specimen (m), A = area of cross-section of specimen (m^2).

Resistor combination rules

Resistors in series The pd across a series combination is equal to the sum of the individual pds. Therefore, for two resistors R_1 and R_2 in series, the combined resistance R is based upon

$IR = IR_1 + IR_2$ since the current I is the same in each resistor. Hence, $R = R_1 + R_2$ gives the total (i.e. combined resistance).

Resistors in parallel The total current is equal to the sum of the individual currents through each resistor. Therefore, for two resistors in parallel (R_1 and R_2), the combined resistance is based upon $V/R = V/R_1 + V/R_2$, since the pd V is the same across each resistor. Hence, the combined resistance R is given by the rule $1/R = 1/R_1 + 1/R_2$. The combined resistance for parallel resistors is always less than the smallest individual resistance value.

Meter conversion

An ammeter or a voltmeter can have its range extended as follows.

Ammeters are converted by connecting a suitable resistance (called a 'shunt') in parallel with

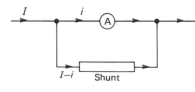

Fig. 7.6 Using a shunt

the ammeter. For an ammeter of resistance r with a full-scale deflection current of i, the value of the shunt resistance which will extend its range to I will be $ir/(I-i)$, since the pd across the shunt will be ir, and the current through the shunt will be $I-i$. Fig. 7.6 shows the basic arrangement.

Voltmeters are converted by connecting a suitable resistance (called a 'multiplier') in series with the voltmeter. For a voltmeter of full-scale deflection voltage v and resistance r, to extend its scale to full-scale deflection voltage V, the multiplier resistance must be $(V-v)r/v$, since $(V-v)$ is the pd across the multiplier, and v/r is the current through the multiplier. See Fig. 7.7.

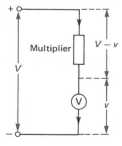

Fig. 7.7 Using a multiplier

Because moving-coil voltmeters require current for their basic operation, the use of a moving-coil voltmeter to measure pd in a circuit will affect the current flow in the circuit being measured. For example, consider the simple circuit shown in Fig. 7.8(a). Clearly the pd across R_2 is 6 V. However, if a voltmeter of resistance 6000 Ω is connected across R_2 to measure the pd across R_2, then the circuit

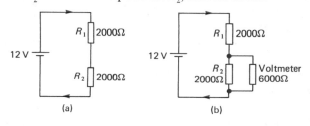

Fig. 7.8 Effect of a low-impedance voltmeter

has been changed to that in Fig. 7.8(b). The total circuit resistance will now be 2000 Ω + 1500 Ω = 3500 Ω (the combined resistance of R_2 and the voltmeter in parallel is 1500 Ω, then add on the resistance of R_1), so that the current from the 12 V battery (negligible internal resistance) will be 12/3500 = 0.0034 A. This means that the pd across R_1 is 2000 × 0.0034 = 6.8 V, which leaves a pd across the parallel combination of 12 − 6.8 = 5.2 V. An alternative way of calculating the pd across the parallel combination is to multiply the current taken by the combination (0.0034 A) by the combined resistance (1500 Ω) to give 5.2 V (5.1 V actually on account of rounding off 0.00342 A too early!). The voltmeter therefore reads 5.2 V, which is less than the pd across R_2 without the voltmeter in circuit.

Because a moving-coil voltmeter takes current, its use to measure the emf of a cell will cause an error if the cell has internal resistance. Consider the arrangement shown in Fig. 7.9 where the cell has emf E and internal resistance r, and the voltmeter has resistance R. As shown, the circuit resistance will be $(R + r)$ and the current I will be $E/(R + r)$. Thus, the pd across

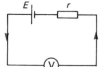

Fig. 7.9 emf measurement with a voltmeter

the voltmeter will be $IR = ER/(R + r)$, and so the pd must always be less than the emf E. As a numerical example, suppose $E = 12$ V, $r = 50$ Ω and a voltmeter of resistance $R = 5000$ Ω is used as in Fig. 7.9. Then, the current will be 12/5050 A giving a pd across the voltmeter of 12 × 5000/5050 = 11.88 V on account of a 'lost voltage' of 0.12 V. The bigger the voltmeter resistance R is (compared with the cell's internal resistance r), the smaller will be the error.

7.4 POTENTIAL DIVIDERS AND POTENTIOMETERS

Potential dividers are used to supply required levels of pd to a circuit from sources of fixed emf. The circuit in Fig. 7.10(a) shows the simplest form of the arrangement with two resistors R_1 and R_2 connected in series. With a battery supplying fixed pd V across the ends as shown, the current taken by the two resistors will be $V/(R_1 + R_2)$; thus, the pd across R_2 will be $VR_2/(R_1 + R_2)$. By choosing suitable values of R_1 and R_2, the pd across R_2 can be made equal to any value from 0 to V.

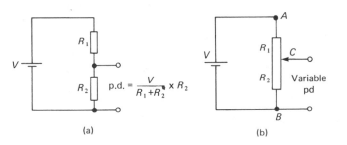

(a) (b)

Fig. 7.10 Potential dividers

A more useful form of the potential divider is shown in Fig. 7.10 (b). This time, R_1 and R_2 are adjacent sections of the same resistor R which has a 'tapping off' point at C. Contact C is a sliding contact, so that the resistance of section R_2 can be made to vary from 0 to R. In this way, the pd across R_2 (i.e. between point B and C) can be made to change from 0 to V by sliding the contact C from end B to end A. Thus, the pd between B and C can be set at any specified value between 0 and V.

The principle of a **potentiometer** is based upon the variable potential divider as in Fig. 7.10(b). A potentiometer is used to measure pd without taking any current (unlike a moving-coil voltmeter), and it does this by balancing up its pd (from a variable potential divider) against the pd to be measured, as in Fig. 7.11. Balance is achieved when no current is detected on the meter M in the circuit diagram.

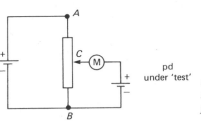

Fig. 7.11 Potentiometer principle

Useful insight into the operation of a potentiometer can be obtained by considering the simple circuits of Fig. 7.12. In circuit (a), battery A has a greater emf than the battery B so current flows from the +ve terminal of A into the +ve terminal of B. In circuit (b), battery A has a smaller emf than battery C so current flows from the +ve terminal of C into the +ve terminal of A. In circuit (c), battery A has the same emf as battery D so no current flows either to or from A; A 'balances' D exactly.

With the potential divider arrangement of circuit Fig. 7.11, the position of the sliding contact C will determine the size and direction of the current through the meter M. When C is at end A, then current will flow through M from left to right; when C is at end B, then current will flow through M from right to left. Clearly, there will be one single position for C at which no current will flow through M. At this position, the pd to be measured will have been balanced out exactly by the pd between C and B. The position of C at balance can then be used to determine the 'unknown' pd.

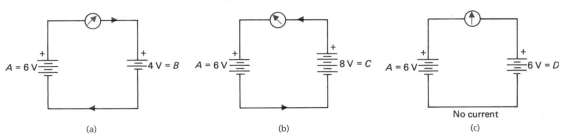

(a) (b) (c)

Fig. 7.12 Current flow in circuits with many cells

Comparison of cell emfs

The emfs of the two cells X and Y may be compared using the 'slide wire' form of the potentiometer shown in Fig. 7.13(a). This form of the potentiometer consists of a length of resistance wire across which a fixed pd is maintained. Contact C is capable of being moved along the wire. Provided that the wire has a uniform cross-section, then the pd between the contact C and one end of the wire is in proportion to the length of wire from that end to the contact (i.e. $V = kl$ where k is a constant). Each of the cells to be compared is connected in turn into the potentiometer circuit. A centre-reading meter must be connected in series with the 'test' cell, as shown by Fig. 7.13(a). By moving contact C along the wire, the point is found at which the meter current is zero (i.e. a 'null' reading). At this point, called the balance point, the pd across the balance length CB ($= l$ in the diagram) must be exactly equal to the emf of the test cell. Thus the test cell emf $E = kl$. By measuring the balance length for cell X and then for cell Y in turn (then recheck for cell X) the emf ratio can be calculated from:

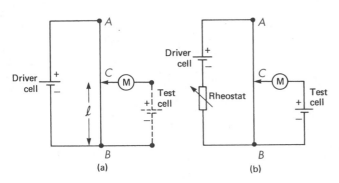

Fig. 7.13 Comparison of emfs

E7.9 $\quad \dfrac{E_x}{E_y} = \dfrac{l_x}{l_y}$

If Y is a standard cell of known emf, then the emf of cell X can be calculated. The key point is that no current is drawn from the test cell when the balance point has been located; hence, there is no loss of pd inside the test cell due to internal resistance. The method is called a 'null' method because the measurement is made at balance when the meter reads zero. The meter is required only to detect current, not to measure it.

Some common errors in setting up potentiometer circuits are:

❶ Wiring the test cell with incorrect polarity when connecting it into the circuit. The test cell polarity must always be such as to oppose the pd along the wire due to the driver cell.

❷ Allowing the driver cell emf to 'run down' during the experiment so that readings with the same test cell at the end will differ from readings taken at the start (with that test cell). A rheostat is sometimes included in series with the driver cell to limit the driver cell current, as in Fig. 7.13(b). Once set, the rheostat must not be adjusted otherwise the pd across the wire will change.

❸ Insufficient pd across the wire to balance up a test cell. The rheostat in point 2 must be set in the first place (if used) to enable the largest test emf to be balanced near the far end of the wire (end A in Fig. 7.13(b)).

7.5 THE WHEATSTONE BRIDGE

The Wheatstone bridge network shown in Fig. 7.14(a) will be balanced (i.e. zero current through M) when the resistance values satisfy the equation:

E7.10 $\quad \dfrac{P}{Q} = \dfrac{R}{S}$

In the metre bridge arrangement shown in Fig. 7.14(b) resistors P and Q together form a metre length of uniform resistance wire AB, resistor S is usually a standard resistor and R is the 'unknown' resistance. Contact C is moved along the wire until the galvanometer current is zero. Then, the length of AB that corresponds to resistor P (length l in the diagram) is measured. R is given by:

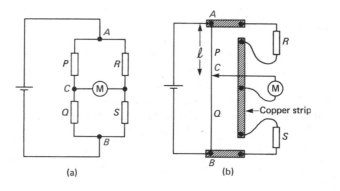

Fig. 7.14 Wheatstone bridge

(a)   (b)

E7.11 $R = \dfrac{lS}{(L-l)}$

where S = standard resistance (Ω), R = unknown resistance (Ω), l = balance length (from the same end as R) (m), L = total length of the wire (m).

Some understanding of the balanced Wheatstone bridge can be achieved by considering Fig. 7.14(b) again. Consider wire AB and sliding contact C as a supplier of variable pd between B and C, and also consider R and S as supplying a fixed pd across S. When at balance, the variable pd between B and C must exactly equal the fixed pd across S.

The Wheatstone network in its metre bridge form is used to measure the resistance of resistance thermometers (see 6.1) and to determine resistivities of wires, etc. To measure the resistivity of a given uniform specimen, the resistance R of the specimen, its area of cross-section A and its length must be determined. Then, resistivity can be calculated from resistance × area of cross-section/length.

7.6 CAPACITORS IN DC CIRCUITS

Capacitance (C) is defined as the charge stored per unit pd:

E7.12 $C = \dfrac{Q}{V}$

where Q = charge stored (C), V = applied pd (V), C = capacitance (farads, F).

Combination rules

In series The combined capacitance (C) of two capacitors in series is given by:

E7.13 $\dfrac{1}{C} = \dfrac{1}{C_1} + \dfrac{1}{C_2}$

Two capacitors in series, as in Fig. 7.15, each store the same charge (Q) so that the pd across C_1 is Q/C_1 and the pd across C_2 is Q/C_2. Thus, the pd across the series combination is $Q/C_1 + Q/C_2$. Since the combined capacitance C = charge stored/total pd, then E7.13 follows. Note that the total charge stored is still Q.

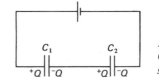

Fig. 7.15 Capacitors in series

In parallel The combined capacitance C is given by:

E7.14 $C = C_1 + C_2$

Capacitors in parallel have the same pd across their terminals (V), so that the charge stored by each capacitor is in proportion to its capacitance (i.e. C_1 stores charge $C_1 V$ and C_2 stores charge C_2V). The total charge stored is thus $C_1V + C_2V$.

When a charged capacitor shares its charge with another capacitor, as in Fig. 7.16 when switch S is disconnected from A and reconnected to B, the final sharing of charge can be determined using two basic rules:

❶ The total final charge = the total initial charge.

❷ The final charge is distributed in proportion to the capacitance (because the two capacitors have the same pd after sharing has taken place).

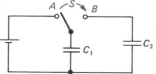

Fig. 7.16 Capacitors in parallel

Discharging a capacitor through a resistor

Initially the discharge current is large because the pd across the resistor (= pd across the capacitor) is at its greatest. Subsequently, the discharge current falls because the capacitor pd falls. A discharge circuit is shown in Fig. 7.17, together with graphs for (i) y = charge, x = time and (ii) y = current, x = time.

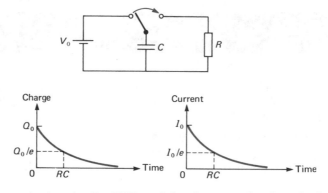

Fig. 7.17 Discharging a capacitor

Since the current is given by $I = V/R$, and the charge on the plates is given by $Q = CV$, it follows that $I = Q/CR$. Because current I is equal to the rate of flow of charge *off* the plates $(\mathrm{d}Q/\mathrm{d}t)$, the differential equation $\mathrm{d}Q/\mathrm{d}t = -Q/RC$ can be used to give a formula for the variation of charge with time:

E7.15 $\quad Q = Q_0 e^{-t/RC}$

where Q = charge at time t (C), Q_0 = initial charge (C), RC = circuit 'time constant'.

The circuit **time constant RC** is a measure of the rate of discharge. Use of E7.15 shows that the capacitor discharges to 37% of its initial charge in a time RC. Note that the discharge current follows the same exponential decay law as the charge; in other words, $I = I_0 e^{-t/RC}$ where I_0 = initial current = Q_0/RC.

Charging a capacitor through a resistor

Initially, a capacitor will charge up with a high rate of flow of charge (i.e. current) onto the plates, but as the charge on the plates builds up, then the charging current becomes less and less. Fig. 7.18 shows a charging circuit, and the graphs show how the charge on the plates and the current through the resistor vary with time. Note that the gradient of the charge curve gives the current curve (since $I = \mathrm{d}Q/\mathrm{d}t$). Also, since the pd across the resistor is equal to IR, and the pd across the capacitor is equal to Q/C, then the pd across each component varies as in Fig. 7.19. The sum of the pds is equal to the battery pd.

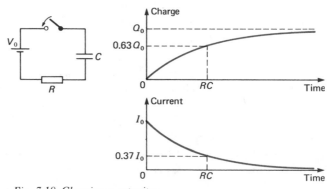

Fig. 7.18 Charging a capacitor

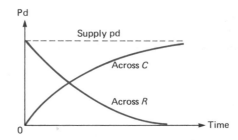

Fig. 7.19 Capacitor and resistor pd on charging

The charge curve of Fig. 7.18 is a 'build-up' exponential (because it 'builds up' to a final level), and it can be shown that the time constant RC represents the speed of charging up in a similar way to the discharge curve where it represents the speed of discharge. In fact, RC for the charging circuit is the time taken for the charge to build up to 63% (i.e. $100 - 37\%$) of its final value.

7.7 MEASURING AC

Frequency is the number of complete cycles per second. The unit is the hertz (Hz). **Amplitude** (or peak value) is the maximum value of an alternating signal. The **root mean square** (rms) value of an alternating current (or pd) is the value of the direct current (or pd) which would give the same power dissipation as the alternating current (or pd), in a given resistor. The term 'alternating' can be taken to include not only sine wave signals but any other regular waveform such as a square wave signal or a triangular waveform, as in Fig. 3.1.

E7.16 rms value $= \dfrac{\text{peak value}}{\sqrt{2}}$ **for sine waves only**

Representing **sine wave** signal can either be by:

❶ graphs of y = instant value, x = time, as in Fig. 7.20(b), or

❷ an equation of the form:

E7.17 $I = I_0 \sin(2\pi f t)$

where I = current at time t (A), I_0 = peak current (A), f = frequency (Hz), or

❸ rotating vectors sometimes called **phasors**, as in Fig. 7.20(a). The vector length is scaled to the peak value, and the vector is considered to rotate at frequency f; thus, the projection of the vector onto a straight line represents the instant value.

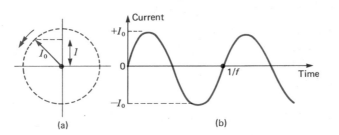

(a)   (b)

Fig. 7.20 Representing a sinusoidal change

Rectifiers convert ac into dc. Fig. 7.21 shows a bridge rectifier with which sine wave ac may be converted into full-wave dc as illustrated by the graphs in the diagram. On one half of each cycle, diodes D_1 and D_3 conduct, whereas on the other half of the cycle diodes D_2 and D_4 conduct. A smoothing capacitor is usually included, as shown on the diagram, so as to give a steadier direct current.

Oscilloscopes are used to display and measure alternating voltage waveforms. With a calibrated time base, the time period of an alternating signal may be measured; with a calibrated Y-scale, the peak value may be measured. Note that the vertical deflection of the spot (or trace)

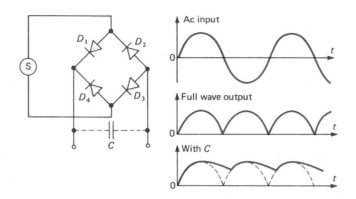

*Fig. 7.21 Full-wave
rectification and smoothing*

is in proportion to the pd between the *Y*-input terminals. Fig. 7.22 shows an example of the use of an oscilloscope to measure peak voltage and frequency. If the time base is set to a rate of 5 ms cm⁻¹, then the time period must be 20 ms since one full cycle takes 4 cm along the *X*-scale; if the *Y*-sensitivity is 0.5 V cm⁻¹, then the peak value (i.e. from centre to top) will be 1.0 V since the trace is 4 cm from top to bottom.

An oscilloscope can be used to display a **Lissajous** figure, as in Fig. 7.23, by disconnecting the time base circuit from the *X*-input plates and connecting an alternating voltage of frequency

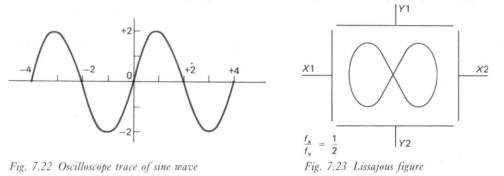

Fig. 7.22 Oscilloscope trace of sine wave   *Fig. 7.23 Lissajous figure*

f instead. Then, a second alternating voltage of the same amplitude and of frequency *nf* (where *n* is an integer) is applied to the *Y*-input terminals. In the diagram, the spot has moved up and down in the same time as it has moved across just once; i.e. $f_Y = 2f_X$.

7.8 AC CIRCUITS

The effect of alternating pds upon individual components must first be understood before dealing with combinations of components. For a **sine wave** pd applied across a single component (i.e. a resistor or a capacitor or an inductor), the current can be calculated from the pd if the following two quantities are known:

❶ the resistance *R* if a resistor, or the reactance *X* if a capacitor or an inductor;

❷ the phase difference between current and pd (i.e. the fraction of a cycle between peak pd and peak current). Remember that one full cycle is 2π radians.

Resistance only

Current is in phase with the applied pd, so that at every instant the current is given by $I = (V_0/R)\sin(2\pi ft)$ for an applied pd $V_0 \sin(2\pi ft)$.

Capacitance only

Current is ahead of the applied pd by $\frac{1}{4}$ cycle. The reason is that with an applied pd $V = V_0 \sin(2\pi ft)$, the charge *Q* at any instant on the plates is given by:

$Q = CV = CV_0 \sin(2\pi ft)$

Since the current $I = \mathrm{d}Q/\mathrm{d}t$, then differentiation gives:

$I = (2\pi fC)V_0 \cos(2\pi ft)$

The peak current I_0 is when $\cos(2\pi ft) = 1$, so giving:

$I_0 = (2\pi fC)V_0$

Since the reactance of a capacitor is defined as (peak pd)/(peak current) then reactance is given in terms of capacitance by:

E7.18 $$X_C = \frac{1}{2\pi fC}$$

where X_C = capacitor reactance (Ω), f = frequency (Hz), C = capacitance (F).

Fig. 7.24 illustrates the circuit involved.

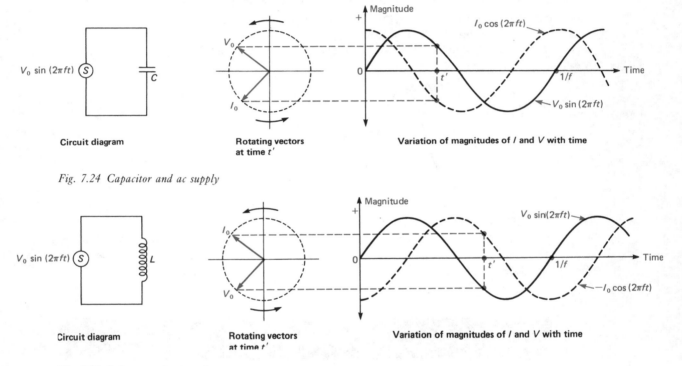

V₀ sin (2πft)
S
C
Circuit diagram
V₀
I₀
Rotating vectors at time t′
Magnitude
+
0
I₀ cos (2πft)
t′
1/f
Time
V₀ sin (2πft)
Variation of magnitudes of I and V with time

Fig. 7.24 Capacitor and ac supply

V₀ sin (2πft)
S
L
Circuit diagram
I₀
V₀
Rotating vectors at time t′
Magnitude
+
0
V₀ sin(2πft)
t′
1/f
Time
−I₀ cos (2πft)
Variation of magnitudes of I and V with time

Fig. 7.25 Inductor and ac supply

Inductance only

Current is behind the applied pd by $\frac{1}{4}$ cycle. The explanation here lies in the fact that the applied pd causes current change, which causes an induced emf to match the applied pd. If $V_0 \sin(2\pi ft)$ is the applied pd, then the induced emf ($= L(\mathrm{d}I/\mathrm{d}t)$) must equal $V_0 \sin(2\pi ft)$. By integration:

$$I = \frac{-V_0}{(2\pi fL)} \cos(2\pi ft)$$

The peak current I_0 is when $\cos(2\pi ft) = 1$, so $I_0 = V_0/(2\pi fL)$. The variation of applied pd, and of current, with time is shown in Fig. 7.25.

E7.19 $X_L = 2\pi fL$

where X_L = reactance of the inductor (Ω), f = frequency (Hz), L = self-inductance (H).

A useful aid for memorising phase differences is **'CIVIL'** (I ahead of V for C; I after V for L).

Filter circuits

A capacitor will tend to block the passage of low-frequency currents because its reactance is high at low frequency. An inductor will allow low-frequency currents to pass with relative ease, but will tend to block the passage of high-frequency currents since its reactance increases with increased frequency. A simple tuning circuit can be made by connecting a capacitor in parallel with an inductor, as in Fig. 7.26. The aerial will pick up many frequencies, but the high-frequency signals will pass to earth via the capacitor whereas the low-frequency signals will pass to earth via the inductor. Frequencies of an intermediate value will not pass easily through either the inductor or the capacitor, and so will pass into the amplifier, etc. If the signal to be detected has a frequency near this intermediate value, then it will pass into the amplifier. The intermediate frequency is then given by equal reactance of the capacitor and inductor (see E7.23).

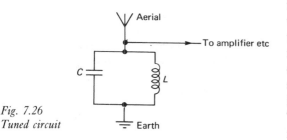

Fig. 7.26
Tuned circuit

Series circuits and sine wave pds

Consider a series *LCR* circuit connected to a sine wave pd, as in Fig. 7.27. There are two key points to use in working out currents and pds:

❶ The current at any instant is the same in each component (because they are in series).

❷ The sum of the pds at any instant is equal to the source pd at that instant.

In this example, currents and pds can best be dealt with by using the rotating vector (phasor) method. Since the current is the same for each component, then the individual pd vectors can be drawn relative to the current vector, as in Fig. 7.27. The sum of the three pd vectors V_R, V_C and V_L is equal to V_s, the supply pd vector, so giving the following equation:

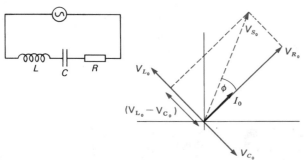

Fig. 7.27 LCR series circuit

E7.20 $\qquad V_{s_0}^2 \ = \ V_{R_0}^2 + (V_{L_0} - V_{C_0})^2$

where V_{s_0} is the peak value of the supply pd, etc. Then, since individual peak pds are related to peak current (I_0) by the equations $V_{R_0} = I_0 R$, $V_{L_0} = I_0(2\pi fL)$ and $V_{C_0} = I_0(1/2\pi fC)$, E7.20 may be written as:

$$V_{s_0}^2 = I_0^2 \left[R^2 + \left(2\pi fL - \frac{1}{2\pi fC} \right)^2 \right]$$

Impedance (Z) of a combination of components is defined as:

$$\frac{\text{peak applied pd } (V_0)}{\text{peak current } (I_0)}$$

so that the impedance of a series *LCR* circuit is given by:

E7.21 $\quad Z = \sqrt{\left[R^2 + \left(2\pi fL - \frac{1}{2\pi fC} \right)^2 \right]}$

where R = resistance (Ω), C = capacitance (F), L = self-inductance (H), f = frequency (Hz), Z = impedance (Ω).

Phase angle φ between the supply pd vector and the current vector for the series *LCR* circuit is given by:

E7.22 $\ \tan \varphi = \frac{1}{R} \left(2\pi fL - \frac{1}{2\pi fC} \right)$

For a series RC circuit or a series RL circuit, the vector method as above applies in a similar way; the final results can be deduced from E7.21 and E7.22 by omitting the reactance term corresponding to the excluded component. For instance, for a series RL circuit, the impedance $Z = \sqrt{[R^2 + (2\pi f L)^2]}$. It is worth noting that the term 'coil' is usually taken to mean a device with inductance *and* resistance, so unless you are given that a particular coil has negligible resistance, you must treat it as an inductance in series with a resistance.

Power dissipation only occurs in the resistances of an ac circuit. The reason is that the pd and current for either a capacitor or an inductor are $\frac{1}{4}$ cycle out of phase, so that power (= pd $\times$ current) averages out at zero over a complete cycle for L or C. The impedance of a capacitor or an inductor is called its **reactance**, a term used to signify that the average power dissipated is zero. The power dissipated by a resistance R which passes ac of peak value I_0 is $\frac{1}{2}I_0^2 R$ which equals $I_{rms}^2 R$ since the rms current $I_{rms} = I_0 / \sqrt{2}$.

Resonance of a series *LCR* circuit

Consider a series LCR circuit with a variable-frequency supply pd. The impedance, as given by E7.21, will be a minimum when the applied frequency is such that the capacitor reactance $(1/2\pi f C)$ is equal to the inductor reactance $(2\pi f L)$. This frequency is known as the 'resonant' frequency and corresponds to equal values of capacitor pd and inductor pd, so the two pds cancel one another out (the vectors V_C and V_L of Fig. 7.27 will be of equal length). At resonance, the current will be a maximum (= V_{s_0}/R) and will be in phase with the supply pd:

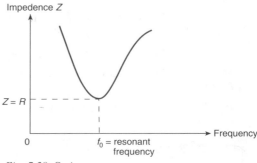

Fig. 7.28 Series resonance

E7.23 $f_0 = \dfrac{1}{2\pi\sqrt{LC}}$

where C = capacitance (F), L = self-inductance (H), f_0 = resonant frequency (Hz).

Fig. 7.28 shows a graph of y = impedance, x = frequency of supply.

7.9 SEMICONDUCTORS

Semiconductors are the basic materials used in the manufacture of transistors and integrated circuit 'chips'. The element silicon (Si) is a widely used semiconductor. Without any atoms of different elements added (i.e. without doping), the semiconductor is known as an **'intrinsic'** semiconductor. At absolute zero, the atoms have loosely attached electrons in their outer shells. At room temperature, these electrons become detached (i.e. become conduction electrons) and so respond to applied pds. The resistance of intrinsic semiconductors falls as the temperature rises because more electrons become detached from 'parent' atoms. In other words, intrinsic semiconductors have a −ve temperature coefficient of resistance (compared with a +ve coefficient for metals). **Thermistors** are temperature-dependent resistors made from semiconducting material.

A **light-dependent resistor (LDR)** contains a semiconducting material which is in a transparent seal and can therefore be illuminated. Light falling on the semiconductor surface gives sufficient energy to loosely attached electrons to allow them to break free and become conduction electrons. Thus the resistance of an LDR falls if the light intensity rises.

Extrinsic semiconductors are made by adding controlled amounts of a different element to an intrinsic semiconductor (i.e. doping). When the added atoms have one more outer-shell electron each than is required for bonding, the surplus electron per added atom then becomes a conduction electron; the semiconductor is then known as an **n-type** (extrinsic) semiconductor. Alternatively, by doping an intrinsic semiconductor with atoms which each have one less outer-shell electron than the 'host' atoms, a **p-type** semiconductor is produced. In p-type material, the electron vacancies are called 'holes', and these are responsible for conduction. Being produced by electron vacancies, the holes are +ve charge carriers, a fact which can be demonstrated by the Hall effect (see 2.11). Fig. 7.29 illustrates the idea of conduction by holes.

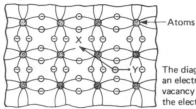

Atoms

The diagram shows that when an electron fills a vacancy, the vacancy moves to the site where the electron moved from.

Fig. 7.29 Movement of a vacancy (hole)

The p-n junction diode consists of a p-type semiconductor in contact with an n-type semiconductor. The arrangement and circuit symbol are shown in Fig. 7.30(a) and Fig. 7.30(b).

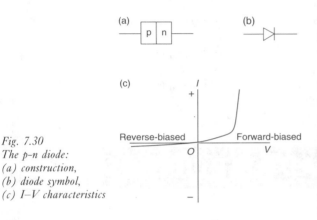

*Fig. 7.30
The p-n diode:
(a) construction,
(b) diode symbol,
(c) I–V characteristics*

Current only passes through a p-n diode in a circuit when the p-type material is positive and the n-type is negative; the diode is said to be **forward-biased** when it conducts.

The reason why a p-n diode only allows current through in one direction is because conduction electrons from the n-type semiconductor spread into the p-type semiconductor and holes transfer in the opposite direction. This process makes the n-type material positive and the p-type material negative, thus creating a *barrier* to the further transfer of charge carriers. The barrier is removed when the p-type material is connected to the positive terminal of a battery and the n-type to the negative; the diode thus conducts. Connecting the battery the other way round increases the barrier so the diode cannot conduct. Fig. 7.30(c) shows how the current varies with applied pd.

Transistor action

The most common form of transistor is the silicon n-p-n junction transistor in which two regions of n-type material are separated from one another by a layer of p-type material. The arrangement and circuit symbol are shown in Fig. 7.31. To understand transistors in action, you should first appreciate that current flow into the collector and out through the emitter

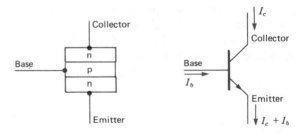

Fig. 7.31 n-p-n transistor

(I_c) is controlled by a much smaller current flow into the base (I_b) (and out at the emitter). In normal operation, the collector current is always determined by, and in proportion to, the current taken by the base.

$$\text{E7.24 Current gain } (\beta) = \frac{\text{change of } I_c}{\text{change of } I_b}$$

where I_c = collector current, I_b = base current.

The characteristics of a junction transistor are represented by the two graphs of Fig. 7.32.

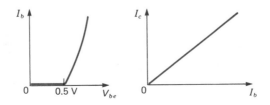

Fig. 7.32 Transistor characteristics: (a) input, (b) transfer

The **input characteristic**, as in Fig. 7.32(a), shows how the base current varies with base–emitter pd. The graph shows that the base–emitter resistance is not constant, and that the base–emitter pd must exceed approximately 0.5 V before the base will take any current (i.e. before the transistor is 'turned on').

The **transfer characteristic**, as in Fig. 7.32(b), shows how the collector current varies with base current. The slope is equal to the current gain (β).

Consider as an example a transistor with a current gain of 150. If the base current is 2 mA, the collector current is therefore $2 \times 150 = 300$ mA. If the base current is zero, the collector current is also zero. This example shows that a transistor is a device in which a small current controls a much larger current.

Fig. 7.33 shows a transistor used in a circuit for a temperature-operated alarm. When the thermistor T becomes hot, its resistance falls and so it allows more current into the base of the transistor. Hence more current passes into the collector of the transistor

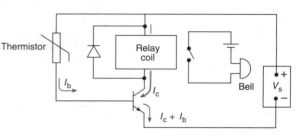

Fig. 7.33 A temperature-operated alarm

through the relay coil. Thus the relay is energised and its switch is closed, turning the alarm bell on. The small current through the thermistor controls a much larger current through the relay coil. Note the reverse-biased diode across the relay coil to protect the transistor from back emfs induced in the relay coil when it is switched off.

7.10 OPERATIONAL AMPLIFIERS

Op-amps in integrated form are widely used in **analogue** circuits where output pds can take any value between the limits of the supply pd. The essential features of an op-amp are (i) a very high gain (typically 10^5), (ii) a very high input resistance (typically $10^{12}\ \Omega$ in modern op-amps). The circuit symbol is shown in Fig. 7.34, and represents an 'open-loop' amplifier. Note that there are two inputs: input P is the inverting input and input Q is the noninverting input. The output pd is proportional to the pd between Q and P provided it does not 'saturate' (i.e. reach the limits set by the supply pd):

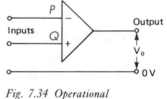

Fig. 7.34 Operational amplifier: basic connections

E7.25 $\quad V_o = A(V_Q - V_P)$

where V_o = output pd (V), V_Q = pd at the noninverting input (V), V_P = pd at the inverting input (V), A = open-loop gain (typically 10^5).

Open-loop voltage comparator

If the input to the noninverting terminal Q is set at fixed pd, and a sine wave input is applied to the inverting input P, as in Fig. 7.35, then the output pd will be $-V_s$ (V_s is supply pd)

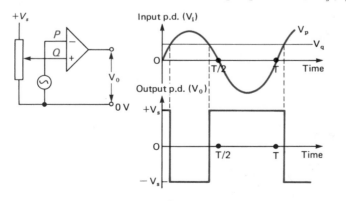

Fig. 7.35 Operational amplifier: comparator

whenever the sine wave pd exceeds the fixed pd at Q. When the sine wave pd is less than the fixed pd, then the output pd will be $+V_s$. The diagram also shows the variation with time of the input and output pds. The high open-loop gain (A) forces the output to **saturation** (i.e. to $+V_s$ or $-V_s$) whenever the pd between the inputs exceeds $\pm V_s/A$. So for a supply pd of ± 15 V with $A = 10^5$, the input pd only needs to exceed ± 150 µV for saturation to occur.

Inverting amplifier

To amplify input pds greater than approximately 150 µV without saturation, a feedback resistor and an input resistor must be added. Fig. 7.36 shows the arrangement for an inverting amplifier (a closed-loop amplifier).
The voltage gain is given by:

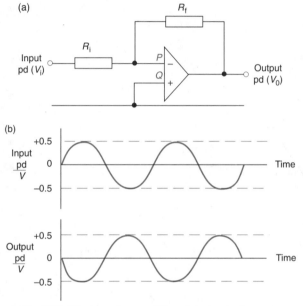

(a)

(b)

E7.26   voltage gain $\dfrac{V_o}{V_i} = -\dfrac{R_f}{R_i}$

where R_f = feedback resistance (Ω), R_i = input resistance (Ω).

This circuit provides an example of 'negative feedback', in which a fraction of the output signal is fed back so as to reduce the gain from the open-loop value. Note that the voltage gain here depends only upon the external resistors. Also in the circuit, the noninverting input is earthed. Provided the output does not saturate, the inverting input will be at 'virtual earth' potential. This is because $V_Q = 0$, and with $A = 10^5$ and $V_o <$ supply pd (say 15 V), then V_p will be less than 150 µV (i.e. $\simeq 0$ V).

Suppose a sine wave of peak value 0.50 V is applied to an inverting amplifier which has a voltage gain of -10. The output pd will therefore be an *inverted* sine wave of peak value 5.0 V$(= -10 \times 0.5$ V), as shown in Fig. 7.36(b).

Fig. 7.36 The inverting amplifier: (a) circuit diagram, (b) input and output waveforms for a voltage gain of -10

The voltage follower

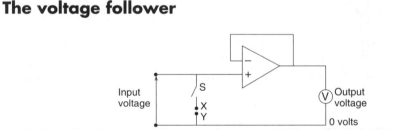

Voltage follower	S open
Nanoammeter	S closed: resistor between X and Y
Nanocoulombmeter	S closed: capacitor between X and Y

Fig. 7.37 The voltage follower

The voltage follower (Fig. 7.37) is a circuit in which the input resistance is very high and output voltage equals the input voltage. The circuit may therefore be used as a buffer or 'impedance converter' to monitor or measure a voltage signal without drawing current.

The circuit consists of an operational amplifier in which the output terminal is connected to the inverting input. The voltage to be measured is applied to the noninverting terminal. Since the inverting input is virtually at the same potential as the noninverting input (assuming the output is not saturated), the output voltage is therefore equal to the input voltage.

The circuit can also be used:

❶ as a nanoammeter, by connecting a high-resistance resistor R between the noninverting terminal and the 0 V line – the current through the resistor is equal to V_{out}/R;

❷ as a nanocoulomb meter, by connecting a suitable capacitor C between the noninverting terminal and the 0 V line – the charge on the capacitor is equal to $C \times V_{out}$.

Black boxes

The op-amp provides a good example of an electronic 'black box' in that only the input/output characteristics need to be known; detailed knowledge of the internal circuit is not essential to use the device. For example, consider a black box with input/output characteristics as shown in Fig. 7.38. If the sine wave pd of peak value 2 V is applied between the input terminals, the output pd can be deduced from the characteristics as follows: (i) $V_i \leq 0.5$ V, so $V_o = -10$ V; (ii) $V_i \geq 1.5$ V, so $V_o = +10$ V; (iii) $0.5 < V_i < 1.5$ V – in this range V_i is amplified, and since the voltage gain (from the gradient of the graph) is $\times 20$, then $V_o = 20(V_i - 1)$ in this example.

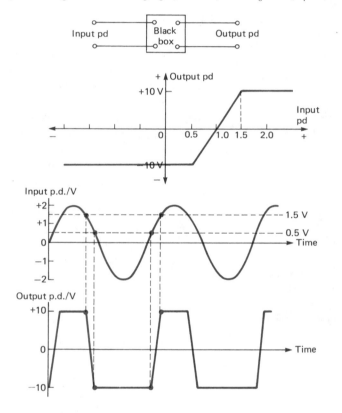

Fig. 7.38 Operational amplifier as a 'black box'

7.11 LOGIC CIRCUITS

Logic circuits give a simple introduction to **digital** electronics in which circuit units can have only two states, so that inputs and outputs are either high (i.e. +ve saturation, termed as '1') or low (i.e. –ve saturation termed as '0').

The circuit in Fig. 7.39 will operate as a **logic switch** if the base resistance is sufficiently small. When the input pd is high, the output pd is low; when the input is low, the output will be high. The circuit is a NOT unit in logic terms; its symbol and 'truth table' relating output to input are shown in Fig. 7.39.

Fig. 7.40 shows several other simple logic units based upon transistors. The key to each is the truth table.

You should be able to work out the truth table of any simple combination of the logic units in Fig. 7.40. For example, consider the combination shown of two NOT units and a NOR

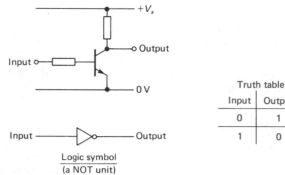

Truth table

Input	Output
0	1
1	0

Logic symbol
(a NOT unit)

Fig. 7.39 Transistor as a logic block

Name	NOR	OR	NAND	AND	AND constructed from 2 NOTs & 1 NOR
Logic symbol					
Truth table — A	0 0 1 1	0 0 1 1	0 0 1 1	0 0 1 1	
B	0 1 0 1	0 1 0 1	0 1 0 1	0 1 0 1	
V_0	1 0 0 0	0 1 1 1	1 1 1 0	0 0 0 1	

Fig. 7.40 Logic gates

unit; the output is 1 only when the two inputs are both 1, so its truth table is the same as that of an AND unit.

A **multivibrator** consists of two logic switches that are coupled via either resistors or capacitors or both. With capacitor coupling, as in Fig. 7.41, the circuit is known as an **astable** multivibrator; its outputs change back and forth automatically between the two logic states such that when one output is high, the other is low. The diagram shows the output variation with time for one of its outputs.

When the output of gate 1, X, goes high, the voltage at Y goes high temporarily. Thus Z and W go low while Y is high, keeping X high. However, as the capacitor between X and Y charges up, the voltage drops at Y. When the voltage at Y decreases below a certain level, gate 2 switches and Z goes high, taking W high temporarily – thus switching X off until the voltage at W decreases below a certain level. The sequence repeats itself automatically.

A **bistable** multivibrator, as in Fig. 7.42, is a combination of logic gates that has two stable states; it can only be in one of these states at any time. An input pulse makes it change from one state to the other.

The bistable unit is the basis of the binary pulse counter, as in Fig. 7.43, in which a series of bistable units are connected with indicator lamps (or light-emitting diodes) to demonstrate the state of each bistable unit.

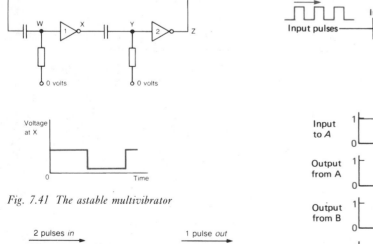

Fig. 7.41 The astable multivibrator

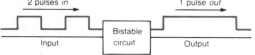

Fig. 7.42 The bistable multivibrator

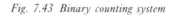

Fig. 7.43 Binary counting system

Fall of the input at *A* causes the input at *B* to change state, and when *B*'s input changes from a 1 to a 0, then *C* changes state, etc. Starting with all outputs at 0, the diagram shows the state of the outputs after *n* input pulses. The output states represent the binary number for *n* in each case; for example, after 4 input pulses (and before the fifth), output *C* is a 1 and all other outputs are 0s. The outputs in order *DCBA* are therefore 0100 which is the binary number for 4.

Digital transmission

Information may be transferred as a sequence of pulses, each pulse caused by switching electrical or electromagnetic carrier waves on and off.

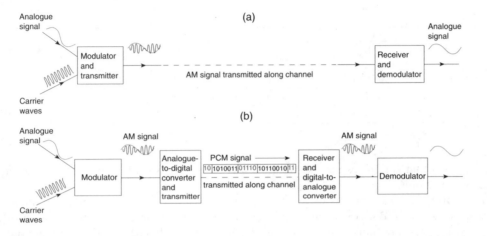

Fig. 7.44 (a) Amplitude modulation, (b) pulse code modulation

Pulse code modulation (PCM) (Fig. 7.44b) involves converting an analogue signal into a string of 1s or 0s representing the signal amplitude in binary form. The pulses may need to be **amplified** and **regenerated** at intervals along the transmission path to remove unwanted signals due to 'noise'. At the receiver, the pulses are used to recreate the original analogue signal. In comparison, **amplitude modulation (AM)** (Fig. 7.44a) is where the analogue signal varies the amplitude of the carrier waves. Noise cannot be removed and amplifiers boost the noise as well as the signal.

Chapter roundup

Be certain that you have a sound grasp of potential difference before you move onto Chapter 8. Make sure that you understand the principles of capacitors and ac circuits in terms of knowledge developed in the early topics of this chapter.

Question bank

1 The diagram shows a piece of pure semiconductor, *S*, in series with a variable resistor, *R*, and a source of constant voltage, *V*. *S* is heated and the current is kept constant by adjustment of *R*. Which of the following factors will decrease during this process?

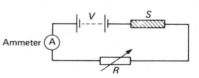

1 The drift velocity of the conduction electrons in *S*.
2 The dc resistance of *S*.
3 The number of conduction electrons in *S*.
Answer: A if 1, 2, 3 correct. B if 1, 2 correct.   C if 2, 3 correct.
   D if 1 only.   E if 3 only.
(London: *and all other Boards except Oxford, SEB***)**

Points

See 7.9 for the behaviour of semiconductors.

2 The current in a copper wire is increased by increasing the potential difference between its ends. Which one of the following statements regarding n, the number of charge carriers per unit volume in the wire, and $\bar{v}$, the drift velocity of the charge carriers, is correct?
  A n is unaltered but $\bar{v}$ is decreased.
  B n is unaltered but $\bar{v}$ is increased.
  C n is increased but $\bar{v}$ is decreased.
  D n is increased but $\bar{v}$ is unaltered.
  E Both n and $\bar{v}$ are increased.

(**Cambridge**: *and all other Boards except Oxford, SEB*)

Points

See 7.1. The situation is not unlike that of water flow from a hosepipe; increased pressure makes the water flow faster but does not change the water density.

3 The diagram shows a 12 V supply of negligible internal resistance connected to two 30 kΩ resistors with a voltmeter of resistance 60 kΩ connected between P and Q.

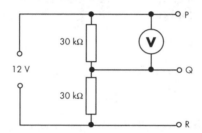

The potential difference, in V, between Q and R is:
A 4.8.   B 6.0.   C 7.2.   D 9.0

(**AEB** June 91: *all Boards*)

Points

Work out the combined resistance of the parallel combination and hence the total resistance between P and R. The current through the resistor between Q and R is the same as the current from the 12 V supply which equals 12/total resistance. Hence work out the pd between Q and R.

4 The diagram shows four identical lamps, J, K, L and M which are all lit.
Lamp K is now removed from the circuit.
Which one of the following statements is *true*?
A J, L and M are now equally bright.
B J is brighter than originally but not as bright as L and M are now.
C J is less bright than originally but not as bright as L and M are now.
D J is less bright than originally but brighter than L and M are now.

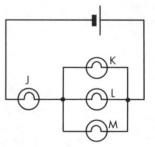

(**AEB** Nov 92: *all Boards*)

Points

Removing K reduces the current through J. How does this affect the pd across L and M and hence the current through L and M?

5 In the circuit diagram below, two resistors X and Y are connected to a 12 V battery of negligible internal resistance. The resistance of X is known to be 6000 ohms. A voltmeter of internal resistance 6000 ohms is then connected across X, and its reading is found to be 9.0 V. What must be the resistance of Y in ohms?

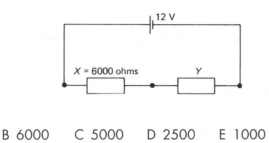

A 7500   B 6000   C 5000   D 2500   E 1000

(–: all Boards)

Points

With 9 V across X when the meter is connected in the circuit, the remainder of the battery pd is dropped across Y. Also, the current through Y = current through X + current through the meter.

6 A galvanometer gives a full-scale deflection when a current of 2mA flows through it and the potential difference across its terminals is 4 mV. Which of the following resistors would be most suitable to convert it to give a full-scale deflection for a current of 1 A?
A 0.004 Ω in series.
B 0.004 Ω in parallel.
C 0.50 Ω in series.
D 500 Ω in series.
E 500 Ω in parallel.

(London: *and all other Boards except O and C, SEB, NICCEA)*

Points

See Fig. 7.6 and the discussion in 7.3 on meter conversion.

7 In the circuit, XY is a potentiometer wire 100 cm long. The circuit is connected up as shown. With switches S_2 and S_3 open a balance point is found at Z. After switch S_1 has remained closed for some time, it is found that the contact at Z must be moved

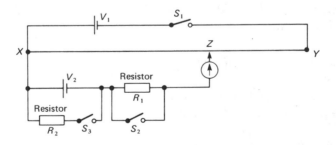

towards Y to maintain a balance. Which of the following is the most likely reason for this?
A The cell V_1 is running down.
B The cell V_2 is running down.
C The wire XZ is getting warm and its resistance is increasing.
D The resistor R_1 is getting warm and increasing in value.
E Polarisation is affecting the emf of V_2.

(London: *and all other Boards except O and C Nuffield, SEB)*

Points

Ignore S_3 and R_2 as S_3 is not closed. Remember that a longer balance length corresponds to either V_1 being smaller or V_2 being larger than at the start. Which is the more likely?

8 The balance point of a slide wire Wheatstone bridge will *not* be changed by increasing:
1 the emf of the driver cell.
2 the resistance of the slide wire.
3 the galvanometer resistance.
A 1, 2, 3 correct. B 1 and 2 correct.
C 1 and 3 correct. D 1 only correct. E 3 only correct.

(–: *NEAB*, Oxford, O and C, SEB*)

Points

What is the equation which represents the balance condition? See E7.10 if necessary. Does that equation contain any of the above factors? For 2, remember that it is the *ratio* of the resistances of the two sections of the slide wire that is important.

9 A box is known to contain three *identical* capacitors wired together in a circuit containing no other components. Two wires lead from this circuit to the outside of the box, and the measured capacitance between these wires is 30 μF. Which one of the following could be the correct capacitance of each capacitor?
A 15μF   B 20 μF   C 40 μF   D 60 μF   E 100 μF

(**London**: *all other Boards except SEB*)

Points

Combination rules for capacitors are given in 7.6. There are four possible arrangements:
❶ All in parallel: will any of the alternatives give 30 μF in total?
❷ All in series: the combined capacitance will be $\frac{1}{3}$ of the individual capacitance for three identical capacitance in series. Will any alternative give 30 μF in total?
❸ Two in parallel wired with one in series: let C be the individual capacitance, so write down the total capacitance in terms of C. Can you now choose a value for C from the alternatives that gives 30 μF in total?
❹ Two in series wired with the third in parallel with the series pair: again let C be the individual capacitance, and use the same approach as in 3.

10 The combined capacitance of the arrangement shown below in μF, is:

A 1   B $\dfrac{18}{11}$ C $\dfrac{30}{11}$ D 4   E 11

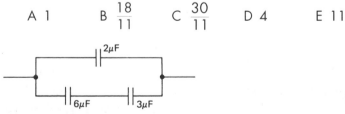

(–: *all Boards except SEB*)

Points

Determine the capacitance of the pair in series firstly (see E7.13) and then consider the parallel combination (see E7.14, if necessary).

11 An alternating current of 1.5 mA rms and angular frequency ω =100 rad s^{-1} flows through a 10 kΩ resistor and a 0.50 μF capacitor in series. The rms pd across the capacitor is:

A 4.8 V   B 15 V   C 30 V   D 34 V   E 190 V
(**Cambridge**: *and Oxford*, all other Boards except SEB*)

Points

The circuit is a series *RC* circuit; calculate the capacitor reactance using E7.18 and referring to the comments in 7.8. Then calculate the rms pd (= rms current × capacitor reactance).

12 In the circuit shown the frequency of the alternating emf is gradually increased from 500 Hz to 5000 Hz, its amplitude remaining constant at 12 V.

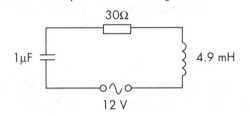

As the frequency is increased, the current flowing in the circuit will:
A increase continually.
B decrease continually.
C increase at first and later decrease.
D decrease at first and later increase.

(**AEB** June 93: *all Boards*)

Points

The current is greatest at the resonant frequency. Use E7.23 to calculate the resonant frequency. Does the resonant frequency lie in the range 500 Hz to 5000 Hz?

13 In a circuit containing a capacitor, an inductor and a resistor in series, V_C, V_L and V_R represent the potential differences across those three components and *I* represents the current through them. Which of the following statements is (are) true?
1 V_C and *I* are 180° out of phase.
2 V_R and *I* are 90° out of phase.
3 V_L and V_C are 180° out of phase.
Answer:
A if 1, 2, 3 are correct.   B if 1, 2 correct.
C if 2, 3 correct.   D if 1 only.   E if 3 only.

(**London**: *all other Boards except SEB*)

Points

See 7.8 if necessary.

14 A moving-coil ammeter is to be adapted to detect small alternating currents. Which of the diagrams shows how a diode could be connected in order to make the conversion?

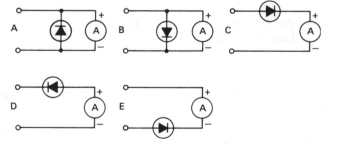

(**London**: *all other Boards except SEB*)

Points

The current must only pass through the meter in the direction + to –. The diodes in A and B will not stop the reverse current, so can be eliminated. Remember that a diode conducts in the direction in which the triangle of the symbol is pointing.

15 In the circuit shown, a capacitor of 10 μF capacitance is connected in parallel with a coil of inductance 1 mH and negligible resistance. The combination is then connected in series with an ac supply of variable frequency and an ac meter.

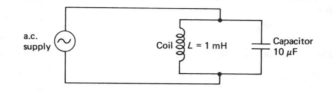

The frequency of the supply is then adjusted to 1600 Hz and the output is adjusted until the meter reads 10 mA. Without changing the output pd of the supply, the meter reading will increase when:
1. the frequency is changed to 2000 Hz.
2. the frequency is decreased to 1200 Hz.
3. the capacitance is decreased to 5 μF.
A 1 only.   B 2 only.   C 1 and 3. D 2 and 3. E 1, 2 and 3.

(–: O and C Nuffield)

Points

The current from the supply is at a minimum at the resonant frequency. Hence the current will drop if any of the changes 1, 2, 3 result in moving closer to resonance. Conversely the current will increase only if the suggested change results in moving further away from the resonance position. E7.23 can be used to determine the resonance frequency; this frequency is the same for 1 and 2 but changes for 3.

16 The overhead cables used in a 132 kV grid system consist of 7 strands of steel wire and 30 strands of aluminium wire. The 7 strands of steel wire have a combined resistance of 3.0 Ω per kilometre and the 30 strands of aluminium wire have a combined resistance of 0.17 Ω per kilometre.
(a) Show that the resistance of the cable is 0.16 Ω per kilometre of cable.
(b) A typical current in the cable is 400 A. Calculate the power loss per kilometre.

(SEB: *all other Boards)*

Points

(a) Each kilometre of cable is equivalent to two resistors of 3 Ω and 0.17 Ω in parallel.
(b) Power loss occurs in the cable since its resistance will cause heating. Since electrical power is given by current $\times$ pd, then the heat produced per second = I^2R.

17 (a) Explain what is meant by the 'electromotive force' and the 'terminal potential difference' of a battery.
(b) A bulb is used in a torch which is powered by two identical cells in series each of emf 1.5 V. The bulb then dissipates power at the rate of 625 mW and the pd across the bulb is 2.5 V. Calculate (i) the internal resistance of each cell and (ii) the energy dissipated in each cell in 1 minute.

(NEAB: *all other Boards)*

Points

(a) See 7.2.

(b) Draw the circuit consisting of the two cells and the torch bulb in series with one another. You should show the internal resistance of each cell clearly. Calculate the current taken by the bulb from the power and pd values given. See E7.4 if necessary. Then, calculate the pd across each internal resistance, using the fact that the total emf is 3.0 V, but only 2.5 V is dropped across the external load (i.e. the bulb). Once you have determined the current in the circuit and the pd across each internal resistance (the lost voltage: see 7.2 if necessary), calculate a value for internal resistance. For (ii), use E7.4 to calculate the power dissipated in each internal resistance (remember to use the 'lost voltage' not the cell emf for power dissipated). Then, since power = energy per second, calculate the energy dissipated in 1 minute in each cell.

18 Find the current I in the circuit shown.

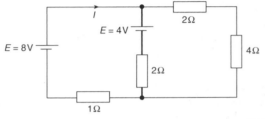

(**WJEC**: *all other Boards*)

Points

Let the current from the 4 V cell be i. Now apply Kirchhoff's second law (see 7.2 if necessary) to a complete loop of the circuit to produce an equation containing I and i. For example, the outer loop consists of a cell of emf 8 V in series with a 1 Ω resistor carrying current I and a 2 Ω resistor and a 4 Ω resistor carrying current $(I + i)$. Hence the equation for this loop is $8 = 1 \times I + 2 \times (I + i) + 4 \times (I + i)$

Now consider a different complete loop and form a second equation containing I and i. Then solve the two equations for I and i.

19 A student wanted to light a lamp labelled 3 V 0.2 A but only had available a 12 V battery of negligible internal resistance. In order to reduce the battery voltage he connected up the circuit shown on the left. He included the voltmeter – using it rather stupidly – so that he could check the voltage before connecting the lamp between A and B. The maximum value of the resistance of the rheostat CD was 1000 Ω.

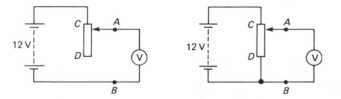

(a) He found that, when the sliding contact of the rheostat was moved down from C to D, the voltmeter reading dropped from 12 V to 11 V. What was the resistance of the voltmeter?

(b) He modified his circuit as shown on the right, using the rheostat as a potentiometer, and was now able to adjust the rheostat to give a meter reading of 3 V. What current would now flow through the voltmeter?

(c) Assuming that this current is negligible compared with the current through the rheostat, how far down from C would the sliding contact have been moved?

(d) The student then removed the voltmeter and connected the lamp in its place, but it did not light. How would you explain this? (The lamp itself was not defective.)

(**O and C Nuffield**: *and all other Boards*)

Points

(a) The rheostat and voltmeter are in series here so they have the same current. Hence $(V/R)_{meter} = (V/R)_{rheostat}$, etc.

(b) Use your value for voltmeter resistance from (a) and the given reading of 3 V to calculate the current through the meter.

(c) See Fig. 7.10(b) and related comments if necessary.

(d) When the lamp replaces the meter, it effectively short-circuits the section of the rheostat from A to B because of the lamp's comparatively low resistance (check the statement using the lamp's rating as given). Thus the circuit resistance is provided by the other section (the upper part) of the rheostat only. How much resistance is in the upper section of the rheostat? What is the maximum battery current with this resistance, and is that amount of current sufficient to light the lamp?

20 Two resistance wires A and B, made of different materials, are connected into a circuit with identical resistors R_1 and R_2 $(R_1 = R_2)$, a sensitive high-resistance galvanometer G, a cell C and a switch S, as shown in the diagram.

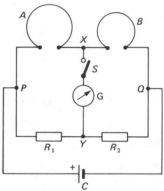

(a) If A and B have equal resistance, no current will flow through G when the switch S is closed even though the cell is still delivering a current. Explain why this is so.

(b) The diameter of A is twice that of B and the resistivity of the material of which B is made is 6×10^{-6} Ω m. It is found that for zero current through G the length of A has to be three times that of B. Calculate, showing your working, the resistivity of the material of which A is made.

(c) If the length of wire B is now reduced by a small amount so that the current through G is no longer zero, say which way the current will flow through G and explain why.

(**O and C Nuffield**: *and NEAB*, Oxford, O and C, SEB*)

Points

(a) See 7.5 for the Wheatstone bridge circuit which is what the above circuit is.

(b) Remember that R_1 equals R_2 so the resistance of A must be equal to the resistance of B. See E7.8 for resistivity; equate $\rho L/A$ for wire A to $\rho L/A$ for wire B, and remember that the area ratio = (diameter ratio)^2.

(c) At balance, the pd from X to Q is equal to that from Y to Q. Suppose S is now opened, and *then* B is shortened; will the change of B's resistance increase or decrease the pd from X to Q, and will X be more positive than Y as a result ... or less positive? Once you have decided, you can then state which way current passes between X and Y when S is closed. The above reasoning should give you the basis of your explanation.

21 In the circuit shown, the parallel-plate capacitors are identical except that the distances apart of the plates, d, are as shown. Find the potential difference across *each* capacitor and the electric field intensity between *each* pair of plates.

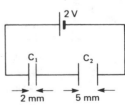

(**WJEC**: *all other Boards except SEB*)

Points

There are two possible methods here.

❶ The formal method: calculate the combined capacitance (see E7.13 if necessary). Then calculate the charge stored by the combination (see E7.12). Since each capacitor stores that amount of charge (because the two are in series), then use E7.12 again to calculate the pd across each capacitor. Once you have calculated the pd across each capacitor, then use E2.9 to calculate E across each gap.

❷ The 'intuitive' approach: the two capacitors have the same charge since they are in series, so the pd ratio is the inverse of the capacitor ratio (because $V = Q/C$). Since the capacitance is inversely proportional to the spacing, then the pd ratio is therefore proportional to the spacing ratio (i.e., 2/5). Since the pds add up to 2 V, then you can calculate the pd across each. Then proceed as in 1 for E.

22 In the circuit shown, the source has negligible internal impedance. Find (a) the rms current in the circuit, (b) the mean rate of heat production.

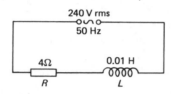

(WJEC: *and Oxford*, O and C*, and all other Boards except SEB)*

Points

(a) Use an appropriately modified form of E7.21 to calculate the circuit impedance. To modify E7.21, leave out the capacitance term as discussed in the text after the equation. The rms current can then be calculated from the impedance value and the rms value of the supply pd.

(b) Heat is only produced in the resistor of the circuit. See the comments on 'power dissipation' after E7.22.

23 Fig. 1 shows how, for a self-contained electronic unit enclosed in a box, the output voltage V_{out} changes when the input voltage V_{in} is varied over a range from −4 V to +4 V.

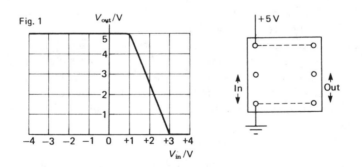

(a) The box, with its input and output terminals marked, is shown in Fig. 1. Add to the drawing of the box input and output circuits which could have been used to obtain the values of V_{in} and V_{out} used in plotting Fig. 1. Label your added circuit components.

(b) Two different inputs to the box are represented on the upper pair of axes as in Fig. 2. On the axes already provided under each input, carefully draw graphs representing the corresponding outputs from the box.

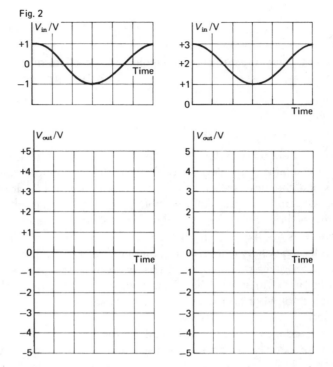

Fig. 2

(**O and C Nuffield**: *and NEAB, Cambridge, O and C, WJEC, SEB**)

Points

(a) To give a variable-input pd, you need to add a potential divider with a battery from −4 V to +4 V across its ends so that any value of pd from −4 V to +4 V can be supplied to the input terminals. Show voltmeters, with ranges, to measure input and output pds.

(b) The first input has a mean value of 0 and a peak value of 1 V. Thus it never exceeds +1 V. Use the input–output graph (Fig. 1)to determine the output in this case. The second input varies from +3 V to + 1 V and back. The simplest way to proceed is to mark the input graph each quarter cycle and read off the value of input pd after each quarter cycle. Then, use the input–output characteristic to determine the corresponding output value each quarter cycle, and plot the value on the output–time axes. Then sketch the waveform over a full cycle, using the plotted points to guide you. Note that since the input varies from +1 V to +3 V, which is exactly the range of the 'sloped section' of the input–output graph, there will not be any 'saturation'; the output should be inverted compared with the input because the slope is negative. You might find it useful to determine the voltage gain from the gradient of the slope (see 7.10 if necessary), and this should tell you how many times 'bigger' the amplitude of the output wave is compared with the input wave.

24 State Ohm's law.

For the circuit shown in the diagram, derive an expression for the potential difference V_1 in terms of R_1, R_2 and V, where the symbols have their customary meanings and the cell has negligible internal resistance.

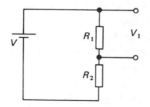

Deduce expressions for the potential differences indicated by a moving-coil voltmeter of resistance R when it is connected (a) across R_1, (b) across R_2.

Calculate the readings on the meter if $V = 2$ V, $R_1 = R_2 = 1 \times 10^3\,\Omega$ and $R = 500\,\Omega$. Comment on the fact that your calculated values do not add to 2 V. Hence, discuss the factors that affect the choice of voltmeters for practical purposes.

(Cambridge: *and all other Boards)*

Points

For Ohm's law, see 7.3.

The circuit here is essentially a potential divider arrangement which is discussed in 7.4.

(a) When the voltmeter is connected across R_1, the circuit is no longer that of the diagram above. Sketch the new circuit. Then, derive an expression for the combined resistance of R_1 and R in parallel, and add on R_2 to give the total circuit resistance (assume the cell has negligible internal resistance). The current from the cell is then V/R_T where R_T is the expression for the total circuit resistance. The pd across the voltmeter is then given by cell current $\times$ combined resistance of R_1 and R in parallel.

(b) Your expression for the pd across R_2 can be derived by the same steps as in (a) using the new circuit with R in parallel with R_2 this time. Since interchange of R_1 and R_2 takes you from (a) to (b), your final expression for (b) should be as for (a) with R_1 and R_2 interchanged.

Use your derived expressions to calculate the readings.

Your comments should be based upon the fact that moving-coil voltmeters take current. See 7.3.

25 (a) The diagrams below show two circuits commonly used to determine the value of an unknown resistance. Both ammeter and voltmeter are moving-coil instruments. State which circuit you would use if the unknown resistance had a value similar to the resistance of the voltmeter. Justify your choice, commenting on the position of the ammeter in each case, and the errors likely to result if the other circuit were used.

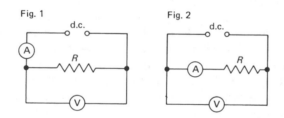

(b) A galvanometer of resistance 10 Ω gives its maximum deflection for a current of 50 mA. (i) What is the maximum pd the galvanometer can measure if it is used as a voltmeter? (ii) How would you convert the voltmeter so that it could read values of pd up to 10 V? A numerical answer is required. (iii) After conversion, find the current in the meter if the scale reading is 2 V.

(SEB: *and all other Boards except O and C and NICCEA)*

Points

(a) In which circuit does the ammeter correctly record the current through R? Remember that R and the voltmeter have similar resistances. The key to your answer lies in the point that in one circuit the ammeter measures only the resistor current, and since the ammeter has low resistance, the voltmeter reading is the same in either case. Use of the other circuit will give the same voltmeter reading but the ammeter will not record only the resistor current.

(b) (i) See E7.7 if necessary. (ii) See 7.3 if necessary. A quick method is to calculate the total resistance which will draw 50 mA from a 10 V supply. Then subtract the meter resistance to give the multiplier resistance. (iii) 10 V gives 50 mA through the meter, so 2 V gives ? mA.

26 (a) Describe qualitatively how the magnitude of the drift velocity of free electrons in a current-carrying metal conductor depends on (i) the temperature of the conductor, (ii) the potential difference between the ends of the conductor. A copper wire of cross-sectional area 1.0 mm² and length 1.5 m carries a current of 2.0 A. Each copper atom contributes one free electron to conduction. Calculate how long it takes for a free electron to drift from one end of the wire to the other. (Electron charge $e = -1.6 \times 10^{-19}$ C, density of copper $= 8.9 \times 10^3$ kg m⁻³, 0.064 kg of copper contains 6.0×10^{23} atoms.)

(b) (i) A resistance network is formed by connecting 12 equal lengths of wire, each of resistance 6 Ω, as the edges of a cube. A current of 3 A enters at S and flows symmetrically through the network, leaving it by the diagonally opposite corner Z. The directions of all the currents and the magnitudes of some are indicated in the diagram. Find the magnitude of the current in each wire. The potential difference between the points S and Z may be

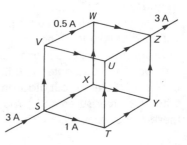

calculated as the algebraic sum of the potential differences along any path taken between the points S and Z. Hence show that the resistance of the network between S and Z is 5 Ω. (ii) If a solid cube of edge 20 mm were made from metal of resistivity 5×10^{-8} Ω m, what would be the resistance between opposite faces of the cube?

(**NICCEA**: and all other Boards except Oxford, SEB)

Points

(a) See 7.1. Use E7.3 to calculate the drift velocity; you will need to do a preliminary calculation of the number of copper atoms (and hence electrons) in the wire, so use its dimensions, density and the given value of Avogadro's number to do this. Remember to put the area into m², and that n in E7.3 is the number of free electrons per unit volume. Lastly, calculate the time taken from the value of drift velocity and the length.

(b) (i) The key lies in the fact that the resistances of each edge are equal, and the current flows symmetrically through the network. Hence the currents in ST, SX and SV will be equal and will also equal the currents in WZ, UZ and YZ. At V, the total current arriving must equal the total current leaving; hence obtain the current in VU. The currents in the remaining conductors may be determined by symmetry (or by applying Kirchhoff's first law to each remaining junction, i.e. total current in = total current out). For the resistance from S to Z, choose a short path from S to Z (e.g. SV, VW, WZ). Apply E7.7 to each conductor in turn, and then add the separate pds to find the total pd between S and Z. Then divide the total pd by the total current taken to give the resistance. (ii) Rearrange the resistivity equation to give $R = \rho L/A$. Use this equation, but take care with units.

27 A series circuit consists of a 100 Ω resistor, a 20 μF capacitor and a 0.2 H inductor driven by an alternating source of frequency f. The potential difference across the inductor, measured with a high-impedance voltmeter, is 50 V rms, and that across the capacitor is 200 V rms. Find f, and the rms current in the circuit.

Draw a vector diagram showing in magnitude and phase the peak potential differences across each of the circuit components (R, L and C) and the peak emf of the source. Find the phase difference between the emf and the current.

Find the resonant frequency for this circuit, and point out with the aid of a sketch how the above vector diagram becomes altered at resonance.

How would you detect resonance experimentally in such a circuit by means of a cathode ray oscilloscope?

(**WJEC**: and Oxford*, O and C* and all other Boards)

Points

The series *LCR* circuit is discussed in 7.8.

For the vector diagram, see Fig. 7.27 if necessary. Before drawing the diagram to scale, you must first calculate the peak pd across the resistor (= peak current × resistance) and the peak pds across L and R from the rms values. See E7.16 for the link between peak and rms values. The phase angle can either be measured off your diagram or calculated from E7.22.

Resonance of a series *LCR* circuit is discussed in 7.8. To calculate the resonant frequency, use E7.23.

At resonance, the circuit impedance is a minimum, so for a fixed supply pd, the current will be a maximum. Since a cro measures pd, the pd across the resistor will give a trace in proportion to the current. You should redraw the circuit showing where you would connect the cro to determine the current and then to check that the supply pd remains unaltered. Refer to your diagram in describing the procedure you would follow, and state what trace measurement(s) you would make and how you would detect resonance from these measurements.

28 This question is about the design of an experiment to investigate the frequency response of an amplifier. In the experiment the gain of an amplifier is to be investigated over a range of frequencies.

The diagram shows the circuit of the amplifier to be investigated. The operational amplifier may be assumed to be ideal. It is to be operated using a −15 V to 0 to +15 V supply which is not shown.

(a) (i) Select from the following list the two resistors you would use for R_1 and R_2 so that the gain of the amplifier at low frequencies would be as near to 30 as is possible.

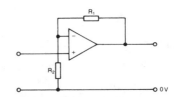

1 kΩ, 2.7 kΩ, 4.7 kΩ, 10 kΩ, 39 kΩ, 150 kΩ

(ii) Calculate the expected low-frequency gain of the amplifier using the resistors you have chosen.

(b) A sinusoidal input signal is to be provided by an uncalibrated oscillator which has a variable frequency with a range 1 Hz to 1 MHz. The output has a peak value which is constant at 2 V for all frequencies.

(i) Explain why this voltage is too large for investigating the frequency response of the amplifier.

(ii) Draw a diagram to show how you would reduce this voltage to a suitable magnitude using components from the list in (a) (i) and calculate the new peak output voltage.

(iii) Explain how you would proceed to measure the dc gain of the amplifier given that a 2 V dc supply is available. Indicate clearly the instrument(s) you would use, where the instrument(s) would be connected and the measurements you would make.

(iv) Describe how you would use an oscilloscope to calibrate the oscillator and use the calibrated oscillator to investigate the frequency response of the amplifier.

(v) Draw a graph indicating the shape of the frequency response graph you would expect.

(AEB June 89: *and NEAB*, O and C* and all other Boards*)

Points

(a) The circuit is a noninverting amplifier. Its voltage gain = $(R_1 + R_2)/R_2$. Choose values from the list to give a voltage gain as near to 30 as possible. Then use the chosen values to work out the voltage gain exactly.

(b) (i) See comments in 7.10 relating to saturation. (ii) A potential divider is needed. See 7.4. (iii) Give a circuit diagram showing a potential divider used to supply a variable input voltage and showing voltmeters to measure the input and output voltages. (iv) and (v) See your textbook if necessary.

29 (a) What do you understand by a logic gate?

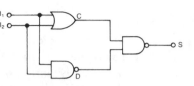

(b) The diagram illustrates a circuit which has inputs I_1 and I_2 and output S.

(i) Identify the logic gates shown and write out their respective truth tables.

(ii) Copy out and complete the truth table shown below for the circuit.

I_1	I_2	C	D	S
0	0			
0	1			
1	0			
1	1			

(iii) Describe in words the logic function of the circuit.

(c) An electric motor is to be controlled by three switches P, Q and R. The motor is to be running (logic state 1) when switches P and Q are in the same state. Whenever P and Q are in different states, the motor is to be controlled by switch R, such that the motor is running when R is in logic state 1.

(i) Write out a truth table for the control circuit.

(ii) Hence, using the circuit given in (b) or otherwise, design a circuit which could be used to control the motor.

(d) In the operational amplifier circuit below, a sinusoidal emf of 2.0 V (rms) and frequency 50 Hz is applied to the noninverting input. The inverting input is at earth potential.

Draw sketch graphs on the same axes to show the variation with time of

(i) the potential at the noninverting input,

(ii) the output potential V_o.

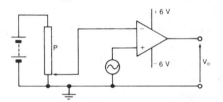

The potential at the inverting input may be made positive with respect to earth by adjustment of the potentiometer P. Draw a sketch graph to show, in detail, how V_o varies with time t when the inverting input is held at +2.0 V.

(**Cambridge**: *and all other Boards except London and O and C*)

Points

(a) and (b) See 7.11. In this circuit, S = 1 whenever the input states are the same. The circuit is therefore a comparator.

(c) (i) With three switches, P, Q and R, there are eight possible input conditions. The output to the motor must be 1 when P is the same as Q or when R = 1. Use this statement to write out the truth table. (ii) The statement in (i) should tell you what to add to the circuit in (b) to make the control circuit.

(d) The circuit is an open-loop op-amp with the output at + or − saturation according to whether the noninverting input voltage is more than or less than the inverting voltage. When P is set to give nonzero voltage, the output is an 'uneven' square wave. See Fig. 7.35 if necessary.

30 (a) The operational amplifier shown in the diagram can be used as a voltage amplifier. Describe how you would determine experimentally its dc input/output characteristic for positive and negative input voltages. Include a labelled circuit diagram showing how the amplifier is connected to a suitable power supply and explain how you would obtain different input voltages using a potential divider. Suggest suitable ranges of voltmeters you would use to measure V_1 and V_o given that the resistance of $R_1 = 10$ kΩ and the resistance of $R_2 = 100$ kΩ.

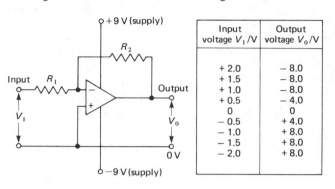

Input voltage V_1/V	Output voltage V_0/V
+ 2.0	− 8.0
+ 1.5	− 8.0
+ 1.0	− 8.0
+ 0.5	− 4.0
0	0
− 0.5	+ 4.0
− 1.0	+ 8.0
− 1.5	+ 8.0
− 2.0	+ 8.0

(b) The table shows typical results for a voltage amplifier similar to that shown above. Draw the input/output characteristic and, by reference to it, explain what is meant by (i) voltage gain, (ii) saturation and (iii) inversion. State the range of input voltages for which the amplifier has a linear response and calculate the voltage gain within this range.

(c) A sinusoidal voltage of frequency 50 Hz is applied to the input terminals of the amplifier described in (b). Sketch graphs on one set of axes showing how the output voltage varies with time when the input voltage is (i) 0.5 V rms, (ii) 1.0 V rms. In each case indicate the peak value of the output voltage and comment on the waveform.

(**NEAB**: *London*, O and C* and all other Boards except AEB*)

Points

(a) The circuit given should be redrawn, and a potential divider added (see Fig. 7.10(b)) to give a variable input pd. The voltage supply for the potential divider should be such that the potential divider output can be varied over the full range of input voltages (i.e. from + to 0 to −). Show clearly on your diagram voltmeters to measure the input and output pds.

(b) Plot y = output pd, x = input pd. See 7.10 for the meaning of the terms in (i), (ii) and (iii). You should explain these terms by reference to your graph. A linear response is where the output is in proportion to the input; use your graph to decide on the input range which gives a linear response.

(c) (i) Calculate the peak input pd; see E7.16 if necessary. Then make a sketch of the input pd (for one complete cycle) against time. Then draw the output pd against time by 'multiplying the input wave by the voltage gain value'; in other words, if the input is a sine wave of peak value V_i, then the output will be a sine wave of peak value $V_i \times$ voltage gain. If the voltage gain is −ve, the output is inverted compared with the input. Then draw the limits of the output pd on your graph. If the output wave exceeds the limit, then you must 'clip' the wave; see 7.10 and Fig. 7.38. (ii) As (i).

Answers: **1** B; **2** B; **3** C; **4** D; **5** E; **6** B; **7** A; **8** A; **9** B; **10** D; **11** C; **12** C; **13** E; **14** C; **15** E; **16**(ii) 25.7 kW; **17**(b)(i) 1 Ω (ii) 3.75 J; **18** 2 A; **19**(a) 11 000 ohms (b) 0.27 mA (c) 3/4 of the way down; **20**(b) 8 × 10⁻⁶ Ω m; **21** $v_1 = 4/7$ V, $v_2 = 10/7$ V, $E_1 = E_2 = 290$ V m⁻¹; **22**(a) 47.2 A (b) 8.91 kW; **24** 0.5 V in both cases; **25**(b)(i) 0.5 V (ii) 190 ohms in series (iii) 10 mA; **26**(a) 1.0×10^4 s (b)(i) 1 A in *ST:SX:SV: YZ:WZ: UZ*, 0.5 A in remainder (ii) 2.5×10^{-6} ohms; **27** 39.8 Hz, 1.0 A; *I* is 56.3° ahead of *V*, 79.6 Hz; **28**(a)(i) 4.7 kΩ, 150 kΩ (ii) 33; **30**(b) Voltage gain = −8

CHAPTER 8

STRUCTURE OF THE ATOM

Units in this chapter

Chapter objectives

After working through the topics appropriate to your syllabus in this chapter, you should be able to:

- describe and carry out simple calculations on the motion of charged particles in uniform electric and magnetic fields
- explain the photoelectric effect and use Einstein's photoelectric equation
- explain excitation and ionisation processes
- work out possible photon frequencies from energy level diagrams
- understand and explain the production of X-rays and X-ray spectra
- state the equations for α and β decay
- describe the main properties of radioactive emissions
- describe the use of a Geiger counter, an ionisation chamber and a cloud chamber
- define activity and half-life and solve simple problems
- explain the release of energy due to fission and fusion processes
- outline the operation of a thermal reactor

8.1 PROPERTIES OF THE ELECTRON

Electrons were originally discovered during experiments involving the passage of electric currents through gases. The direction of electron travel (cathode to anode) coupled with information from their deflection by electric and magnetic fields proved that electrons must carry a negative charge.

The charge on an electron: Millikan's experiment

A charged oil drop is kept in static equilibrium. The downward force upon the drop (its weight $= mg$) is balanced by an equal and opposite upward force due to the uniform electric field (force $= qV/d$, see E2.9). Equating the forces gives:

E8.1 $q = \dfrac{mg}{V/d}$

where q = total charge upon oil drop (C), m = mass of the oil drop (kg), g = acceleration due to gravity (m s^{-2}), V = pd between parallel plates (V), d = separation of the parallel plates (m).

The mass of the drop is obtained by measuring its terminal velocity when falling through air in the absence of the electric field. Remember that q is the charge upon the oil drop and not the charge on an electron. However, this charge q is found to be an integral (whole number) multiple of a value e (e.g. $3e$, $1e$, $6e$), but never a fractional multiple of e (e.g. $1.5e$, $7.8e$). Electric charge is hence quantised (see 1.10) in equal-sized quanta of size e (-1.6×10^{-19} C). So the smallest amount of charge available is e; this is assumed to be the charge on one electron.

The charge/mass ratio of an electron

There are two methods commonly used to obtain a value for e/m_e (the charge/mass ratio, or specific charge, of an electron). The first method is normally carried out using a **fine beam tube** and involves bending an electron beam around in a circle under the influence of the uniform magnetic field provided by a pair of Helmholtz coils. The force Bev on each electron (see E2.16) always acts at right angles to the direction of travel of the electrons. If this force can be kept constant (uniform magnetic field and electrons travelling at constant speed), then the electrons must travel in a circle (see 1.7). The centripetal force required (mv^2/r, see E1.12) is supplied by the magnetic field effect (Bev) so:

E8.2 $\dfrac{e}{m_e} = \dfrac{v}{Br}$

where e/m_e = charge/mass ratio of an electron (C kg^{-1}), v = speed of the electrons (m s^{-1}) (see E8.4), B = strength of the magnetic field (T), r = radius of the circle in which the electrons move (m).

Beware of mistakenly putting a voltage value into E8.2 instead of the speed of the electrons v. You will often be given the accelerating voltage V_A to use in E8.4 to calculate the electron speed v. It is often necessary to combine E8.4 and E8.2 to obtain the charge/mass ratio. Fig. 8.1 illustrates the experiment; use the left-hand rule to check on the direction in which the electron beam 'bends' at C and D. Remember that you do not see the electrons in the experiment! The light is emitted by atoms that have been excited (see 8.4) by the electron beam passing through them.

The second method of measuring e/m_e is a **'null' method** involving the cancelling of two equal and opposite effects (other examples of null methods include using potentiometers to measure voltages and Millikan's method as already described). The experiment is carried out using a **deflection tube** and requires the deflection in one direction by an electric field to be cancelled by the use of a uniform magnetic field that will deflect the beam in the

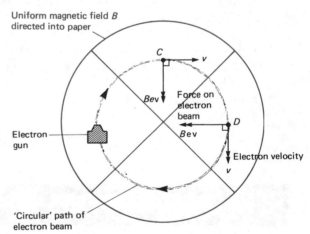

Fig. 8.1 Electron path in a uniform magnetic field

opposite direction. Fig. 8.2 shows the electron beam deflected upwards by an electric field; the left-hand rule shows that the magnetic field must be directed into the paper to deflect the

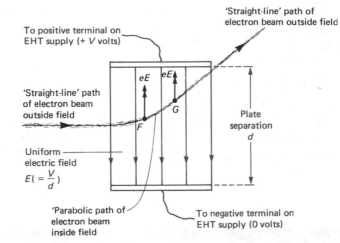

To positive terminal on EHT supply (+ V volts)

'Straight-line' path of electron beam outside field

eE eE

G

'Straight-line' path of electron beam outside field

F

Plate separation d

Uniform electric field $E\left(=\dfrac{V}{d}\right)$

'Parabolic path of electron beam inside field

To negative terminal on EHT supply (0 volts)

Fig. 8.2 Electron path in a uniform electric field

beam back down. For an undeflected beam the force due to the magnetic field (Bev) must equal the electric field force (eV/d) so:

E8.3 $v = \dfrac{V}{Bd} = \dfrac{E}{B}$

where v = speed of the electrons (m s^{-1}), V = pd used to deflect the electrons (V), B = strength of the magnetic field (T), d = separation of the deflecting plates (m), $E = V/d$ = strength of the electric field (V m^{-1} or N C^{-1}).

e/m_e is calculated by combining the above formula with E8.4. Often the same pd (V) is used both to accelerate the electrons and to deflect them.

In both experiments for e/m_e the electron beam is often found to 'spread' considerably as the electrons acquire a range of speeds owing to collisions with gas atoms inside the tubes. The formulae are only correct for the fastest moving electrons (i.e. those that have not lost kinetic energy in collisions). This can be allowed for by using the largest radius in the first method and the largest magnetic field that still leaves part of the beam undeflected in the second method.

Mass, speed and energy of electrons: velocity selector and electron volts

The result of Millikan's experiment (e) can be combined with the value for the charge/mass ratio of the electron (e/m_e) to enable a value for the mass of the electron to be obtained (m_e = 9.1×10^{-31} kg).

If an electron with no kinetic energy is accelerated by a pd (V_A) the work done on the electron ($V_A e$) supplies ke ($\frac{1}{2} m_e v^2$) hence:

E8.4 $v = \sqrt{\dfrac{e}{m_e} 2V_A}$

where v = speed of the electron (m s^{-1}), e/m_e = charge/mass ratio of the electron (C kg^{-1}), V_A = pd used to accelerate the electron (V).

It is also possible to measure the speed of the electrons (or any beam of charged particles) using the experiment summarised by E8.3. This equation has far wider applications as it demonstrates that the charged particles left undeflected by a combination of electric and magnetic fields will all have the same velocity *regardless* of the mass or charge on any of the particles. Hence a beam of charged particles all travelling at the same velocity can be obtained from a beam of assorted particles with many different velocities. The chosen velocity is selected by the value of E/B and the device is known as a **velocity selector**. Such a device is very useful in mass spectrometers.

Care must be taken when using E8.3 and E8.4 to calculate the speeds of electrons. These formulae are based on classical physics and cannot be applied to electrons travelling near the speed of light (see 1.10), when relativistic effects become important.

In atomic physics it is common for electrons to be accelerated or decelerated by the use of pds. The change in energy of an electron as it moves through a pd of 1 V is 1.6×10^{-19} J (energy = charge × pd). This amount of energy is called **one electron volt** (1 eV).

The following conversions are important:

❶ To convert from J to eV: divide by 1.6×10^{-19}.

❷ To convert from eV to J: multiply by 1.6×10^{-19}.

Remember that the value in J will be a much smaller quantity than the same energy expressed in eV. Note that any energy can be expressed in eV; this method of measuring energies is not only applied to electrons, e.g. alpha-particle and gamma-ray energies are usually measured in millions of electron volts (MeV).

8.2 FREE ELECTRONS IN METALS

Thermionic emission

Free electrons are those electrons that can be 'moved' by electric or magnetic fields, i.e. those not bound into atoms. There are a large number of free electrons present in all metals (about one free electron per atom on average). These free electrons are the charge carriers of electric current in metals and are also referred to as **'conduction' electrons**. It is possible to remove these free electrons from the surface of the metal; the energy that must be given to a free electron to make this happen is called the **work function** (φ).

Removing electrons from a metal in this way is very similar to the process of evaporation in liquids. A liquid molecule needs energy to enable it to escape from the surface of the liquid to become a vapour molecule. Molecules in a liquid can collide with one another and as a result have a wide range of energies. Similarly, free electrons in a metal acquire a wide range of energies as a result of collisions with each other and the metal atoms. In a liquid if the molecules with high energies reach the surface they can escape; this process is known as evaporation. In a metal at normal temperatures the free electrons almost never acquire enough energy to be 'emitted' in a similar fashion. However, if the metal is heated to a sufficiently high temperature the atoms and electrons have far more ke and free electrons can escape from the metal surface. This process is known as **thermionic emission**. Whenever a beam of electrons is to be produced (e.g. in a cathode ray tube, electronic valve or X-ray tube) it is usually generated by thermionic emission, the number of electrons per second (tube current) being controlled by the temperature of the thermionic emitter, which is heated electrically.

Photoelectric effect

Another method of supplying free electrons near the surface of a metal with sufficient energy for them to escape is by transferring energy from a photon (of electromagnetic radiation) to the free electron. This is known as the **photoelectric effect**. It is essential that each photon has at least enough energy to enable an electron to escape; this minimum energy is of course the work function (φ). Since the energy of a photon (hf) depends upon its frequency (f), there must be a minimum frequency of radiation needed to cause the photoelectric effect. This frequency is known as the **threshold frequency** (f_0) and no beam of radiation at a lower frequency than this, however intense, can cause photoelectric emission to occur.

E8.5 $hf_0 = \varphi$

where f_0 = threshold frequency of the metal (Hz), h = Planck's constant (J s), φ = work function of the metal (J).

The number of electrons emitted per second determines the **photoelectric current** and clearly is proportional to the number of photons per second hitting the metal surface (assuming that each photon is above the threshold frequency). The photoelectric effect provides important evidence of the need for a quantum theory in physics, as the photon model of electromagnetic radiation is essential to explain photoelectricity. Occasionally a photon may deliver all its energy to a single electron; conservation of energy requires that the photon energy should equal the kinetic energy of the emerging electron added to the energy needed to extract the electron from the metal, giving **Einstein's equation** for the photoelectric effect (E8.6). This equation makes it possible to calculate the maximum ke with which an electron may be emitted from the **photocathode**. This maximum energy can also be measured experimentally. Normally the electrons pass through the photocell from the negative photocathode to the positive anode. However, if the supply voltage is reversed, the electric field will actually repel the electrons and cause a decrease in the photoelectric current. As this reverse voltage is increased the photoelectric current will reduce to the point where it becomes zero. At this point the potential difference across the cell will just be sufficient to stop the fastest electron (i.e. those with maximum ke); this potential difference is called the **stopping voltage** (V_s). The work done by these 'fastest' electrons against the pd they have travelled through is eV_s and must equal the kinetic energy they have lost giving:

E8.6   Maximum ke of electrons $= eV_s$
$$= hf - \varphi$$
$$= hf - hf_0$$

where V_s = stopping voltage (V), e = charge on electron (C), h = Planck's constant (J s), φ = work function (J), f = photon frequency (Hz), f_0 = threshold frequency (Hz).

From E8.6 it can be seen that a graph of the stopping voltage plotted against the frequency of the photons incident upon the photoelectric material will be as shown in Fig. 8.3: slope, h/e; area, no physical significance; intercept, the threshold frequency (f_0).

Few materials exist with a threshold frequency that lies in the visible region of the electromagnetic spectrum; most materials require higher frequencies (ultraviolet radiation is particularly useful for photoelectric work).

It is possible for the reader to use Fig. 8.3 to confirm that the slope of the graph is equal to h/e and that the threshold frequency of the chosen photoelectric material is unusually low (it corresponds to red light of wavelength 750 nm).

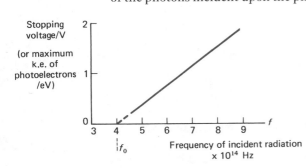

Fig. 8.3 Stopping voltage plotted against frequency

8.3 ELECTRONS WITHIN ATOMS

Energy levels, ionisation and excitation

In metals there are some 'free electrons', but the majority of electrons in metals and all materials are 'bound' to individual atoms. These electrons are 'held in the atom' by the force of attraction that exists between their negative charge and the positive charge on the nucleus. Hence each 'bound' electron must have electrical potential energy (see 2.2) and the work that must be done to remove an individual electron from the atom (i.e. to 'infinity') is called the **ionisation energy** of that electron. The energy needed to ionise atoms can be supplied in many ways, e.g. by radioactivity, heat, electron collision, X-rays.

Experiments (notably by Franck and Hertz) showed that each atom exhibits its own unique pattern of ionisation energies. Hence atoms can be identified from their ionisation energies and electrons in any atom can only have specific values of pe, i.e. electron energies are quantised (see 1.10). These specific values are referred to as **energy levels** and Fig. 8.4 shows four of the energy levels (−19, −12, −7 and −3 eV) for an imaginary atom There are many more energy levels available in the shaded region but they have been left out of the diagram.

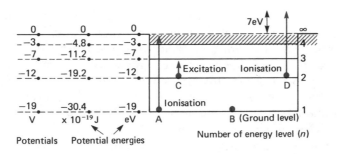

Fig. 8.4 Electron energy levels in an imaginary atom

The pd needed to give a free electron sufficient energy to enable it to ionise an atom is called the **ionisation potential**. For electron A in Fig. 8.4 this is 19 V, which corresponds to an ionisation energy of 19 eV (or 30.4×10^{-19} J). If, on the other hand, 19 eV of energy were given to electron D it would have 7 eV more energy than it needed for ionisation (12 eV); hence it would emerge from the atom with 7 eV of ke. It is possible to give an electron extra energy that is not enough to remove it from the atom; in this case the electron must be given precisely enough energy to move it to a new level within the atom. This process is called **excitation**. It is possible to excite electron C as shown with precisely an extra 5 eV of energy to move it up to the next level. However, the electron would not accept 4 eV or 6 eV of extra energy, yet it would accept 9 eV of extra energy which would move it up to level 4.

It can be seen from the diagram that the energy levels are numbered (n), starting from the lowest potential energy level ($n = 1$). This lowest level is often referred to as the **ground level**.

Energy levels in hydrogen

An equation for calculating the pe levels (in J) of the hydrogen atoms was developed by Balmer:

E8.7 $$E_n = -\frac{21.8 \times 10^{-19}}{n^2}$$

where E_n = pe of the nth level in the hydrogen atom (J), n = the number of the level – see Fig. 8.4.

As hydrogen only has one electron in its atom, which will usually be found in the ground level ($n = 1$), it should be possible for you to predict that the ionisation potential of an unexcited hydrogen atom is 13.6 V (using E8.7).

Pauli exclusion principle

It is a general rule in science that all systems will try to adopt their position of minimum pe. If this were to happen to electrons within atoms we would find all the electrons on the $n = 1$ level (in the ground level); this does not happen! There is a maximum number of electrons allowed on each level, as shown in Fig. 8.5.

Number of level (n)	1	2	3	4	etc.
Maximum number of electrons allowed on the level:	2	8	18	32	etc.
Letter code for level:	K	L	M	N	etc.

Fig. 8.5 Electron shells

The set of rules that predicts Fig. 8.5 is collectively referred to as the Pauli exclusion principle. Basically, the total energy of each electron in an atom is defined by a set of **quantum numbers**; the level number n is one of these quantum numbers. The exclusion principle states that no two electrons in an atom may have the same set of quantum numbers; hence a limited number of electrons can be found on each level (where the quantum number n stays fixed but other quantum numbers may alter). Energy levels are sometimes referred to as **shells** and these are given code letters as shown in Fig. 8.5. Hence a K shell electron will be found in the ground level ($n = 1$) of an atom, etc.

8.4 ENERGY EMITTED BY ELECTRONS

Electrons can lose energy in two distinct ways:

❶ The energy is transferred to other particles or bodies (e.g. electrons striking a target will heat it up as electron ke is converted into molecular ke in the target).

❷ The energy is converted into a photon of electromagnetic radiation (e.g. an electron changing to a lower energy level in an atom emits a photon to get rid of the energy it must lose).

Electrons changing energy levels

Fig. 8.4 shows the four electrons in an imaginary atom occupying the lowest levels that they can according to the exclusion principle (see Fig. 8.5 – only two electrons are allowed in the ground level). This atom is said to be in the **ground state** or **unexcited**. Fig. 8.6 shows the same atom in an **excited** state (i.e. it is now possible for one or more of the electrons to move to lower energy levels).

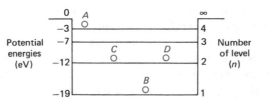

Fig. 8.6 Imaginary atom in an excited state

It can be seen in Fig. 8.6 that electron A has been excited to the $n = 4$ level (N shell) by receiving 16 eV of energy. This excited atom will spontaneously (i.e. of its own accord) return to the ground state (its position of minimum allowable potential energy). However, for Fig. 8.6 there are four different methods by which the electrons can move levels to achieve this, as shown in Fig. 8.7.

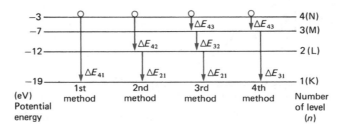

Fig. 8.7 Transitions from $n = 4$ to $n = 1$ level

Any electron on a level may be involved in a transition. For example, in the second method in Fig. 8.7 after the ΔE_{42} transition, electrons A, C and D will all be on level $n = 2$ and any one of them could then carry out the ΔE_{21} transition.

An electron involved in a downward transition must lose energy, e.g. a transition from level $n = 3$ to level $n = 2$ involves an electron energy loss (ΔE_{32}) of 5 eV. This energy is emitted as a photon of electromagnetic radiation giving:

E8.8 $\Delta E = hf = hc/\lambda$

where ΔE = energy difference between the levels (J), h = Planck's constant (J s), f = frequency of photon emitted (Hz), c = speed of light (m s^{-1}), λ = wavelength of photon emitted (m).

The wavelengths of photons emitted from atoms can vary from infrared right through the entire visible spectrum and ultraviolet to X-rays. It should be pointed out that the energy needed for X-rays is so large (of the order of keV) that only transitions between the lower levels (small n and large ΔE) in larger atoms can produce X-ray emissions. The light from excited atoms is used in neon and sodium lamps and the ultraviolet emissions from excited mercury vapour provide the energy needed for fluorescence in fluorescent lights. An **emission**

spectrum of the radiation given out by excited atoms can be plotted. Part of the emission spectrum from our imaginary atom is plotted in Fig. 8.8.

Fig. 8.8 is called a **line spectrum** (compared with a **continuous spectrum**, e.g. Fig. 8.9) because only certain frequencies are emitted as defined by E8.8 ,which gives rise to a spectrum of 'vertical lines'. The wavelengths most frequently emitted give 'higher' lines. The six wavelengths of emission in Fig. 8.8 correspond to the six different transitions (ΔE_{41}, ΔE_{42}, ΔE_{21}, ΔE_{43}, ΔE_{32}, ΔE_{31}) of Fig. 8.7. You can use E8.8 to check that this is the case. Emission spectra can be used to identify atoms as each atom emits a unique line spectrum; obviously the spectra can also be used to deduce information about the energy levels of the electrons. It is also

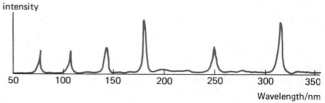

Fig. 8.8 Part of emission spectrum from an imaginary atom

possible to study atoms by making them absorb energy (i.e. by exciting the atoms) and studying the associated **absorption spectrum**; atomic absorption will also produce a line spectrum.

Photons created by very-high-energy electrons: X-rays

X-rays are the photons produced (energy 1 keV or greater) when very-high-energy electrons suddenly lose energy. They are produced in an X-ray tube when electrons of energy 10 keV or more are 'fired' into a dense target (frequently tungsten). Well over 90% of the ke lost by the electrons as they hit the target is converted into heat energy (ke of the target's molecules). However, some of the electron energy is converted into high-energy photons or X-rays. This can happen in two distinct ways:

❶ The electrons lose energy in inelastic collisions with nuclei of the target atoms. The lost ke is emitted as X-ray photons. The electrons are slowed down in this process and hence the X-rays are also called **bremsstrahlung** – this is German for 'braking radiation' and refers to the way in which they are made. Electrons can lose any fraction of their energy in a collision with a nucleus – this means photons of a continuous range of wavelengths are produced, resulting in a **continuous spectrum** for bremsstrahlung, see Fig. 8.9.

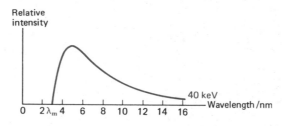
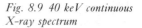

Fig. 8.9 40 keV continuous X-ray spectrum

❷ The electrons lose energy in collision with electrons in the target atoms. The target atoms become excited and on returning to the ground state emit photons of well-defined energies producing **X-ray line spectra**.

The shape in the continuous bremsstrahlung spectrum of Fig. 8.9 depends only upon the energy of the electrons and not upon the nature of the target. There will always be a clearly defined minimum wavelength of emission (λ_m) which corresponds to the largest energy photon that can be emitted. This must refer to a collision in which a maximum energy electron converts all its energy into a photon (40 keV in Fig. 8.9). You can use E8.8 to confirm that λ_m in Fig. 8.9 is equivalent to a photon energy of 40 keV.

If the X-ray spectrum for 40 keV electrons is examined it will show both types of spectra superimposed on top of each other. Fig. 8.10 shows the full X-ray spectrum for 40 keV and 80 keV electrons 'fired' at the same target element.

The wavelengths defined by the line spectrum in Fig. 8.10 depend only on the nature of the target, i.e. the X-ray line spectrum will identify the element from which the target is made. The extra lines appear in the 80 keV X-ray spectrum only because 40 keV electrons do not have enough energy to excite the electrons that are causing these lines.

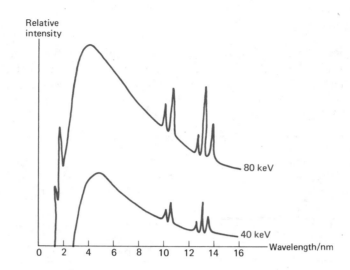

Fig. 8.10 40/80 keV X-ray spectra (same target element)

8.5 PROPERTIES OF THE NUCLEUS

The nuclear model: neutrons, protons and isotopes

Under the direction of Lord Rutherford, alpha-particle scattering experiments that proved the existence of a nucleus within the atom were carried out by Geiger and Marsden in 1909. In a typical medium-sized atom (a sphere of diameter of about 10^{-9} m) the nucleus (a sphere of diameter of about 10^{-14} m) occupies a minute fraction of the total volume of the atom (which is mostly 'empty space'). Yet as virtually all the mass of the atom is concentrated in the nucleus, the nuclear density has a phenomenally high value (about 10^{15} kg m^{-3}). The nucleus consists of **nucleons** (neutrons and protons) bound together by a very strong (but very short range) 'nuclear field force' that exists between two or more nucleons. The basic properties of the nucleons are summarised in Fig. 8.11 where an important comparison is made with the properties of electrons.

Nucleon	charge		mass	
PROTON	$+1.6 \times 10^{-19}$	$+1(e)$	$M_p = 1.673 \times 10^{-27}$ kg	about $1800m_e$
NEUTRON	Zero	$0(e)$	$M_n = 1.675 \times 10^{-27}$ kg	about $1800m_e$

Fig. 8.11 Properties of nucleons

A nucleus is identified as being of a particular element by its number of protons: the number of protons in a nucleus is called the **atomic number** (Z). For example, a nucleus containing 1 proton ($Z = 1$) is hydrogen, a nucleus containing 92 protons ($Z = 92$) is uranium. The atomic number hence defines the positive charge carried by a nucleus. The total number of nucleons (neutrons + protons) in a nucleus is referred to as the **mass number** (A).

It can be seen that the **neutron number** (N) is defined by:

E8.9 $N = A - Z$

where N = neutron number, A = mass number = total number of nucleons, Z = atomic number = total number of protons.

It was discovered using mass spectrometers that nuclei of the same element can have different masses as they can contain different numbers of neutrons; each different nucleus is said to be an **isotope** of that element. Two important isotopes of uranium are uranium-235 and uranium-238; both isotopes contain 92 protons but one has 143 and the other 146 neutrons. A graph of N plotted against Z (an N–Z plot, see Fig. 8.14) has several important interpretations in nuclear physics.

Nuclear equations

In nuclear physics a particle is identified by its mass number (i.e. how many nucleons it contains) and its charge compared with the charge on a proton (for any nucleus this is also its atomic number). The particle itself can be labelled by its chemical symbol (e.g. He for helium nuclei), by a capital letter (e.g. X, D, E) or by a 'code symbol' (e.g. e for electron, α for alpha particle). Some examples are given in Fig. 8.12.

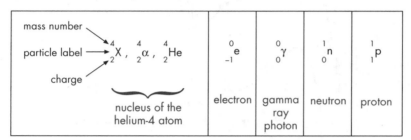

Fig. 8.12 Notation of particles

In any nuclear physics reaction both the total number of nucleons and the total charge are conserved. Consider the following nuclear interaction of an alpha-particle colliding with the nucleus of a nitrogen atom:

E8.10 $^{4}_{2}\alpha + ^{14}_{7}N \rightarrow ^{17}_{8}O + ^{?}_{?}X$

On the left-hand side of the equation the total mass number is 18; hence the mass number of X is 1 to ensure that there are also 18 nucleons on the right-hand side of the equation. On the left-hand side of the equation the total charge is +9; hence X must carry a charge of +1 (no sign on the charge means that it is positive). Hence as X is one nucleon ($A = 1$) carrying the charge of a proton ($Z = 1$) it must be a proton ($^{1}_{1}p$)! This is the nuclear interaction observed by Rutherford when he discovered the **proton** in 1919.

In nuclear physics it is important to realise that although charge and the total number of nucleons will be conserved in a reaction, neither mass nor energy need be conserved as one can be converted into the other (see 1.10 and 8.7).

8.6 RADIOACTIVITY

If an isotope is **stable** it will never spontaneously (of its own accord) emit energy. However, there are many unstable isotopes which will eventually emit energy *spontaneously*; such nuclei are said to be **radioactive**. The three best-known radioactive energy emissions (alpha α, beta β and gamma γ) are summarised in Fig. 8.13.

Radioactive emissions occur as a result of the basic principle that any system will adopt a state of lower total potential energy if possible. The pe of the **parent** nucleus must be higher than that of the **daughter** nucleus that will be formed after the radioactive decay. The energy difference must be sufficient to permit the radioactive emission to take place. For large radioactive nuclei ($Z >$ about 80) alpha or beta emissions are possible, but alpha particles are not emitted from smaller nuclei (i.e. with $Z < 60$).

In an **alpha-particle emission** two protons and two neutrons are ejected as one particle (the alpha particle – a helium-4 nucleus). So if D is the parent nucleus and E the daughter:

E8.11 $^{A}_{Z}D \rightarrow ^{A-4}_{Z-2}E + ^{4}_{2}\alpha$ alpha-particle emission

e.g. $^{226}_{88}Ra \rightarrow ^{222}_{86}Rn + ^{4}_{2}\alpha$

The arrow indicates that the process is 'one-way' i.e. not reversible.

In a **beta-particle emission** a neutron in the parent nucleus (D) is turned into a proton and an electron (the beta particle) so producing a new nucleus, the daughter (E):

E8.12 $^{1}_{0}n \rightarrow ^{1}_{1}p + ^{0}_{-1}\beta$ beta-particle emission

Type	α	β	γ
Nature	Helium-4 nucleus (2 neutrons and 2 protons)	Electron	Photon of electromagnetic radiation *emitted from excited nuclei*
Charge, e	+2	-1	0
Mass, kg	6.7×10^{-27}	9.1×10^{-31}	0
Mass, u	4	1/1836	0
Energy, MeV	0.5–1.0	0.01–10	0.01–10
Ionising power	Very good ioniser	About 1/10 of α	About 1/10 000 of α
Speed	Up to $0.01c$	$(0.01–0.9)c$	c
Absorbed by (range)	Thin paper	mm of aluminium	Partially absorbed by cm of lead
Detected by	Thin-window Geiger tube Solid-state detector Cloud/bubble chambers Ionisation chamber Photographic film Scintillation counter Spark counter	Geiger tube Solid-state detector Cloud bubble chambers Photographic film Scintillation counter	Geiger tube Solid-state detector Photographic film Scintillation counter
Uses/applications	Artificial disintegration of nuclei Luminous paint	Radioactive dating Thickness gauges Medical tracing	Gamma-ray photography through metals Thickness gauges Medical tracing

Fig. 8.13 Properties of radioactivity

or $\quad {}^{A}_{Z}D \rightarrow {}^{A}_{Z+1}E + {}^{0}_{-1}\beta$

e.g. $\quad {}^{90}_{38}\text{Sr} \rightarrow {}^{90}_{39}Y + {}^{0}_{-1}\beta$

All beta-particle emissions are accompanied by the emission of extra energy which is carried away by a particle of no mass and no charge called a **neutrino** ($ {}^{0}_{0}v $). Beta particles are electrons and are no different to any other free electrons of the same energy; the term 'beta particle' refers to the way in which the electron was created, i.e. by radioactive emission. Although electrons can be ejected from nuclei (by neutrons 'turning into' protons), it is impossible to find an electron within a nucleus. The wave properties of such an electron (see 1.10 for particle/wave equivalence) would result in energy of the order of GeV (10^9 eV) for the electron; this energy would shatter any nucleus!

A key feature of all alpha and beta decays is the fact that the daughter is a different element to the parent; radioactivity hence enables one element to spontaneously turn into another! In E8.11 radium (Ra) turns into radon (Rn) as a result of an alpha decay, and in E8.12 strontium (Sr) turns into yttrium (Y) as a result of a beta decay. Most daughter nuclei created by radioactive decay are radioactive themselves and undergo further decays. It can be possible to trace a whole sequence of radioactive decays from a single parent; a complete sequence of possible decays is called a **radioactive series**.

All unstable nuclei will eventually undergo radioactive decay. It is possible to predict likely radioactive decays by examining a plot of the stable nuclei; Fig. 8.14 shows such a plot. The **N–Z plot**, as it is called, has the neutron number (N) plotted on the y-axis and the atomic number (Z) plotted on the x-axis. It is not necessary for A level to know the position of any points that represent the stable nuclei, so instead the 'locus' of these points is shown (shaded area on graph).

If a radioactive isotope (i.e. an unstable nucleus) is plotted on Fig. 8.14 it is possible to make a good prediction of its likely radioactive decay from its position on the plot in relation to the stable nuclei. To illustrate this more clearly, Fig. 8.15 shows an enlargement of a portion of the plot as defined by the coloured rectangle on Fig. 8.14.

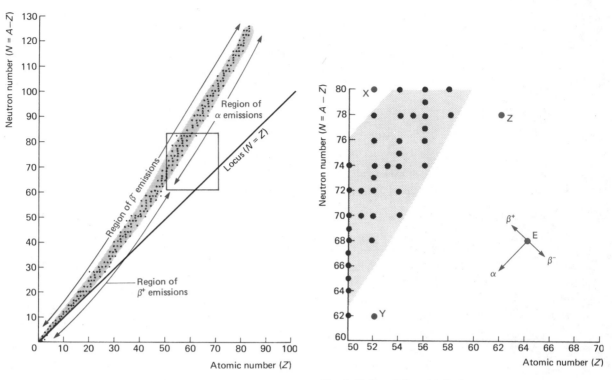

Fig. 8.14 N–Z plot for stable isotopes

Fig. 8.15 Detail from N–Z plot

In Fig. 8.15 the stable nuclei are shown as black dots and the radioactive nuclei (X, Y, Z and E) are plotted in colour. Arrows show the effect of three basic radioactive decays on E. The position of the daughter nucleus on the plot in relation to the parent nucleus can be summarised as follows: for an alpha decay (N and Z reduced by two) the nucleus moves at 45° down to the left, for a β^- decay (Z up one, N down one) the nucleus moves at 45° to the right, and for a β^+ decay (N up one, Z down one) the nucleus moves at 45° up to the left. In a β^+ decay a proton is turned into a neutron by emission of a positive electron (**positron**) or by the nucleus capturing an electron; an understanding of these decays is not required for most A-level courses. For X, Y and Z it can be seen that the nuclei can be moved closer to the zone of stability in the following ways: X down to the right (β^-), Y up to the left (β^+) and Z down to the left (α). Clear examples have been chosen, but sometimes the appropriate beta decay or an alpha decay will seem to move a nucleus nearer to stability; indeed some nuclei are found to undergo either type of radioactive decay! An important point about the N–Z plot is that for 'small' and 'medium' sized nuclei the plot follows the line defined by $N = Z$ (see Fig. 8.14). An alpha decay will move a nucleus parallel to this line and hence no nearer stability. This means that alpha decay occurs only for large nuclei ($Z >$ about 60) in the region marked on the plot. Smaller nuclei below the region of stability tend to undergo β^+ emissions. β^- emitters can be found anywhere above the region of stability.

In the same way that excited atoms (see 8.4) lose energy as electrons drop into lower energy levels, an **excited nucleus** can lose energy as the nucleons carry out a similar process. In each case the energy is emitted as photons of electromagnetic radiation. However, the photon energies of nuclear emissions (MeV) are much higher than those from the atomic electrons (eV to keV). Photons of electromagnetic radiation emitted from excited nuclei are called **gamma rays**. On the completion of an alpha or beta decay most nuclei are left in an excited state. This means that most alpha and beta decays are also accompanied by gamma emissions. Excited nuclei are indicated with an asterisk (*) hence:

E8.13 $^A_Z X^* \rightarrow\ ^A_Z X\ +\ ^0_0 \gamma$ gamma-ray emission

It is possible to get X-rays with the same wavelength as gamma rays. They are of course identical and are only given different names as they have been produced in different ways.

All three basic types of radioactivity have sufficient energy to ionise atoms (see Fig. 8.13) and it is their ionising properties that enable them to be detected. Detectors that work on this principle include cloud and bubble chambers, Geiger–Müller tubes, spark counters, ionisation chambers, solid-state detectors and scintillation counters. In scintillation devices the electrons produced by ionisation (due to the radioactivity) then excite other atoms which proceed to emit light. Photographic emulsions that use the energy of the radioactivity to produce a chemical effect are one of the few detectors not to rely on ionisation. It is possible to distinguish between the three types of radiation by their different ionising powers (i.e. thickness of tracks in cloud and bubble chambers) or by using absorbers to classify them (see Fig. 8.13). As gamma rays undergo negligible absorption and travel in straight lines in air, they will obey an inverse square law:

E8.14 $\quad I_r \propto \dfrac{1}{r^2}$

where I_r = intensity of gamma-ray source a distance r away (W or photons per second; per square metre), r = distance of detector from source (m).

As both alpha and beta particles undergo significant absorption in air, their intensity drops off at an even greater rate than that predicted by E8.14 for gamma rays. It is important to remember that the two most vital concepts in radioactive safety are time of exposure and distance from the source; a surface 10 cm away from a source will be exposed to at least 100 times the dosage that would be received 1 m away over the same period of time.

Mathematical model of radioactivity: decay laws and half-life

If it were possible to watch a group of identical radioactive nuclei, the time that would pass before any particular nucleus carried out its radioactive emission would appear to be random. Such is the random nature of radioactivity. However, if the average time were taken for a large number of the same nuclei to decay, this time would have a consistent value. Large numbers of objects that individually behave randomly can follow very predictable patterns, e.g. the toss of a coin is random (head or tail), but the result of a large number of tosses (half heads, half tails) can be reliably predicted (see also 6.10). The number of radioactive particles emitted per second from a particular type of nucleus, the **activity**, is the same quantity as the rate of change of the number of that nucleus actually present (dN/dt). Furthermore, the number of radioactive emissions per second must be proportional to the number of radioactive nuclei actually present, hence:

E8.15 $\quad \dfrac{dN}{dt} = -\lambda N$

where dN/dt = number of radioactive emissions per second (s^{-1}), λ = radioactive decay constant (s^{-1}), N = number of radioactive nuclei present (no units).

This equation defines the **radioactive decay constant** (λ). The negative sign indicates that the number of radioactive nuclei present is being reduced (not increased) and hence as time progresses the decay rate drops. E8.15 can be represented graphically by plotting the number of radioactive nuclei present (N) against time (t) as shown in Fig. 8.16. In Fig. 8.16: slope, decay rate (dN/dt) (s^{-1}); area, no physical significance; intercepts, etc., N_0 is the number of radioactive nuclei initially present – after sufficient time the graph asymptotically approaches the t axis. The graph has a constant half-life ($T_{\frac{1}{2}}$).

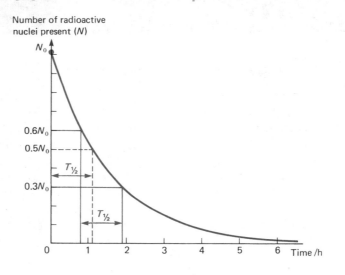

Fig. 8.16 Theoretical curve of radioactive decay ($T_{\frac{1}{2}}$ = 1.1 h)

Like all exponential graphs Fig. 8.16 has a constant ratio property. In radioactivity this property is defined by measuring the time taken for the number of radioactive nuclei present to be halved; this time is called the **half-life** ($T_{\frac{1}{2}}$) of the radioactive isotope. On the graph two half-lives have been measured: one from $N = N_0$ to when $N = \frac{1}{2}N_0$, the other from when $N = 0.6N_0$ to when $N = 0.3N_0$. You can measure some more half-lives from the graph; like the two shown they will all be 1.1 hours. Half-lives for radioactive isotopes can vary from millions of years to millionths of seconds. The half-life and decay constant are linked:

E8.16 $T_{\frac{1}{2}} = (\log_e 2)/\lambda$

where λ = decay constant (s^{-1}), $T_{\frac{1}{2}}$ = half-life (s).

The numerical value of $\log_e 2$ is 0.693. It is possible to use E8.15 and E8.16 to help find the number of radioactive nuclei present after a time t has elapsed when N_0 radioactive nuclei were originally present. The equation needed is obtained by integrating E8.15:

E8.17 $N_t = N_0 e^{-\lambda t}$

where N_t = number of radioactive nuclei present at time t (no units), N_0 = number of radioactive nuclei originally present, i.e. at time $t = 0$ (no units), λ = decay constant (s^{-1}), t = time that has elapsed from when $N = N_0$ (s).

The numerical value of e is 2.718. It is possible to use E8.15 and E8.17 in several different forms by replacing N, N_0 and N_t with quantities that are proportional to them, e.g. R, R_0 and R_t, ratemeter readings (count rates), or I, I_0 and I_t, ionisation chamber currents, etc. Similarly the vertical axis of Fig. 8.16 may be relabelled with these quantities, though an actual experiment will produce the graph shown as Fig. 8.17 rather than the theoretical result of Fig. 8.16. In Fig. 8.17: slope, rate of change of count rate (s^{-1}); area, total number of radioactive particles detected; intercepts, etc., C_0 is the count rate at the start of the experiment, C_B is the count rate due to the background radiation.

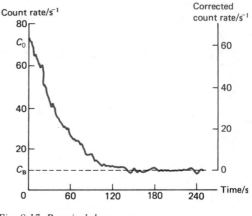

Fig. 8.17 Practical decay curve

The graph can be seen to show random fluctuations typical of the radioactive process but it still adopts the theoretical shape defined by Fig. 8.16. An important further difference is that when the radioactive decay has ceased there is still a 'steady' count rate (C_B) due to the ever-present **background radiation** that is found on the Earth. Before the radioactive decay can be analysed it is necessary to subtract the background count from all the readings; this can simply be done by rescaling the vertical axis as shown by the numbers in colour on the right-hand side of the graph. You can use this scale to check that the half-life is 30 s.

8.7 INDUCED EMISSION OF NUCLEAR ENERGY

Measuring the energy of nuclear interactions

Any unstable nucleus will eventually emit energy (radioactivity). It is also possible to *induce* a stable nucleus to emit energy. This is usually done by causing an interaction to occur between two nuclear particles; E8.10 is an example of an induced release of nuclear energy (the alpha particle induces a stable nitrogen-14 nucleus to form an oxygen-17 nucleus and the energy released is carried away as ke of the proton and oxygen-17 nucleus).

The nucleons in a nucleus are 'held together' by the short-range attractive nuclear force (which is far stronger than the electric field repulsive force between the protons). Thus to remove a nucleon from the nucleus requires work to be done against the attractive force so that the pe of the nucleon is increased; this is a similar concept to the work done against the attractive force of gravity as the pe of a rocket is increased as it leaves the surface of the Earth. The total energy needed to remove a nucleon from a nucleus is called the **binding energy of the nucleon**, i.e. the energy with which it was originally bound into the nucleus. The total energy needed to separate all the nucleons from a nucleus so that they are all 'free' is called the **binding energy of the nucleus**. The lower the potential energy of a nucleon, the greater will be its binding energy; in much the same way, the lower the potential energy of a rocket (i.e. the nearer it is to the surface of the Earth), the greater will be the energy needed to free it from the Earth's gravitational field.

Nuclear interactions do not obey the normal (classical mechanics) laws of conservation of energy and of mass (see 1.10) as it is possible for energy to be converted into mass and vice versa (see E1.17 for the conversion formula). So as mass and energy are 'equivalent' in nuclear physics, it is the total of 'mass + energy' that is conserved. This leads to a simple method of analysing nuclear interactions; consider an interaction in which particles A and B interact to create particles C and D:

$$A + B \longrightarrow C + D$$

If mass of A + mass of B is greater than mass of C + mass of D

then the missing mass on the right-hand side will have been turned into energy and the interaction will release this amount of energy. But:

If mass of A + mass of B is less than mass of C + mass of D

then to make the equation balance the missing mass must be supplied to the left-hand side of the equation in the form of energy, i.e. this interaction needs an energy supply if it is to happen.

The 'missing mass' in a nuclear interaction measures either the energy released in the interaction or the energy needed to make it happen. It is a vital property of any nuclear interaction.

The measurement of mass is of such importance in nuclear physics that a new unit is defined for greater convenience: the **atomic mass unit** (abbreviation amu) is a unit mass equal to 1/12th of the mass of an atom of carbon-12. The units of amu (u) can be converted into mass (in kg) or energy (in MeV or J):

E8.18 1 atomic mass unit $= 1\ u$
$$= 1.66 \times 10^{-27}\ \text{kg}$$
$$\equiv 931\ \text{MeV} \equiv 1.49 \times 10^{-10}\ \text{J}$$

It can be seen that each nucleon corresponds to nearly 1 u of mass. The following example illustrates how to calculate the energy involved in an interaction:

E8.19 $^{2}_{1}\text{H} + {}^{2}_{1}\text{H} \rightarrow {}^{3}_{2}\text{He} + {}^{1}_{0}\text{n}$

Total mass of $({}^{2}_{1}\text{H} + {}^{2}_{1}\text{H}) = (2.015 + 2.015) = 4.030\ u$

Total mass of $({}^{3}_{2}\text{He} + {}^{1}_{0}\text{n}) = (3.017 + 1.009) = 4.026\ u$

Hence 0.004 u of mass 'goes missing' and is converted into the 3.7 MeV of energy released in this process. You can calculate the energy released when an alpha particle is formed from the following information:

Mass of proton = 1.0076 u

Mass of neutron = 1.0090 u

Mass of alpha particle = 4.0028 u

The energy released should be 28.3 MeV, which is therefore the binding energy of the alpha particle. It is possible to calculate the binding energy of any other nucleus in a similar fashion; the 'missing mass' difference between the mass of the nucleus and the mass of the 'free' nucleons from which it is made up is called the **mass defect**. If we divide the binding energy of a nucleus by the number of nucleons it contains we have measured the average **binding energy per nucleon** ($\bar{B}$). A graph of this quantity plotted against mass number (Fig. 8.18) is one of the most important graphs in nuclear physics.

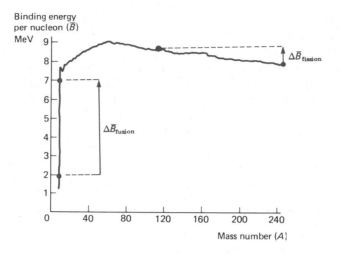

Fig. 8.18 Average binding energy per nucleon against mass number

FISSION AND FUSION

Such a phenomenal amount of energy can be generated from mass (10^{17} J kg^{-1}, see E1.17) that extracting energy from the nucleus in this fashion has been the subject of over 40 years of concentrated research. Two features of nuclear interactions are of key importance in designing nuclear energy systems (bombs or reactors):

❶ Only a small proportion of the mass is converted into energy in most nuclear reactions.

❷ It is very difficult to create the experimental conditions that will cause nuclear reactions to take place sufficiently often to generate significant quantities of energy.

The amount of energy released in a single nuclear reaction is equal to the increase in binding energy (reduction in mass) that occurs. Hence the amount of energy released per unit mass of 'fuel' is equal to the increase in binding energy per unit mass; this can be expressed as the increase in binding energy per nucleon ($\bar{B}$) which can be determined from Fig. 8.18. The nuclear reactions which will be most efficient at generating energy will be those that result in the biggest increase in $\bar{B}$. Two basic reactions are suggested by the graph.

❶ **Fusion**: very small nuclei ($A \sim 2$, $\bar{B} \sim 2$ MeV) are fused (joined together) to form a larger nucleus ($A \sim 4$, $\bar{B} \sim 7$ MeV) creating an increase in binding energy per nucleon (or energy release per nucleon) of $\Delta \bar{B}_{\text{fusion}} \sim 5$ MeV.

❷ **Fission**: very large nuclei ($A \sim 240$, $\bar{B} \sim 7.5$ MeV) undergo fission (splitting) creating two smaller nuclei ($A \sim 115$, $\bar{B} \sim 8.5$ MeV) resulting in a release of energy per nucleon of $\Delta \bar{B}_{\text{fission}} \sim 1$ MeV.

A typical fusion reaction has already been analysed (see E8.19). Fuel for fusion is plentiful (there is an almost limitless supply of hydrogen on the Earth in the form of water). However, fusion reactions only take place at extraordinarily high temperatures (typically 10^6 K) from which their alternative name of **thermonuclear reactions** is derived. It is possible to produce such temperatures for uncontrolled reactions (hydrogen bombs), but scientists have yet to build a successful thermonuclear reactor for power generation.

Despite the fact that fission generates less energy (per kg of fuel) than fusion and that fission fuel (e.g. uranium and plutonium) is far less plentiful and more difficult to process than fusion fuel, fission reactors remain the only viable means of producing controlled nuclear power. However, atomic bombs (fission) have in the main been superseded by the more powerful thermonuclear (fusion) weapons. Fission processes do not need the same high temperatures as fusion; a typical fission reaction illustrates that the process can be self-sustaining:

E8.20 $^{235}_{92}\text{U} + {}^{1}_{0}\text{n} \rightarrow {}^{148}_{57}\text{La} + {}^{85}_{35}\text{Br} + 3{}^{1}_{0}\text{n}$

The energy released by a nucleon when it changes from being a member of a U^{235} nucleus to being a member of a nucleus of mass $\cong \frac{1}{2} \times \text{U}^{235}$ mass is about 0.8 MeV. Hence the energy released by the fission of E8.20 is about 185 MeV ($\cong 0.8 \times 235$). The fission is induced by a neutron colliding with the parent nucleus of U^{235}. The fission creates several further free neutrons (in E8.20, three further free neutrons are released). If at least one of these neutrons goes on to create a further fission, then a **chain reaction** of continual fissions will be underway.

In the 'thermal' type of **nuclear fission reactor**, fission neutrons (produced with energies $\cong$ MeV) must be slowed down to 'thermal' energies (of the order of eV) to produce further fission of U^{235}; otherwise, the fast neutrons released after fission will be absorbed by U^{238} atoms (since the fuel is mostly U^{238}) without further fission. The slowing down of fast neutrons to become thermal neutrons is achieved by setting the fuel pins into a **moderator** block, as in Fig. 8.19. Here fast neutrons leave the fuel pins after fission and enter the moderator where collisions with moderator atoms (usually light atoms, such as carbon for maximum energy transfer) cause the neutrons from the fuel pins to lose energy quickly. Eventually the slow (by now) neutrons in the moderator re-enter the fuel pins so producing further fission. In this way, the fission energy is transferred to heat energy of the moderator which is then removed by a **coolant** fluid. To prevent the chain reaction from proceeding too rapidly, **control rods** of high neutron-absorbing power can be lowered into the 'core' to maintain a constant neutron flux in the core. In this way, a steady chain reaction is maintained.

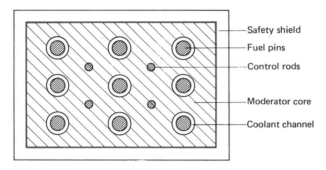

Fig. 8.19 Core of a fission reactor

A rough estimate of the energy per unit mass of fuel for a thermal reactor can be made from the following figures:

Assume fuel is 1% U^{235}, and 200 MeV of energy are released from every U^{235} fission. For complete fission of U^{235} in 1 kg of fuel, 10 g of U^{235} containing $(10/235) \times N_A$ atoms will be fissioned (N_A = Avogadro's number). Thus, the energy released from 1 kg of fuel will be $(10/235) \times 6 \times 10^{23} \times 200 \times 1.6 \times 10^{-13}$ J, which is approximately 10^{12} J kg^{-1}.

It is possible to release nuclear energy with the type of reaction illustrated in E8.10. However, the energy released per unit mass in these reactions tends to be far less than in fission or fusion. Further, these reactions are very rare and cannot be induced frequently enough to be viable energy producers.

8.8 PARTICLE PHYSICS AND RELATIVITY

Special relativity

Einstein based his Special Theory of Relativity on two postulates:

❶ Physical laws have the same form in all inertial frames of reference (i.e. frames in which Newton's first law is obeyed).

❷ The speed of light in free space, c, is invariant (i.e. always the same, regardless of the motion of the source or the observer).

Using these two postulates, Einstein showed that space and time are not absolute. In particular, he showed that:

- no material object can travel faster than the speed of light,
- moving clocks run more slowly than stationary clocks (i.e. there is time dilation),
- moving rods contract in the direction of motion due to the motion (length contraction),
- the mass of a particle increases as the particle moves faster.

In addition to the famous formula $E = mc^2$, he proved the following formulae concerning the above points:

E8.21 $\quad t = \dfrac{t_0}{\left(1 - v^2 / c^2\right)^{\frac{1}{2}}}$

where t_0 = the proper time for an event (i.e. as measured by an observer at rest relative to the event), and t = the time for an event as measured by an observer moving at velocity v relative to the event.

E8.22 $l = l_0\left(1 - v^2 / c^2\right)^{\frac{1}{2}}$

where l_0 = length of a rod measured by an observer at rest relative to the rod, and l = length of the rod measured by an observer moving at velocity v in the direction of the rod.

E8.23 $m = \dfrac{m_0}{\left(1 - v^2 / c^2\right)^{\frac{1}{2}}}$

where m_0 = the rest mass of a particle, and m = the mass of the particle at speed v. Experimental evidence from particle beam experiments has confirmed all these formulae.

Particle physics

For every particle, there is a corresponding **antiparticle**. The existence of antimatter particles, or antiparticles, was predicted in 1928 by Paul Dirac. General properties of particles and antiparticles include:

❶ A charged antiparticle has the same mass and equal and opposite charge to the corresponding particle.

❷ When a particle meets its antiparticle, they annihilate each other, creating electromagnetic radiation.

❸ A gamma photon of sufficiently high energy is capable of creating a particle and its antiparticle. This is called pair production (Fig. 8.20).

Particles and antiparticles are created in high-energy accelerators by collisions between

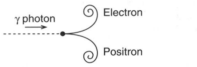

Fig. 8.20 Pair production in a magnetic field

particles and/or antiparticles. The rules for creation or annihilation are based on the conservation laws for energy, momentum and charge. For example, to create an electron and its antiparticle, the positron, a gamma photon passing through matter suddenly ceases to exist and its energy is used to create the electron and the positron. This process can occur only if the gamma photon's energy is at least equal to $2\,m_0 c^2$, where m_0 is the rest mass of the electron.

E8.24 $hf \geq 2m_0c^2$

where f = the gamma photon frequency, c = speed of light in vacuo, m_0 = electron rest mass, h = the Planck constant.

Note that particle mass and momentum values are often expressed in units of GeV/c^2 and GeV/c (or MeV/c^2 and MeV/c) respectively.

The positron was discovered in 1932 from cloud chamber photographs which showed two particle tracks curving in a magnetic field in opposite directions from the same point. Antiparticles such as the antiproton can be created in high-energy particle collisions.

Particle accelerators are used to make charged particles collide at high energies:

❶ The **Van de Graaff accelerator** is used to accelerate protons and ions to energies up to 10 MeV or more. The potential of a Van de Graaff dome is limited by insulation failure and corona discharge.

❷ The **linear accelerator** is used to accelerate charged particles to energies up to 50 GeV. A high-frequency alternating pd is applied between adjacent electrodes so a charged particle is accelerated from one electrode to the next, to the next, etc.

❸ The **cyclotron** (Fig. 8.21) is used to accelerate charged particles to energies up to 20 MeV. The particles spiral out from the centre owing to the combined effect of the

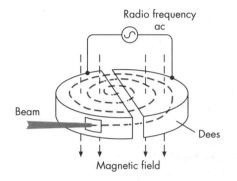

Fig. 8.21 The cyclotron

magnetic field and the alternating pd between the 'dees'. Each time the particle moves around a semicircle inside a dee, the alternating pd reverses polarity so that the particle is attracted to the other dee. For a charged particle of mass m and charge q, the magnetic force Bqv provides the centripetal acceleration v^2/r. Hence $Bqv = mv^2/r$, thus the radius of orbit $r = mv/Bq$. Hence the frequency of the alternating pd must be equal to $v/2\pi r$ $= Bq/2\pi m$.

E8.25 $f = Bq/2\pi m$

where f = the frequency of the alternating pd, v = the particle speed, B = magnetic flux density.

As the speed increases, the radius of orbit increases until the particle leaves the cyclotron. The relativistic increase of mass with speed causes the alternating pd to become more and more out of phase with the frequency of rotation, which limits the maximum ke of the particles.

❹ The **synchrotron** (Fig. 8.22) is a ring of electromagnets which confines charged particles in an evacuated tube to an orbital path. Radio frequency cavities along the path accelerate the charged particles to higher and higher energies. The magnetic flux density is increased to compensate for the relativistic increase of mass. The kinetic energy gained by a charged particle is limited by **synchrotron radiation**, which is electromagnetic radiation emitted by high-energy charged particles forced to move around a curved path.

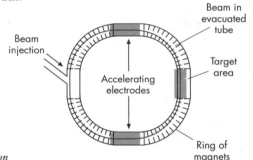

Fig. 8.22 The synchrotron

The synchrotron can cause high-energy particles to collide with other particles in a fixed target, or with antiparticles moving in the opposite direction. A collision between a high-energy particle and a particle in a fixed target cannot use all the energy of the moving particle to create new particles because some kinetic energy must remain to conserve momentum. However, in a 'head-on' collision between a particle and an antiparticle moving in opposite directions the total momentum is zero, hence all the kinetic energy can be used to create new particles.

The **quark–lepton model** was developed to explain the huge number of unstable particles and antiparticles which have been created in particle beam experiments. In addition to properties such as charge, new properties such as **strangeness** and **charm** (Fig. 8.23) have been invented to describe the formation of these particles.

Quark	Charge	Mass
Up	$+^2/_3\,e$	
Down	$-^1/_3\,e$	
Strange	$-^1/_3\,e$	INCREASING
Charm	$+^2/_3\,e$	
Top	$+^2/_3\,e$	
Bottom	$-^1/_3\,e$	

Fig. 8.23 The quark family

In the quark–lepton model:

❶ there are six different quarks, each with a corresponding antiquark,

❷ quarks (or antiquarks) combine in threes forming a class of particles called **baryons** (or antibaryons),

❸ a quark and an antiquark can combine to form a class of particle called a **meson**,

❹ electrons, positrons and neutrinos belong to a class of particles called **leptons**. Like quarks, these are elementary particles, but do not combine with each other. There are thought to be six different leptons.

Particles interact with each other through force fields which are quantised. The quantum of energy exchanged is referred to as the **exchange carrier** of the field.

❶ The electromagnetic force between charged particles is caused by the exchange of photons, referred to as **virtual photons** – virtual because they are undetectable; if they could be detected, the interaction would cease.

❷ The force holding quarks together is caused by **gluon exchange**. The **strong nuclear force** holding protons and neutrons together in a nucleus is the residual effect of this exchange force.

❸ The **weak nuclear force** is responsible for beta decay when a proton changes into a neutron or vice versa. This happens because an up quark changes to a down quark (or vice versa). A particle called a **W-boson** is emitted when a quark changes from one type to another; the W-boson is unstable and decays into two leptons (e.g. an electron and an antineutrino) (Fig. 8.24). The W-boson is the exchange carrier of the weak nuclear force. It has been created artificially in high-energy electron–positron and proton–antiproton collisions.

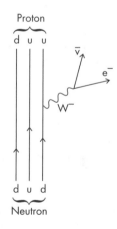

Fig. 8.24 Role of the W-boson in beta decay

Chapter roundup

Before leaving this chapter, make sure you can see the links between each topic and the knowledge that has been brought in from other chapters.

Question Bank

1 The diagram shows a charged oil drop between two horizontal plates connected to a high voltage source *V*. Which of the following statements is/are correct?

1 If the drop is stationary, then it must carry a +ve charge.
2 If the drop is falling at 'terminal velocity', then a resultant force must be acting on it.
3 If the drop is stationary and then suddenly starts to rise, it must have gained extra electrons.

A 1 only.   B 2 only.   C 3 only.   D 1 and 2.   E 2 and 3.

(–: NEAB*, and all other Boards except SEB)

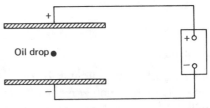

Oil drop ●

Points

For 1, see 8.1 if necessary. For 2, remember Newton's first law: see 1.3 if necessary. For 3, a sudden rise means that the drop's attraction to the top plate must have suddenly become greater.

2 An electron is moving with velocity *v*. It enters a region of uniform magnetic flux density *B* and uniform electric field, *E*. *E* and *B* are mutually perpendicular. The velocity *v* will not remain constant unless *v* is:

A perpendicular to *E* and parallel to *B*.
B perpendicular to *B*.
C parallel to *E*.
D perpendicular to *E* and to *B* and of magnitude *B/E*.
E perpendicular to *E* and to *B* and of magnitude *E/B*.

(London: *all other Boards except SEB)*

Points

See 8.1 and apply E8.3. Also, think of how you arrange your fingers for Fleming's left-hand rule!

3 An electron of mass *m* travelling with a speed *u* collides with an atom and its speed is reduced to *v*. The speed of the atom is unaltered, but one of its electrons is excited to a higher energy level and then returns to its original state, emitting a photon of radiation. If *h* is the Planck constant, the frequency of the radiation is:

A $\dfrac{m(u^2 - v^2)}{2h}$ B $\dfrac{m(v^2 - u^2)}{2h}$ C $\dfrac{m(v^2 + u^2)}{2h}$

D $\dfrac{mu^2}{2h}$ E $\dfrac{mv^2}{2h}$

(London: *all other Boards)*

Points

Apply E8.8 where ΔE will be the change in ke of the electron.

4 An ultraviolet light source causes the emission of photoelectrons from a zinc plate. A more intense source of the same wavelength would give

	Maximum energy/electron	No. of electrons/second
A	More	The same
B	The same	More
C	The same	The same
D	More	More
E	Less	More

(**SEB**: *all other Boards*)

Points

Intensity of incident light is the energy per second per unit area of the beam. Increased intensity of the same wavelength means more photons of the same energy. What effect will this have on the emission of electrons? See 8.2 if necessary.

5 In a series of photoelectric emission experiments a number of different metals were illuminated with monochromatic light of a number of different photon energies and of different intensities. The variables in the experiments were thus:
1 work function of the metals.
2 photon energy.
3 light intensity.
It was found that, for each experiment, the emitted electrons emerged with a spread of kes up to a certain maximum value. This maximum ke depends on:
A 1 only.   B 1 and 2 only.   C 1 and 3 only.
D 2 and 3 only.   E 1, 2 and 3.

(**NICCEA**: *all other Boards*)

Points

Note that an increase of light intensity causes more electrons to be emitted, but their individual kinetic energies are not increased by the increase of light intensity. See 8.2 if necessary.

6 The energy levels of an atom of an element *X* are shown in the following diagram, with values given in eV. Which electron transition A–E will produce photons of wavelength = 620 nm?

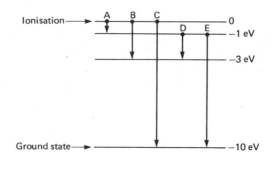

(–: *all Boards*)

Points

Use E8.8 to calculate the photon energy in J. Then convert to eV, remembering that 1 eV = 1.6×10^{-19} J. Assume that the speed of light $c = 3 \times 10^8$ m s^{-1} and that Planck's constant $h = 6.6 \times 10^{-34}$ J s.

7 In an atom containing its normal complement of electrons:
1 the ground state of the atom is the state in which all its electrons are in their lowest possible energy levels.

2 the gaps between excited energy levels are evenly spaced as the atom approaches ionisation.

3 an electron while in an excited energy level is continuously radiating energy in multiples of h, where h is Planck's constant.

Answer:

A if 1, 2, 3 correct.   B if 1, 2 only.   C if 2, 3 only.

D if 1 only.   E if 3 only.

(**London**: *all other Boards*)

Points

For 1, see 8.4. For 2, see Fig. 8.4. For 3, see E8.8 (noting the use of the word *transition*!).

8 Both X-rays and γ-rays:

1 might have wavelengths of the same order.

2 can be detected by a Geiger counter.

3 originate in the nucleus of an atom.

Answer:

A if 1, 2, 3 correct.   B if 1, 2 only.   C if 2, 3 only.

D if 1 only.   E if 3 only.

(**London**: *all other Boards except O and C and SEB*)

Points

See 8.6, E8.13 and also Fig. 8.13.

9 The half-life of a certain radioactive element is such that 7/8 of a given quantity decays in 12 days. What fraction remains undecayed after 24 days?

A 0   B $\dfrac{1}{128}$ C $\dfrac{1}{64}$ D $\dfrac{1}{32}$ E $\dfrac{1}{16}$

(**London**: *all other Boards except SEB*)

Points

If 7/8 decays in 12 days, then 1/8 remains undecayed after 12 days. Hence after another 12 days (24 days in total) then …

10 Two radioactive isotopes P and Q have half-lives of 10 minutes and 15 minutes, respectively. Freshly prepared samples of each isotope initially contain the same number of atoms as one another. After 30 minutes, the ratio

$$\dfrac{\text{the number of atoms of } P}{\text{the number of atoms of } Q} \text{ will be:}$$

A 0.5   B 2.0   C 1.0   D 3.0   E 0.25

(–: *all Boards*)

Points

30 minutes is 3 half-lives for P and 2 half-lives for Z.

11 A radioactive source contains two materials. One decays by the emission of an α particle with a half-life of 4 days while the other emits β particles with a half-life of 3 days. The initial corrected count rate is 176 s^{-1} but this becomes 80 s^{-1} when a piece of tissue paper is placed between the source and the detector. What will be the corrected count rate after 12 days without the tissue paper present?

A 5 s^{-1} B 11 s^{-1} C 12 s^{-1} D 17 s^{-1} E 22 s^{-1}

(**SEB**: *and all other Boards*)

Points

Which component of the radiation, α or β, does the tissue paper absorb? See Fig. 8.13 if necessary. Hence decide what contribution to the initial corrected count rate of 1.76 s^{-1} is made by α particles after 12 days.

Now consider the contribution to the initial corrected count rate made by β particles. What contribution will β particles make after 12 days? Then add the two contributions (of each component after 12 days) to give the final answer.

It is worth noting and using the fact that 12 days is a whole number of half-lives in both cases: after one half-life, the contribution drops to 50%; after two half-lives, the contribution drops to 25% of the initial value; etc.

12 A radioactive atom X emits a beta particle to produce an atom Y which then emits an alpha particle to give an atom Z:
1 The atomic number of X is less than that of Z.
2 The atomic number of Y is less than that of Z.
3 The mass number of X is the same as that of Y.
A 1, 2, 3 correct.   B 1, 2 correct.   C 2, 3 correct.
D 1 correct.   E 3 correct.

(**NICCEA**: *all other Boards*)

Points

It is probably easier to consider the last two statements before the first one since they both involve comparison between an atom and *its* daughter product. Can you make your choice by considering 2 and 3 only? See 8.6 for the basic ideas, if necessary.

13 As a result of a number of successive decay processes in a radioactive series the nucleon number decreases by 4 while the proton number is unchanged. The particles emitted are:
A 1 alpha and 1 beta particle.
B 1 alpha and 2 beta particles.
C 2 alpha and 1 beta particle.
D 2 alpha and 2 beta particles.
E 4 alpha particles.

(**London**: *all other Boards*)

Points

Remember that the nucleon number is the number of neutrons *and* protons, and that the nucleon number is unchanged by beta decay. See E8.11 and E8.12 if necessary.

14 The deviation of α particles by thin metal foils through angles that range from 0° to 180° can be explained by:
A scattering from free electrons.
B scattering from bound electrons.
C diffuse reflection from the metal surface.
D scattering from small but heavy regions of positive charge.
E diffraction from the crystal lattice.

(**Cambridge**: *all other Boards except Oxford and O and C*)

Points

See 8.5 if necessary.

15 Alpha particles and protons of the same initial *velocity* are fired in turn at nuclei of carbon-12. Which of the following statements is/are correct?

1 The alpha particles have a greater initial ke than the protons.
2 The alpha particles are able to approach closer to the carbon-12 nuclei than the protons can (assuming 'head-on' collision in both cases).
3 Each incident particle which is deflected by a carbon-12 nucleus has a greater speed after deflection than before.
A 1 only.   B 2 only.   C 1 and 2.   D 1 and 3.   E 2 and 3.

(–: *all Boards except SEB*)

Points

For 2, the closest distance (*d*) is given by equating the initial ke of the 'projectile' to the pe at the closest distance of approach. E2.3 gives the electrostatic pe.

16 The plates of a horizontal parallel-plate capacitor are 1 cm apart and the potential difference is 10 kV. It is found that an oil drop of mass 1.96×10^{-13} kg remains stationary between the plates. How many electronic charges does the oil drop carry? Indicate briefly how the mass of the drop could be determined experimentally assuming its charge is not known. (Electronic charge = 1.6×10^{-19} C.)

(**WJEC**: *NEAB* and all other Boards except SEB*)

Points

To calculate the number of electronic charges carried by the drop, you must first calculate its charge in C. Use E8.1 to determine the drop charge. Assume $g = 9.81$ m s^{-2}.

To determine the drop mass, the drop must be timed over a measured distance while it falls in zero electric field (i.e. with zero pd across the plates). Because of air resistance, the drop falls at a steady velocity and by measuring this terminal velocity, the drop mass can be calculated. Use your textbook to find further details.

17 When a beam of electrons moving with speed 8.8×10^5 m s^{-1} was subjected to a uniform magnetic field of flux density 5 mT directed at right angles to the original beam, it was found that the electrons described a circular path of radius 1 mm. Find the charge-to-mass ratio for the electron.

(**WJEC**: *all other Boards except SEB*)

Points

Use E8.2.

18 When a light of frequency 5.4×10^{14} Hz is shone on to a metal surface the maximum energy of the electrons emitted is 1.2×10^{-19} J. If the same surface is illuminated with light of frequency 6.6×10^{14} Hz the maximum energy of the electrons emitted is 2×10^{-19} J. Use this data to calculate a value for the Planck constant.

(**London**: *and all other Boards*)

Points

Use E8.6, substituting the appropriate values to give two equations, each with two unknowns. Subtract the two equations to eliminate the work function and then solve for *h* (remembering that *h* has units!)

19 The accelerating voltage across an X-ray tube is 33.0 kV. Explain why the frequency of the X-radiation cannot exceed a certain value and calculate this maximum frequency. (The Planck constant = 6.6×10^{-34} J s; the charge on an electron = 1.6×10^{-19} C.)

(**AEB**: *NEAB, WJEC and NICCEA*)

Points

Your explanation should make it clear that:
❶ The amount of ke gained by a 'beam' electron in an X-ray tube is determined by the accelerating voltage.
❷ The ke of the beam electrons is released as photons when the electrons strike the tube. A single electron may produce one or more photons; if only one photon is produced by a given electron, then all the electron ke becomes energy of a single photon.
❸ The link between energy and frequency for a photon is given by E1.16. Hence show that there is a maximum frequency value for a given tube voltage V.
In the calculation, remember that the energy gained by an electron accelerated through a pd V volts is Ve joules.

20 Radon-222 is a radioactive gas for which the decay constant is 2.1×10^{-6} s^{-1}. If the initial decay rate is 5.6×10^{10} s^{-1} calculate:
(a) the initial number of radioactive atoms present, and
(b) the time which will elapse before the activity is reduced to one-quarter of its initial value (ln 2 = 0.693).

(AEB June 81: all other Boards except SEB)

Points

(a) Use E8.15.
(b) Use E8.16 to find the half-life of the radon. To calculate the time, remember that the activity halves for every half-life (i.e. time equal to one half-life) which passes.

21 A scientist wished to find the age of a sample of rock in which he knew that radioactive potassium ($^{40}_{19}$K) decayed to give the stable isotope argon ($^{40}_{18}$Ar). He started by making the following measurements:
Decay rate of the potassium in the sample = 0.16 disintegrations/second
Mass of the potassium in the sample = 0.6×10^{-6} g
Mass of the argon in the sample = 4.2×10^{-6} g
(a) Show how he could then calculate that for the potassium (i) the decay constant (λ) was 1.8×10^{-17} s^{-1}, (ii) the half-life was 1.2×10^{9} years.
(b) Calculate the age of the rock, assuming that originally there would have been no argon in the sample. Show the steps in your calculation.
(c) Identify and explain a difficulty involved in measuring the decay rate of 0.16 s^{-1} given earlier in the question.

(O and C Nuffield: all other Boards except SEB)

Points

(a) (i) Use E8.15. The decay rate is given in the question, but it is necessary to calculate the number of radioactive nuclei present (i.e. the number of potassium-40 nuclei in 0.6×10^{-6} g). There are two methods, both of which will incur a very slight error (owing to the mass defect of the nucleus – see 8.7). The front of the exam paper will give Avogadro's number and the masses of neutrons and protons. The first method involves using Avogadro's number and the fact that a mole of potassium-40 nuclei would weigh 40 g. The second method involves calculating the mass of an individual potassium-40 nucleus from the neutron and proton masses. (ii) Use E8.16 and the answer to (a)(i).
(b) Use E8.17. However, it is not necessary to calculate the number of nuclei; the masses can be used instead of N_0 and N_t as only a ratio is required. If the mass lost in decay is assumed negligible then originally there must have been 4.8×10^{-6} g of potassium. Time can be saved by noticing that as the potassium has decayed to an 'eighth' of the original quantity, an exact number of half-lives must have passed.

(c) The decay rate is so small that it may be necessary to count for many hours to get a reasonable total count. However, the major difficulty is that it will be extremely difficult to detect the potassium decays; see C_B on Fig. 8.17.

22 When a deuteron of mass 2.0141u and negligible ke is absorbed by a lithium nucleus of mass 6.0155 u, the compound nucleus disintegrates spontaneously into two alpha particles, each of mass 4.0026 u. Calculate the energy, in joules, given to each alpha particle. (1 u = 1.66 × 10^{-27} kg, speed of light in a vacuum = 3.00 × 10^8 m s^{-1}.)

(**London**: *NEAB* and all other Boards except AEB*)

Points

Write the interaction in the form of E8.19 and evaluate the mass difference. Convert this into energy by using $E = mc^2$ (having put m into kg). By conservation of momentum, the alpha particles must have velocities that are equal in magnitude and opposite in direction, and so they will share the energy equally between them.

23 (a) Describe and explain how to obtain a stream of fast-moving electrons. (Such a stream of electrons must also be focused if a beam is required, but you are not asked to describe the focusing system.) Briefly describe one method of detecting or tracking the path of an electron beam.
(b) Describe an experiment which can be used to measure the specific charge of the electron (e/m_e) and explain how the result is obtained from the observations.
(c) An electron beam, in which the electrons are travelling at 1.0×10^7 m s^{-1}, enters a magnetic field in a direction perpendicular to the field direction. It is found that the beam can pass through without change of speed or direction if an electric field of strength 1.1×10^4 V m^{-1} is applied in the same region at a suitable orientation. (i) Calculate the strength of the magnetic field. (ii) If the electric field were switched off, what would be the radius of curvature of the electron path? (Electron charge $e = -1.6 \times 10^{-19}$ C; electron mass $m_e = 9.1 \times 10^{-31}$ kg.)

(**NICCEA**: *and Oxford* and all other Boards except SEB*)

Points

(a) Describe the 'electron gun' part of a cathode ray tube. See your textbook for details. To 'track' the path, the beam must pass over a suitably positioned sheet of fluorescent material, or a low pressure of gas must be maintained in the tube so that ionisation caused by the beam makes its path visible in a darkened room. Describe one of these methods – see your textbook for details.
(b) See 8.1 for the principles of two methods, and use your textbook to find details of the apparatus.
(c) (i) Use E8.3. Remember to give the units in which your answer is expressed.
(ii) Use E8.2.

24 (a) When atoms absorb energy by colliding with moving electrons, light or X-radiation may subsequently be emitted. For each type of radiation, state typical values of the energy per atom which must be absorbed and explain in atomic terms how each type of radiation is emitted.
(b) State **one** similarity and **two** differences between optical atomic emission spectra and X-ray emission spectra produced in this way.
(c) Electrons are accelerated from rest through a potential difference of 10 000 V in an X-ray tube. Calculate (i) the resultant energy of the electrons in eV, (ii) the wavelength of the associated electron waves and (iii) the maximum energy and the minimum wavelength of the X-radiation generated.

(Charge on electron = 1.6×10^{-19} C, mass of electron = 9.11×10^{-31} kg, Planck's constant = 6.62×10^{-34} J s, speed of electromagnetic radiation in vacuo = 3.00×10^8 m s^{-1}.)

(**NEAB**: *and WJEC, NICCEA*)

Points

(a) See 8.4.

(b) You must discuss similarities and differences of the *spectra*, so think in terms of the presence of a continuous background, of lines and of the wavelengths (or energies) involved.

(c) (i) See 8.4. (ii) Use the De Broglie equation E1.18. (iii) See 8.4 and Fig. 8.9 for the maximum photon energy. Use E8.8 to calculate the minimum wavelength; remember that minimum wavelength corresponds to maximum frequency (and hence maximum energy).

25 (a) (i) Explain what is meant by *photoelectric* emission. (ii) Briefly describe a simple experiment to demonstrate this effect qualitatively.

(b) Write down Einstein's photoelectric equation and explain the meaning of each term in it.

(c) In an experiment in photoelectricity, the maximum kinetic energy of the photoelectrons was determined for different wavelengths of the incident radiation. The following results were obtained:

Wavelength/nm	300	375	500
Maximum kinetic energy/eV	2.03	1.20	0.36

Use the results to determine:
(i) the work function for the metal,
(ii) a value for Planck's constant.

(d) (i) Describe how a photocell functions.
(ii) Describe **one** practical application of a photocell.
($c = 3 \times 10^8$ m s^{-1}, $e = 1.6 \times 10^{-19}$ C.)

(**WJEC**: *all other Boards*)

Points

(a) and (b) See 8.2 if necessary. (c) Work out the frequency for each wavelength value then plot the maximum ke against frequency as in Fig. 8.3. You can then work out (i) and (ii) from the graph. See 8.2 if necessary.

26 When the stable isotope of manganese $^{55}_{25}$Mn is irradiated in a nuclear reactor the radioactive isotope $^{56}_{25}$Mn is produced. $^{56}_{25}$Mn decays to a stable isotope $^{56}_{26}$Fe. Write equations for these nuclear transformations.

$^{56}_{25}$Mn has a half-life of a few hours. Give a full account of an experiment to measure the half-life of an isotope such as $^{56}_{25}$Mn. What factors affect the precision of the value you obtain?

Calculate the difference in mass between the particles present before and after the decay of $^{56}_{25}$Mn, given that the total energy liberated in the decay of a $^{56}_{25}$Mn nucleus is 5.9×10^{-13} J.

When $^{56}_{25}$Mn decays to $^{56}_{26}$Fe the iron nucleus is left in an excited state. The iron nucleus radiates this excitation energy as a single γ-ray of wavelength 1.47×10^{-12} m. Calculate the energy carried by the γ-ray photon, and hence the maximum kinetic energy carried by the other decay products.

($h = 6.6 \times 10^{-34}$ J s, $m_e = 9.1 \times 10^{-31}$ kg.)

(**O and C**: *NEAB*, and all other Boards except SEB and AEB*)

Points

There are two nuclear transformations that you must write equations for:

1 $^{55}_{25}\text{Mn} + X \rightarrow ^{56}_{25}\text{Mn}$

2 $^{56}_{25}\text{Mn} \rightarrow ^{56}_{26}\text{Fe} + Y$

Use your knowledge of nuclear structure, in terms of neutrons and protons, to work out what the particles X and Y above ought to be. See 8.5 for the meanings of atomic number and mass in terms of neutrons and protons if necessary.

For experimental determination of the half-life, see your textbook.

The formula for the scale of conversion of mass into energy (and vice versa) is given by E1.17. Assume the total energy liberated has been produced by the conversion of mass into energy.

To calculate the energy of a photon, use E1.16. You will first need to calculate the photon frequency from its wavelength value and the value of the speed of light (3×10^8 m s^{-1}). See E3.7 if necessary.

The maximum ke carried away by the other decay products can be calculated from the difference between the total energy liberated and the energy of the γ-ray photon.

27 (a) (i) Use the following values of masses to calculate the binding energy per nucleon of a ^4_2He nucleus. Give your answer in MeV.
Mass of a neutron = 1.0087 u, mass of a proton = 1.0073 u.
Mass of a ^4_2He nucleus = 4.0015 u.
(ii) Sketch a graph showing the dependence of the binding energy per nucleon of a nucleus on its mass number. State approximate values for, or indicate on your graph, the maximum binding energy per nucleon and the mass number at which it occurs. With reference to this graph, show how energy may be released through nuclear fission, and how energy may be released through nuclear fusion.

(b) (i) Describe briefly how the fission of uranium–235 can lead to a chain reaction. Your answer may be in the form of a simple, carefully labelled diagram. What single condition must apply to achieve a sustained, controlled chain reaction?
(ii) Draw a simple labelled sketch to illustrate the main components of a fission reactor used in electrical power production. Comment briefly on the function of each component of the reactor, and name a suitable material for each component.

(c) Give a brief account of one industrial application of radiation from a radioactive source.

(**NICCEA**: *all other Boards except AEB*)

Points

(a) See 8.7 if necessary. Remember that 1 u is equivalent to 931 MeV.
(b) (i) To sustain the chain reaction, one fission neutron per fission must produce a further fission reaction. See 8.7 if necessary. (ii) See your textbook if necessary for the sketch.

Answers: **1** C; **2** E; **3** A; **4** B; **5** B; **6** D; **7** D; **8** B; **9** C; **10** A; **11** D **12** E; **13** B; **14** D; **15** C; **16** 12; **17** 1.76×10^{11} C kg^{-1}; **18** 6.7×10^{-34} J s; **19** 8.0×10^{18} Hz; **20**(a) 2.67×10^{16} (b) 6.6×10^5 s; **21**(b) 1.08×10^{17} s; **22** 1.82×10^{-12} J; **23**(c)(i) 1.1×10^{-3} T (ii) 0.0517 m; **24**(c)(i) 1.0×10^4 eV (ii) 1.23×10^{-11} m (iii) 1.6×10^{-15} J, 1.24×10^{-10} m; **25**(c) 3.45×10^{-19} J, 6.7×10^{-34} J s; **27** 6.55×10^{-30} kg; photon energy = 1.35×10^{-13} J, max. ke = 4.55×10^{-13} J; **28**(a)(i) 28.4 MeV

DATA ANALYSIS

Units in this chapter

Chapter objectives

After working through the topics appropriate to your syllabus in this chapter, you should be able to:
- relate derived and base units
- check the dimensions of formulae and equations by using base units
- know and use the equation of a straight line and be able to deduce equations from graphs
- know the graph shapes and equations for inverse, inverse square, trigonometric, parabolic and exponential functions
- know the difference between random and systematic errors
- estimate experimental errors and accuracy

9.1 UNITS AND DIMENSIONS

Base units, dimensions and derived units

The system of scientific units (SI, Système International) is founded upon seven internation-ally agreed **base units**. All units for scientific quantities can be derived from the base units, which are hence called 'dimensionally independent' as no one base unit depends upon any other. The base units supply the seven **dimensions** that may be used for dimensional analysis; however, in A-level Physics only the dimensions of mass [M], length [L] and time [T] (these define all the mechanical quantities) are normally required. Note that dimensions are conventionally written in square brackets. Occasionally questions are set involving the dimensions of electric current [A] or temperature [θ]. Table 9.1 lists the base units. The units of any physical quantity that does not appear in Table 9.1 can be derived from the base units; such units are called **derived units** and some are shown in Table 9.2.

Writing units

Although physical quantities could have their units expressed in terms of the base units, this would lead to very lengthy sets of units for many quantities. Hence many derived units (and

Table 9.1 *Base units of SI*

Base quantity	Abbreviation of unit	Name of unit	Dimension
Length	m	metre	$[L]$
Mass	kg	kilogram	$[M]$
Time	s	second	$[T]$
Electric current	A	ampere	$[A]$
Thermodynamic temperature	K	kelvin	$[\theta]$
Amount of substance	mol	mole	Not used at A Level
Luminous intensity*	cd	candela*	

* The 'candela' does not feature on most A-level courses

Table 9.2 *Some derived units of SI*

Derived quantity	Name of unit	Abbreviation of unit	Other common forms of unit	Base units
Frequency	hertz	Hz	–	s^{-1}
Force	newton	N	–	$kg\ m\ s^{-2}$
Energy	joule	J	N m	$kg\ m^2\ s^{-2}$
Power	watt	W	$J\ s^{-1}$	$kg\ m^2\ s^{-3}$
Pressure	pascal	Pa	$N\ m^{-2}$	$kg\ m^{-1}\ s^{-2}$
Electric charge	coulomb	C	–	A s
Electric pd	volt	V	$J\ C^{-1}$	$kg\ m^2\ A^{-1}\ s^{-3}$
Electric resistance	ohm	Ω	–	$kg\ m^2\ A^{-2}\ s^{-3}$
Electric capacitance	farad	F	$C\ V^{-1}$	$A^2\ s^4\ kg^{-1}\ m^{-2}$
Magnetic flux	weber	Wb	$T\ m^2$	$kg\ m^2\ A^{-1}\ s^{-2}$
Magnetic field strength (or magnetic flux density)	tesla	T	$Wb\ m^{-2}$ or $N\ A^{-1}\ m^{-1}$	$kg\ A^{-1}\ s^{-2}$
Speed or velocity	–	–	–	$m\ s^{-1}$
Temperature gradient	–	–	–	$K\ m^{-1}$

two of the base units – A and K) are abbreviated by being named after scientists. Such units are easily recognised as they are written as capital letters; Table 9.2 includes some examples.

To help reduce the length of very large or very small numbers that often have to be written in front of units, a system of abbreviations has been devised. This system is summarised in Table 9.3.

Two 'oddities' of the SI system are worth making clear:

❶ Although mass is measured in kg, fractions and multiples of kg tend to be expressed in terms of g (g = gram = 10^{-3} kg). Hence: 1 Mg = 1000 kg (1 tonne), 1 g = 10^{-3} kg, 1 mg = 10^{-6} kg, 1 μg = 10^{-9} kg, etc.

❷ Although mass is measured in kg, the other base unit relating to amount of substance (the mole – *mol*) is based upon the 'mole' consisting of a fixed number (Avogadro's number) of particles.

The table of constants and numerical values (pp. 298–301) gives many more examples of quantities and their units as well as making use of the prefixes given in Table 9.3.

Table 9.3 *Prefixes used with SI units*

Factor	Name of prefix	Symbol
10^{12}	tera-	T
10^{9}	giga-	G
10^{6}	mega-	M
10^{3}	kilo-	k
10^{-3}	milli-	m
10^{-6}	micro-	μ
10^{-9}	nano-	n
10^{-12}	pico-	p
10^{-15}	femto-	f
10^{-18}	atto-	a

Dimensional analysis (balancing units)

In any scientific equation the units on each side of the equation must be the same. For example, in the equation $E = mc^2$ the units of E must be the same as those of mc^2. However, if this method is always to work correctly then it is essential to reduce the units on each side of the equation to base units (Table 9.1); checking equations by balancing their base units is called the method of **dimensional analysis**. This is not required by all examination boards. Some boards only expect students to balance units in equations without being concerned over whether they are using base or derived units; however, the technique to be used is the same in each case. Most A-level 'dimensional analysis' problems are linked only to mechanical quantities (i.e. dimensions of mass [M], length [L] and time [T]); however, the balancing of units often includes problems involving thermal, optical and electrical quantities, etc.

Dimensional analysis is a powerful technique and can assist in all the following situations:

❶ testing the correctness of equations,

❷ assisting recall of important formulae,

❸ helping to solve physical problems theoretically,

❹ suggesting relationships between fundamental constants.

For example, a student without the knowledge of E3.13 considers that the velocity of transverse waves along a stretched string (v) may be related to the tension in the string (T), the mass of the string (m) and the length of the string (l). First a relationship linking the quantities is suggested (note that this must be written as a proportionality):

$$v \propto T^a m^b l^c$$

where a, b and c are constants to be determined. The next step is to write the units of each term in the base units of mass (kg), length (m) and time (s). These are then converted into the dimensions of mass [M], length [L] and time [T]:

units of v = m s^{-1} so base units of v = m s^{-1}
so dimensions of v = $[LT^{-1}]$

units of T = N   so base units of T = kg m s^{-2}
so dimensions of T = $[MLT^{-2}]$

units of m = kg   so base units of m = kg
so dimensions of m = $[M]$

units of l = m   so base units of l = m
so dimensions of l = $[L]$

These steps were easy except in the case of the units of T, which being a force is measured in N (newtons) – a derived unit. It is not necessary to remember the base units of all derived quantities; usually any equation involving the quantity will enable the base units to be worked out. The easiest equation involving force (force = mass × acceleration) makes it easy to find the base units of force. Now all the terms in the equation are replaced by their dimensions giving:

$$[LT^{-1}] \propto [MLT^{-2}]^a [M]^b [L]^c$$

$$\propto [M]^{a+b} [L]^{a+c} [T]^{-2a}$$

Now each dimension is considered in turn; its power on the left-hand side of the equation being compared to its power on the right-hand side of the equation:

Considering dimensions of M: LHS $\quad 0 = a + b$ RHS

Considering dimensions of L: LHS $\quad 1 = a + c$ RHS

Considering dimensions of T: LHS $\quad -1 = 2a \quad$ RHS

Solving these simultaneous equations gives $a = \frac{1}{2}$, $b = -\frac{1}{2}$ and $c = \frac{1}{2}$. This gives a possible relationship of $v \propto \sqrt{Tl / m}$ (see E3.13).

The following important limitations of the method of dimensional analysis should be appreciated:

❶ It is only possible to 'suggest' relationships using this method. The relationship must then be checked by experiment or further theoretical methods.

❷ Using only the three dimensions ([M], [L], [T]) means that only three unknowns (a, b, c) can be found.

❸ It is impossible to check for dimensionless quantities (i.e. those with no units, such as strain, π, etc.) so any relationship suggested by dimensional analysis can only be written as a proportionality, not as an equation.

Students not required to understand dimensional analysis are still encouraged to study the method, but to apply it to units without worrying about differences between base units, derived units and dimensions. However, it is often necessary to 'juggle' units to balance an equation; an example follows.

It is required to show that the units of the expressions $T = CR$ are consistent (i.e. that they balance). In the equation T represents a time constant (s), C represents capacitance (F) and R represents resistance (Ω).

Units of LHS = units of T = s

Units of RHS = units of CR = FΩ

At this stage the expression must be inspected for methods of simplifying it. The LHS is already in a base unit so the units of the RHS must be reduced to base units. From $Q = CV$ it can be seen that units of capacitance (F) can be written as C V^{-1}. From $V = IR$ it can be seen that units of resistance (Ω) can be written as V A^{-1}. This gives:

Units of RHS = FΩ = CV^{-1}VA^{-1} = CA^{-1}

Now the expression has one base unit (A) but still includes a derived unit (C). Coulombs (C) are units of charge Q so using $Q = It$ we can rewrite the units of charge as A s. This gives:

Units of RHS = CA^{-1} = AsA^{-1} = s

Hence it has been shown that both sides of the equation have the same units.

Table 9.4 *Some of the confusing physical quantities, units and terms*

Physical quantity 'tension' – T	[T] – dimensions of 'time'
Physical quantity 'electric current' – I	A – units of electric current
Units of length – m	[M]– dimensions of mass
Physical quantity 'mass' — M (there are more like this – G, T, etc.)	M – for 'mega-' denoting 10^6
Physical quantity 'inductance' – L	[L] – dimensions of 'length'
Physical quantity 'capacitance' – C	C – units of 'electric charge' (coulombs)
Units of temperature 'kelvin' – K	k – for 'kilo' denoting 10^3
Physical quantity 'spring constant' – k	k – physical constant 'Boltzmann's constant'
Physical quantity 'acceleration due to gravity' – g	g – unit of mass (grams)

When handling physical quantities, units and dimensions, beware of the 'confusing' terms given in Table 9.4. You should add to this any set of terms that cause you problems.

9.2 GRAPHS

Plotting graphs

The first decision to be made in graph plotting is the choice of axes. It is a general rule that the quantity that is 'set' (**independent variable**) is plotted on the x-axis (horizontal axis or **abscissa** – hint for mnemonics, $a$bscissa goes $a$cross) and the quantity that is 'measured' (**dependent variable**) is plotted on the y-axis (vertical axis or **ordinate**). This rule is occasionally broken when the graph will be easier to interpret if the axes are interchanged; for example, if graphs involve 'time' or 'length' these quantities are usually plotted on the x-axis. Hence an experiment could involve a 'set' force (F– the independent variable) which is used to stretch the sample of rubber to produce a 'measured' extension (Δl– the dependent variable); the general rule (F on the x-axis, ΔL on the y-axis) is broken in this case as it is easier to interpret the 'length' on the x-axis and the graph is plotted as shown in Fig. 9.1. When talking about

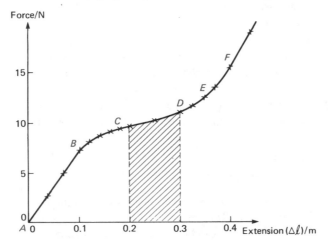

Fig. 9.1 Force (F) against extension (Δl) for rubber

a graph the quantity on the y-axis is mentioned first, so Fig. 9.1 is a graph of the force used to stretch a rubber sample (y-axis) plotted against the extension (x-axis). In questions on graphs this convention also helps in making the choice of axes.

Choosing a sensible scale normally enables the graph to 'cover' as much of the graph paper as possible; it may not be necessary or desirable to include the **origin** ($x = 0$, $y = 0$). It should be possible to see the points on the graph, and the recommended plotting methods include 'crosses' ($\times$, $+$) and 'points' with circles, triangles or squares around them ($\square\triangle\odot$). Some teachers have a distinct preference for 'crosses' on the basis that it is easier to define the plotted point by intersecting lines rather than dots. When plotting several different sets of data onto the same axes it is important to use a different type of point for each set. A change of colour can be used in addition to the five methods already described. The scale should be marked along each axis and the axes clearly labelled with the quantity plotted followed by a stroke (/) and its associated units, e.g. 'F/N' or 'Force used to stretch rubber/N'.

For straight-line graphs five or six points can be sufficient; they should cover as wide a range of x and y values as possible. When curves 'bend' considerably (e.g. from B to C and from D to E in Fig. 9.1) more points are needed to define the shape. It becomes vital to plot extra points when trying to locate a **maximum** or **minimum** ('peak' or 'trough'). Remember to ensure that a graph passes through the origin if this is known to be the case!

Interpreting graphs

When examining graphs, apart from the general trends illustrated by the shape, the physical significance of the following quantities should be considered: slope (gradient), area under the graph, intercepts, other key points (maxima, minima, etc.).

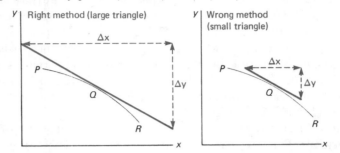

Fig. 9.2 Measuring slope at Q on curve PQR

Fig. 9.2 shows the right (large triangle) and wrong (small triangle) methods of measuring the **slope** at a point Q on curve PQR. The slope ($\Delta y/\Delta x$) in Fig. 9.2 will have a negative value as an increase in y means a decrease in x (i.e. the graph goes 'down'). Remember that horizontal lines have 'zero' slope and that vertical lines have 'infinite' slope. Tangents must be drawn to points on curves to define the slope at that point (see Fig. 9.2). To decide upon the physical significance of a slope, divide the quantity on the y-axis by the quantity on the x-axis and consider the resulting units. For Fig. 9.1 the slope is force/extension measured in unit of N m^{-1}; over the straight line region (AB) where the rubber obeys Hooke's law, the slope represents the spring constant (k) of the rubber band (the force needed to produce an extension of 1 m).

The physical significance of the **area** under the graph can usually by found by multiplying together the quantities plotted on each axis and considering the resulting units. For Fig. 9.1 this gives force $\times$ extension measured in N m (or J); this represents the work done in stretching the rubber. Hence the blue shaded area shown on the graph is the energy (about 1 J) needed to stretch the rubber from an extension of 0.2 m to 0.3 m; this energy is stored as extra potential energy in the stretched rubber. If the equation that defines a graph is known it is possible to measure the area mathematically by the process of **integration** (likewise the slope can be mathematically measured by **differentiation**).

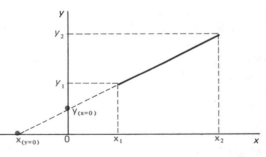

Fig. 9.3 Intercepts on a straight-line graph

Intercepts are the points where a graph crosses the axes. The x intercept ($x_{y=0}$ where y is zero) and the y intercept ($y_{x=0}$ where x is zero) can be read directly off the graph (see Fig. 9.3). However, sometimes the axes chosen will not permit this; then the intercepts can be estimated in the case of a curve or calculated (from two points) in the case of a straight line using E9.1, E9.2 and E9.3:

E9.1 $y = mx + c$
E9.2 $y_{x=0} = c$
E9.3 $x_{y=0} = -c/m$

where x and y are the coordinates of a point on the straight line, m = slope of the straight line, c = value of intercept on the y-axis, $y_{x=0}$ = value of intercept on the y-axis, $x_{y=0}$ = value of intercept on the x-axis.

The coordinates of two points are sufficient to enable m and c to be found from substitution in E9.1, or the slope m can be measured conventionally (see Fig. 9.2) and c found by substitution. When calculating intercepts remember the negative sign in E9.3. Important intercepts in A-level Physics include: Fig. 1.2, x intercept defines range; Fig. 2.5(a), x intercept gives equilibrium separation; Fig. 8.16, y intercept gives the original number of radioactive nuclei. Occasionally a graph fails to pass through the origin as expected and the resulting intercepts can be helpful in assessing the cause; an example to follow up in a textbook is the effect of 'stray capacitance' in the reed switch experiment to measure capacitance.

Other key points can also be found in graphs. Examples of important **maxima** and **minima** include Fig. 1.10, Fig. 2.5(b), Fig. 3.2, Fig. 4.9, Fig. 5.2 – point B, Fig. 6.10, Fig. 7.28. **Asymptotic** behaviour occurs when a graph gradually approaches a fixed value on one axis without theoretically ever reaching that value. Examples of important asympotes can be found on Fig. 2.5 and Fig. 2.6, Fig. 7.18 and Fig. 7.19, Fig. 8.17. **Discontinuities** occur when a graph 'unexpectedly' changes shape. Important examples of discontinuities include Fig. 5.2 – Y_1, etc., Fig. 6.14, Fig. 7.32(a), Fig. 8.10 – the line spectra.

The ability to 'read' graphs is becoming an increasingly more important part of all A-level Physics examinations. This is reflected by the treatment of the graphs featured in this book. A final point of caution should be made; there is not always some sensible physical significance of the area and gradient (etc.) of every graph!

Deducing equations from graphs

Although the need actually to deduce equations from graphs is limited mainly to those involved with project work, the recognition of graph shapes and the physical situations to which they apply is a skill demanded of all A-level students. There are six basic graph shapes that frequently occur in A-level Physics and these are shown in Fig. 9.4. Once a 'shape' has been recognised it may be necessary to try each of the equations (E9.1, E9.4 to E9.9) that might describe the shape. There are two unknowns in each equation which can be found by substituting the coordinates of two points from the graph. The equation can then be checked by making sure that other points from the graph 'fit' it. When 'spotting' graph shapes it is important to look for the effect of moving the origin, making x or y negative, or interchanging x and y. The effects of these are shown in Fig. 9.5 for the graph whose equation is $y = Ax^k$ (see Fig. 9.5 and E9.4).

The 'guessing' method of deducing equations from graphs is poor if the points exhibit any scatter (i.e. errors) from their true position, as is frequently the case when plotting experimental data. Finding an equation that will be a 'good fit' to scattered points is made much easier if the graph can be converted to a straight line. There are two important techniques that enable this to be done.

It is possible to convert all graphs of the form $y = Ax^k$ (i.e. E9.4 and E9.6 where k is negative) to a straight line by taking **logarithms** of both sides. For simplicity logarithms to the base 10 (logs) will be used:

$$y = +Ax^k$$
$$\text{so} \quad \log(y) = \log(Ax^k)$$
$$\text{E9.10} \quad \log(y) = k\log(x) + \log(A)$$

Hence if a graph of $\log(y)$ (y-axis) is plotted against $\log(x)$ (x-axis) a straight line should be obtained of slope k and y-intercept $\log(A)$.

It is possible to convert all graphs of the form $y = Ae^{kx}$ (i.e. E9.5 and E9.7 where k is negative) to a straight line by taking logarithms of both sides once more. In this exponential (i.e. involving e) case, for simplicity, a logarithm to the base e (ln) is used [n.b. $\ln(y) = \log_e(y)$]:

$$y = +Ae^{kx}$$
$$\text{so} \quad \ln(y) = \ln(Ae^{kx})$$
$$\text{E9.11} \quad \ln(y) = kx + \ln(A)$$

Hence if a graph of $\ln(y)$ (y-axis) is plotted against x (x-axis) a straight line should be obtained of slope k and y-intercept $\ln(A)$.

You can try out these two techniques. If E9.10 is used to plot a new version of Fig. 2.4(a), a slight modification is needed; treat the field strength values as positive (you cannot take logs of negative numbers!). k should be found to be -2 and $\log(A) = -11.0$. It is possible to analyse Fig. 8.17 using E9.11; take care to use the 'corrected count rate' for y. In this case $k = 0.012$ and $\ln(A) = +4.1$. It is left to you to construct the two equations and (most importantly) to

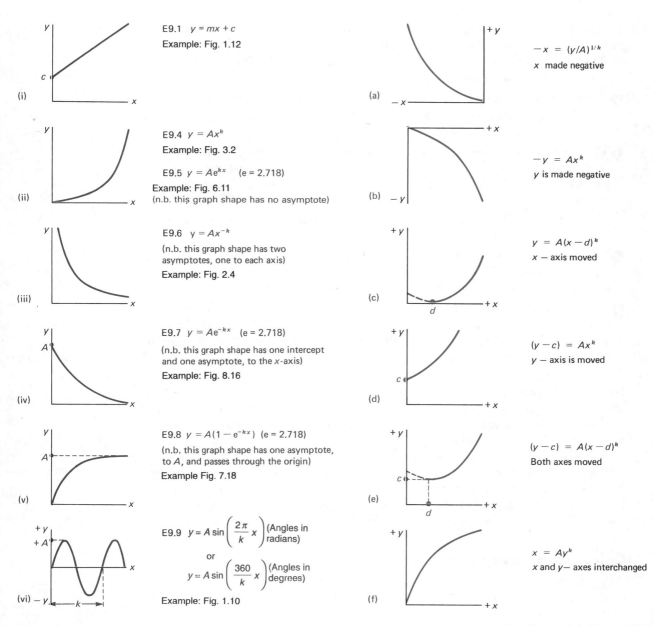

Fig. 9.4 Basic graph shapes and equations

Fig. 9.5 Variations on shape of graph $y = Ax^k$ (Fig. 9.4(ii))

ascertain the units of k and A in each case. Some research in 2.5 and 8.6 will enable the answers to be checked.

9.3 ERRORS

Types of error

There are two methods of quoting the error in a value; they are to state the **absolute error** and to state the **percentage error**. To illustrate the two methods consider the case of a voltmeter observed to read 5.0 V when the error is assessed to be such that it is considered that the reading might reasonably represent a value between 4.9 V and 5.1 V. The reading can be quoted as (5.0 ± 0.1) V (absolute error quoted) or as 5.0 V $\pm 2\%$ (percentage error quoted).

When a measurement is made it is possible to describe two fundamental types of error that may occur. **Random error** is said to occur when repeated measurements of the same quantity can give rise to different values. **Systematic error** refers to an effect that influences all measurements of a particular quantity equally.

Random errors can be demonstrated using the example of the voltmeter where they can arise from two obvious causes:

❶ Friction in the bearings of the system (analogue meter) and deficiencies in its mechanics can cause the pointer to settle in slightly different positions, even though the meter might be repeatedly used to measure the same pd.

❷ The interpretation of the actual position of the pointer is a skill that is subject to random error; this error is less if the observer is aided by a finely marked scale and a mirror to reduce parallax error. Skilled observation can help minimise (but never eliminate) this type of random error.

Systematic errors of both an absolute and percentage nature might arise when using a voltmeter:

❶ If the voltmeter has a 'zero error' then a systematic absolute error will occur. For example, if a 'disconnected' voltmeter reads +0.2 V, then readings will be 0.2 V 'too large', i.e. an absolute systematic error of + 0.2 V will occur in all measurements.

❷ If the voltmeter is incorrectly calibrated than a systematic percentage error will occur. For example, if full-scale deflection (fsd) of the voltmeter is actually 10 V when the meter scale indicates 9 V, then all readings taken with the meter on that range will be 10% too small, i.e. they will have a percentage systematic error of -10%.

A key difference exists when dealing with results that contain random or systematic errors. If a reading is thought to be subject to random error, repeating the measurement several times and taking an average (mean) can improve confidence in the reading (i.e. reduce the likely random error in it). However, this technique is useless for dealing with systematic error which will affect all the observations equally.

Systematic errors are notoriously difficult to detect (because they affect all results equally) and can usually only be found by checking the instrument in which the error is suspected against a known reliable instrument. For our voltmeter example, the zero error is easily found by comparing with a known value of 'zero volts' by disconnecting the meter, but to show up a faulty calibration requires comparison with another voltmeter of known accuracy.

Calculations involving errors

To assess the total error in a calculation it is necessary to evaluate the **likely** error in all the values involved in that calculation. Then the calculation can proceed using the following three rules:

❶ When two quantities are to be **added** or **subtracted**, then **add together their absolute errors** to obtain the absolute error in the answer.

❷ When two quantities are to be **multiplied** or **divided** then **add together their percentage errors** to obtain the percentage error in the answer.

❸ When a quantity is to be **raised to the power** n then **multiply the percentage error by** n to obtain the percentage error in the answer.

It should be pointed out that a full treatment of error calculation goes far beyond this simple scheme and far beyond the scope of any A-level course. An example can illustrate the three simple A-level rules for calculating errors.

A value for the mass of a large ball is required. An experiment to measure the density of the ball yields the result (300 ± 8) kg m^{-3}. The ball is placed upon a metre rule in order to assess its diameter. An observer notes that one end of the diameter is opposite the 35 cm mark and the other end is at 78 cm; the likely error in each of these measurements is assessed to be 1 cm.

D = diameter of ball = $(78 \pm 1) - (35 \pm 1)$ cm = (43 ± 2) cm (using rule 1)

D = (0.43 ± 0.02) m (absolute error)

D = 0.43 m $\pm 4.7\%$ (percentage error)

V = volume of ball = $\frac{4}{3}\pi\frac{D^3}{8}$ = 0.0416 m³ ± 14.1% (using rule 3, we need to treble the percentage error in D)

V = 0.0416 m³ ± 14.1% (percentage error)

ρ = density of ball = (300 ± 8) kg m³ (absolute error)

ρ = 300 kg m⁻³ ± 2.7% (percentage error)

M = mass of the ball = ρV = 12.48 kg ± 16.8% (using rule 2, we need to add the percentage errors in ρ and V)

Finally, it is important to remember never to quote an answer to any more significant figures than the error assessment can justify, so the final value for the mass of the ball is quoted as 12 kg ± 17% or in absolute terms as (12 ± 2) kg.

Experiment design and errors

In an experiment to measure the mass per unit length of a metal rod a student measured the length with vernier callipers (accuracy within 1%) and the mass using a spring balance (accuracy about 10%). This is a classic example of bad experiment design, because the 11% error in the final answer is almost entirely due to the error in one of the two measurements (10% in the mass). If the student had measured the mass using a chemical balance (accuracy within 1%) the result would have been within 2%, and the experiment would have been well designed with each measurement contributing equally to the error in the calculation.

It is a general rule when experimenting that all measurements should contribute equally to the error that will be obtained in the final answer. This normally means that equipment should be chosen to enable all measurement to be made to roughly the same percentage error. An important exception to this suggestions occurs when a measurement is to be raised to a power n in subsequent calculation. The following example will clarify this.

In the sample error calculation the total error in the answer for the mass of the ball was 17%. The error in each measurement was similar, 2.7% for the density and 4.7% for the diameter. Unfortunately the value of the diameter is raised to the power 3 (cubed) in the calculation, which means that the error in the volume is 14.1% (three times the error in the diameter). So most of the error in the answer for the mass arises from the measurement of the diameter. However, if the diameter had been measured to about 1%, then the error in the answer would have been about 6%, with a 3% contribution from each measurement; this would have been much more satisfactory. On the other hand, it can be seen that a measurement that is to be 'square rooted' will have its percentage error halved in the calculation, and hence its value need not be known as accurately as others used in the calculation.

Good experiment design involves realising that the required accuracy of measurements is governed by how the results are to be processed in further calculations.

Assessing errors

A systematic error can be removed by modifying all the observations to remove the error. The difficulty is suspecting that systematic errors might exist! They are detected as a consequence of the analysis of the observations against 'expected' results and the discovery of a consistent error (percentage or absolute). Checking can be done sometimes by repeating the suspect measurement using a standard instrument. Often systematic errors are found by graphs behaving in an unexpected fashion; two good examples from A-level Physics are 'stray capacitance' in the reed switch experiment (see your textbook) and the background count in radioactivity (see Fig. 8.17).

Random errors can be assessed by repeating the same measurement (ideally using different observers) and examining the fluctuations in the results. The following example will illustrate this.

Several observers were asked to time ten oscillations of the same simple pendulum and the following results were obtained: 20.0 s, 20.2 s, 20.1 s, 19.8 s, 20.0 s, 20.0 s, 18.1 s. First, the decision should be made to ignore the last observation as it is obvious from the consistency of the other results that this observer has only timed nine oscillations. (Results can only be ignored if it can be clearly shown that the observations have been wrongly made because their error is far greater than it ought to be for the particular observations.) The average of the rest of the observations is 20.017 s. The largest deviation from this average is 0.2 s (20.2 and 19.8 s); however, to quote this as the 'likely error' is rather exaggerated, as most of the results are within ± 0.1 s. (Students of statistics may chose to quote the standard deviation as the 'likely error', especially as modern scientific calculators have statistical facilities.) Quoting the answer to the correct number of significant figures gives a final value of (20.0 ± 0.1) s. Note that one of the most common examiners' complaints is that students quote answers to calculations to ridiculous numbers of significant figures, so don't copy down a mass of numbers from your calculator but quote just enough significant figures to ensure that the error only affects the last figure!

The above estimate of random error assumes that the error is mainly due to the observer rather than inaccuracy in the instrument itself. This is true in many cases, but an awareness of the manufacturer's specification for the instrument can be important. For example, most dc electrometers have a random error of about 10% in the system, far greater than the error incurred by reading the meter. A sensible balance between the two effects must always be maintained when assessing random errors. This becomes very important when it is impracticable to take an average of several readings to help assess the random error. A useful rule of thumb for estimating random error effects is to evaluate the biggest possible error and then quote half of it as the likely error. One final example will illustrate this.

A manufacturer claims that a voltmeter will give readings accurate to within 1% of its fsd. A meter on fsd of 10 V will hence be subject to a random error of 0.1 V (1% of 10 V). The observer estimates that the meter can be read to an accuracy of ± 0.1 V. Hence if both errors add together a maximum error of 0.2 V will occur in any reading. However, it is unreasonable to consider that the errors will always add up (sometimes they will cancel each other). A sensible quote of likely error is half the maximum error, i.e. ± 0.1 V.

If a straight-line graph is to be plotted, a good estimate of the likely random error can be obtained from the scatter of the points. A common technique is to draw the 'best' line (solid coloured line in Fig. 9.6) and the two acceptable straight lines that deviate most from the best line (dotted coloured lines in Fig. 9.6). The error in the slopes or intercepts (etc.) given by these error lines is used as a basis for the quote of likely error in measurements involving the 'best' line.

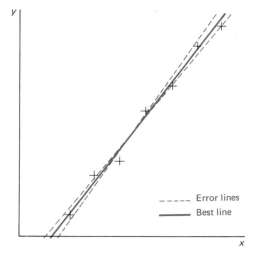

Fig. 9.6 Drawing error lines on a straight-line graph

Chapter roundup

All A-level and AS-level syllabuses test the topics in this chapter, either directly or indirectly. Make sure you have not neglected any of the topics that feature in your syllabus.

Question Bank

1 If p is the momentum of an object of mass m, then the expression p^2/m has the dimensions of:
A Power.   B Impulse.   C Force.   D Acceleration.   E Energy.

(**SEB**: *and all other Boards*)

Points

Obtain the dimensions of momentum (1.3) and work out the dimensions of p^2/m. Then you will have to work out the dimensions of A–E in turn.

Alternatively write momentum as mv so p^2/m becomes $(mv)^2/m$ or mv^2. Putting in a constant without dimensions $(\frac{1}{2})$ gives $\frac{1}{2}mv^2$; this should make it easy to recognise the answer. Remember that 'dimensions' are the same as 'base units' (see 9.1).

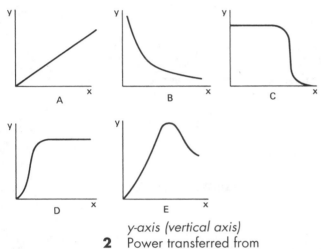

2 and **3**
Five graphs showing the relationship between a physical quantity y and another physical quantity x are lettered A to E. Select from these graphs the one which best represents the relationship between the following pairs of physical quantities. Each graph may be used once, more than once, or not at all.

	y-axis (vertical axis)	*x-axis (horizontal axis)*
2	Power transferred from a battery of finite internal resistance to a resistor connected across it.	The resistance of the load resistor.
3	Flux density along the axis of a long current-carrying solenoid.	The distance from the centre of the solenoid along the axis.

(**NICCEA**: *all other Boards except SEB*)

Points

For 2 firstly consider the point $x = 0$ (i.e. when the load resistance is zero). Here the pd across the load resistor must be zero and all the power supplied by the battery is dissipated in its internal resistance. Thus the graph passes through the origin, limiting the answer to A, D or E. Then consider a second point at 'large x' so that the load resistance is very large and the current will tend towards zero; once again the power dissipated in the load (IV) must tend towards zero. Only one graph fulfils both criteria.

To solve 3 it is necessary to know that the field inside a solenoid is uniform giving a constant y value until on emerging from the solenoid (large x) the field rapidly drops off towards zero. See 2.10.

4 and **5**
The diagram below shows axes which represent the 'logarithms' of pairs of quantities y and x listed below.

log (Y)

log (X)

When the graphs of log (y) against log (x) are drawn, the slopes are among the numbers listed below under the letters A to E:

A –1   B –$\frac{1}{2}$ C $\frac{1}{2}$ D 1   E 2

For each of the pairs of quantities in questions 4 and 5, choose the slope that would be obtained. Each response may be used more than once, once, or not at all.

	y Quantity	*x Quantity*
4	Frequency of small vertical oscillations of a loaded spring.	The mass of the load.
5	Energy stored in a given capacitor.	Pd across the capacitor.

(**NICCEA**: *all other Boards except SEB*)

Points

To answer each question, it is necessary to remember the equation that links the two physical quantities and then use the analysis of E9.10. For 4, use E3.4. For 5, use E2.14; remember that the capacitance in this question is fixed so use the $\frac{1}{2}CV^2$ version.

6   If P is the pressure of a gas and V is its volume, in what unit could the quantity PV be measured?

A Newton.   B Watt.   C Newton/metre.

D Newton second.   E Joule.

(**SEB**: *all other Boards*)

Points

Pressure is force/area and has units N m^{-2}; volume has units m^3. Hence the units of PV are N m. These are the units of force × distance; you should be able to relate them to one of the quantities above. Note that with the knowledge that newton metres are the required units, answers A, C and D are eliminated!

7   In an experiment, the external diameter d_1 and the internal diameter d_2 of a metal tube are found to be (64 ± 2) mm and (47 ± 1) mm, respectively. The percentage error in $(d_1 - d_2)$ expected from these readings is at most:

A 0.3%   B 1%   C 5%   D 6%   E 18%

(–: *all Boards except SEB*)

Points

The question clearly asks for the 'worst possible' error, which must be 3 mm in the measurement of 17 mm. It is easy now to find out the percentage error that this represents. (See 9.3 if necessary.)

8 and 9

The following are five relationships between quantities x and y, k being constant:

A $x = y + k$.   B $x = ky$.   C $x = k/y$.   D $x = k/y^2$.   E $x = ky^2$.

Which of these relationships applies to the following?

8 x is the pd across the plates of Millikan's apparatus needed to hold a certain oil drop stationary, and y is the distance between the plates.

9 x is the speed of an electron as it moves in a vacuum from rest near a negatively charged plate towards a parallel positively charged plate, and y is the time.

(**London**: *Oxford* and all other Boards*)

Points

For 8, rewrite E8.1 with x for V and y for d and rearrange the equation. For 9, note that between parallel plates the electric field is uniform and the electron will hence experience a constant force (see E2.9). A constant force means a constant acceleration; use E1.1 with x for v and y for t. Remember that $u = 0$ since the electron starts from rest.

10 The period T of vertical oscillations of a mass M suspended by a spiral spring is given by

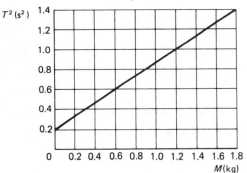

$$T^2 = \frac{A}{g}M + \frac{A}{3g}m$$

where A is a constant depending on the stiffness of the spring and m is the mass of the spring itself.

The graph shows the results of measurements of the period for various values of M. Use the graph to determine the constant A and the mass m of the spring. What are the dimensions of A?

(**SUJB**: *all Boards*)

Points

The graph shows a straight line whose equation is given above. Which part of the equation gives the slope? Hence calculate a value for A, assuming $g = 10$ m s^{-2}. Which part of the equation represents the y-intercept? Hence calculate m, using the previously determined value for A/g.

Each term of the equation has units of s^2, so use the dimensions of g and M to determine the dimensions of A.

11 (30 minutes)

(a) Distinguish between a 'systematic' and a 'random' error in the measurement of a physical quantity.

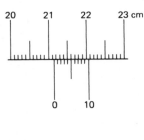

A travelling microscope fitted with a vernier scale is used to measure the internal diameter of a capillary tube. The figures show the vernier when the microscope is adjusted so that the cross-wires are aligned at opposite ends of a diameter.

(i) Write down the two vernier readings.

(ii) What is the maximum uncertainty in a single reading of the vernier?

(iii) Hence find the maximum percentage uncertainty in the area of cross section of the capillary that could arise if it were calculated from these two readings.

(iv) Explain why taking the mean of several microscope readings of the diameter tends to reduce random error.

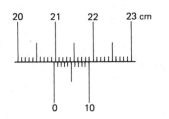

(b) (i) How do you check a formula for its dimensional consistency? Why does this method of checking not give definite confirmation that an equation is correct?

(ii) Express the unit of force and of charge in terms of the SI base units kilogram, metre, second and ampere. Hence by reference to Coulomb's law

$$F = \frac{1}{4\pi\varepsilon_0} \frac{Q_1 Q_2}{r^2}$$

express the unit of ε_0, the permittivity of the vacuum, in terms of these base units.

A unit for μ_0, the permeability of a vacuum, is kg m s^{-2} A^{-2}. Use this unit and your unit for ε_0 to decide which one of the following relations between ε_0, μ_0 and c, the speed of light in a vacuum, is dimensionally consistent

$$\varepsilon_0 \mu_0 = c^2; \quad \varepsilon_0 \mu_0 = c; \quad \varepsilon_0 \mu_0 = c^{-1}; \quad \varepsilon_0 \mu_0 = c^{-2}$$

(**Cambridge**: *all other Boards*)

Points

(a) See 9.3.

(i) and (ii) Each reading can be determined to within one division either way on the lower scale, so convert this into a value in mm since the ten divisions of the lower scale 'correspond' to 1 mm.

(iii) From (ii), calculate the absolute error of the diameter's value. Then, convert this into a % error for the diameter and hence for the area. See 9.3 if necessary.

(iv) See 9.3.

(b) (i) See 9.1.

(ii) Make ε_0 the subject of the equation, then replace the symbols for force, distance and charge with the appropriate combination of base units in each case. Hence express ε_0 in the base units.

Multiply the base units for ε_0 by those for μ_0 to give the base units of $\varepsilon_0 \mu_0$. Hence choose by considering the base units of the right-hand side of each relationship in turn.

12 (45 minutes) The results in the table give corresponding values of the emf, E, mV, of a thermocouple and the temperature, θ, in °C, of the hot junction when the cold junction is maintained at 0 °C. The emf is related to the temperature by the expression $E = b\theta + c\theta^2$ where b and c are constants.

E/m V	θ/°C
5.5	50
10.0	100
13.5	150
16.0	200
17.5	250
18.0	300
17.5	350
16.0	400

(a) Plot a graph of E (y-axis) against θ (x-axis).

(b) Find the slope P of the curve at points $\theta = 50, 100, 150, 200$ and 250 °C. Record your results in a suitable table.

(c) Plot a graph of P (y-axis) against θ (x-axis).

(d) P is known as the thermoelectric power and is related to θ by the expression $P = b + 2c\theta$. From your graph determine the value of b and the temperature at which P is zero.

(**AEB**: *all other Boards*)

Points

(a) Provide a title for your graph and label the axes with the appropriate quantities and their units. Use a sharp pencil!

(b) Draw your tangents as accurately as possble. In measuring the x- and y-values of your gradient triangles make sure that you use the correct units (i.e. do not simply count the graph paper divisions). When recording your values of P remember to state the units.

(c) The same comments apply here as for (a).

(d) The gradient of your graph of P against θ is $2c$, and its intercept on the x-axis (i.e. θ axis) is the value of θ for which $P = 0$ (i.e. when $0 = b + 2c\theta$). Thus $b = -2c\theta_{(P=0)}$.

Answers: 1 E; 2 E; 3 C; 4 B; 5 E; 6 E; 7 E; 8 B; 9 B; 10 A = 6.67 m kg^{-1}, m = 0.9 kg; 11(a)(i) 21.14 cm, 20.97 cm (ii) ±0.01 cm (iii) 24%; 12(d) b = 0.12 mV °C^{-1}, $\theta_{p=0}$ = 300 °C

TEST RUN

In this section:

Test Your Knowledge Quiz

Test Your Knowledge Quiz Answers

Progress Analysis

Mock Exam Questions

Mock Exam Suggested Answers

■ To prepare for module tests during your course, use the practice questions at the end of the relevant chapters, as well as relevant questions from the Mock Exam Questions in this section.

■ To prepare for terminal examinations or final module tests, this section should be tackled towards the end of your revision programme, when you have covered all your syllabus topics, and attempted the practice questions at the end of the relevant chapters.

■ The Test Your Knowledge Quiz contains multiple-choice questions on a wide range of syllabus topics. You should attempt it without reference to the text.

■ Check your answers against the Test Your Knowledge Quiz Answers. If you are not sure why you got an answer wrong, read the points next to the answer and look up any reference to the main text.

■ Enter your marks in the Progress Analysis chart. The notes below will suggest a further revision strategy, based on your performance in the quiz. Only when you have done the extra work suggested should you go on to the final test.

■ The Mock Exam Questions provide a wide spread of question styles and topics, drawn from various examination boards. You should attempt these questions under examination conditions in the time allowed, and without reference to the text.

■ Compare your answers to our Mock Exam Suggested Answers. We have provided tutorial notes to each, showing why we answered the question as we did and indicating where your answer may have differed from ours.

TEST YOUR KNOWLEDGE QUIZ

There are 40 questions to be done in $1\frac{1}{4}$ hours.

1 A constant pd is maintained across a piece of intrinsic semiconductor. When the semiconductor is heated, the current through the semiconductor increases because:
A the atoms vibrate more.
B the conduction electrons move faster.
C the number of conduction electrons increases.
D the cross-sectional area of the conductor increases.
E the electric field across the semiconductor increases.

(−: NICCEA and all other Boards except Oxford, SEB)

2

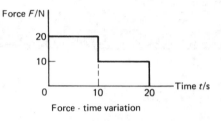

Force - time variation

The resultant force (F) acting upon a 20 kg mass varies with time (t) as shown by the diagram above. At $t = 0$, the speed of the mass is zero. Which of the graphs A–E shows the variation of speed with time?

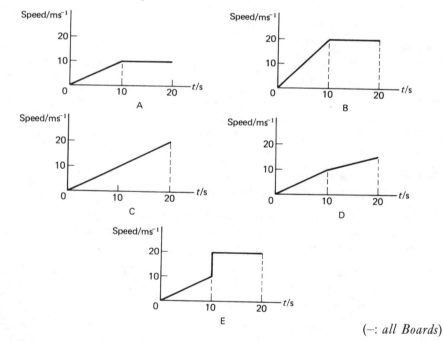

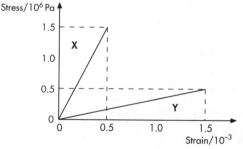

(−: all Boards)

3 The diagram shows the stress–strain graphs for two wires X and Y up to the points at which they break.

Which of the following statements is/are correct?
1 The Young modulus of material X is 3 times that of Y.
2 The ultimate tensile stress of X is 3 times that of Y.
3 At the respective breaking points the energy stored per unit volume of X is the same as that of Y.
A 1, 2 only correct.   B 2, 3 only correct.
C 1 only correct.   D 3 only correct.

(**AEB** June 92: *all Boards*)

4 There is a pd of 6 V between the ends *L* and *M* of the conductor shown. Then:
1 if there is a current of 2 A between *L* and *M*, 3 J of energy will be dissipated in each second as heat in the conductor.
2 electrons in drifting from *L* to *M* lose, on average, 6 eV of energy each.
3 in order that 1 C of charge can flow between *L* and *M*, the conductor must have been supplied with 6 J of energy.
Answer:
A if 1, 2, 3 correct.
B if 1, 2 correct.
C if 2, 3 correct.
D if 1 only.
E if 3 only.

L ▨▨▨▨▨▨▨ M

(**London**: *and all other Boards*)

5 Monochromatic light from a narrow-slit source passes through two parallel slits and forms an interference pattern on a screen. The fringes will be closer together if:
1 the wavelength of the light is reduced.
2 the distance between the slits is reduced.
3 the distance from the source to the slits is reduced.
A 1, 2 only correct.   B 2, 3 only correct.
C 1 only correct.   D 3 only correct.

(**AEB** June 92: *all Boards*)

6 When a two-slit arrangement was set up to produce interference fringes on a screen using a monochromatic source of green light, the fringes were found to be too close together for convenient observation. It would be possible to increase the separation of the fringes by:
A decreasing the distance between the slits and screen.
B increasing the distance between the source and slits.
C increasing the distance between the two slits.
D increasing the width of each slit.
E replacing the light source with a monochromatic source of red light.

(**London**: *and all other Boards*)

7 Consider the arrangement shown in the diagram. At a suitable ac frequency, the wire vibrates as shown (its fundamental mode). If the tension in the wire is now increased considerably, which of the following steps will bring the wire back into vibration in its fundamental mode?

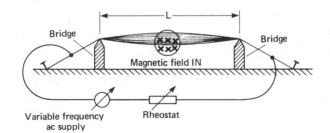

265

1 Increase the frequency.
2 Increase the current.
3 Increase the length L of the wire between the two 'bridges'.
 A 1 only.   B 1 and 2.   C 1 and 3.   D 2 and 3.   E 3 only.

(–: *all Boards except SEB*)

8   Two small objects X and Y, of mass M and $2M$, respectively, are released from rest at heights of 10 m for X and 5 m for Y above level ground. Which of the following properties is the same for X and Y? (Ignore effects of air resistance.)
 A Impact speed.   D Accelerating force just before impact.
 B Impact ke.   E Momentum just before impact.
 C The time taken to fall to the ground.

(–: *all Boards*)

9   A rigid body is in equilibrium due to forces of 2 N, 3 N and 4 N acting upon it. If the 2 N force is suddenly removed, the resultant force at the instant of removal is, in N:
 A 7   B 5   C 4   D 3   E 2

(–: *all Boards*)

10   The Earth is a charged sphere. Near the surface of the Earth, the electric field strength has a constant value of 150 N C^{-1}. The work done, in J, in moving a charge of 2 μC a distance of 5 m at a constant height of 10 m above the surface of the Earth is:
 A 0.0150   B 0.0030   C 0.0015   D 0

(**AEB** June 92: *all Boards*)

11   A beam of monochromatic light of wavelength λ falls normally on a diffraction grating of line spacing d. The angle θ between the 'second'–order diffracted beam and the direction of the incident light is given by:
 A $\sin\theta = \lambda/d$ B $\sin\theta = d/\lambda$ C $\sin\theta = 2\lambda/d$
 D $\sin\theta = 2d/\lambda$ E $\sin\theta = d/2\lambda$

(**NICCEA**: *all other Boards*)

12   The meter in the circuit shown has an uncalibrated linear scale. With the circuit as shown, the scale reading is 20. When another 2000 ohm resistor is connected across XY, the scale reading is:
 A 10
 B 16
 C 25
 D 28
 E 40

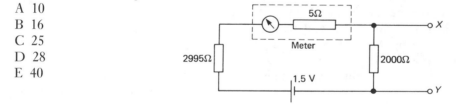

(**Cambridge**: *all other Boards*)

13   Two pupils are asked to find the value of an unknown resistance. They each use the same equipment – a dc supply of negligible internal resistance, an ammeter and a voltmeter. Jane uses circuit J and Margaret uses circuit M.

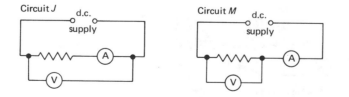

Jane obtained the values: reading on V = 5 volts; reading on A = 1 mA. Margaret's results would be:

Reading on V (volts)	Reading A (mA)
A more than 5	more than 1
B more than 5	less than 1
C 5	1
D less than 5	more than 1
E less than 5	less than 1

(**SEB**: *and all other Boards*)

14 The diffraction pattern produced by a single slit may be demonstrated by illuminating the slit with plane waves of monochromatic light and observing the pattern on a screen *SS'* some distance from the slit.

If the slit is wide compared with the wavelength of the light used, which of the sketches A–E best represents the pattern seen on the screen?

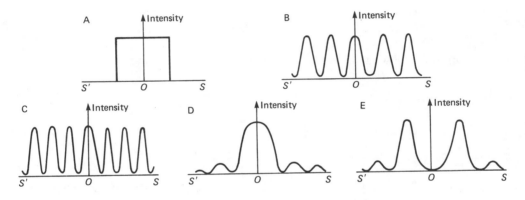

(**NICCEA**: *NEAB* and all other Boards except WJEC, SEB*)

15 The maximum speed of electrons emitted from a given metal surface when illuminated by suitable radiation depends on the:
1 intensity of the radiation.
2 frequency of the radiation.
3 work function of the metal.
Answer:
A if 1, 2, 3 correct.   B if 1, 2 only.   C if 2, 3 only.
D if 1 only.   E if 3 only.

(**London**: *and all other Boards*)

16 Which one of the following experimental phenomena provides evidence for discrete electron energy levels in atoms?
A the spectrum of a tungsten filament lamp.
B the spectrum of a sodium discharge lamp.
C the photoelectric effect.
D the emission of β particles by radioactive atoms.
E the emission of γ rays by radioactive atoms.

(**Cambridge**: *and all other Boards*)

17   A spiral spring was hung vertically with one end attached to a fixed support. A mass of 0.20 kg was hung from the other end, and then made to oscillate vertically with an amplitude of 0.04 m. Its time period was measured and found to be 2.0 s. To increase the time period to 4.0 s with the same spring, it is only necessary to increase:
A the mass to 0.80 kg.   B the mass to 0.40 kg.
C the amplitude to 0.16 m.   D the amplitude to 0.08 m.
E the mass to 0.40 kg and the amplitude to 0.08 m.

(–: *all Boards except SEB*)

18   A disc is rotating about an axis through its centre and perpendicular to its plane. A point P on the disc is twice as far from the axis as a point Q. At a given instant what is the value of the ratio of

$$\frac{\text{the linear velocity of } P}{\text{the linear velocity of } Q}?$$

A 4   B 2   C 1   D $\frac{1}{2}$ E $\frac{1}{4}$

(**London**: *all other Boards except SEB*)

19   A wire that obeys Hooke's law is of length l_1 when it is in equilibrium under a tension F_1. Its length becomes l_2 when the tension is increased to F_2. The energy stored in the wire during this process is:
A $(F_2 - F_1)(l_2 - l_1)$.   B $\frac{1}{4}(F_2 + F_1)(l_2 + l_1)$
C $\frac{1}{4}(F_2 + F_1)(l_2 - l_1)$.   D $\frac{1}{2}(F_2 + F_1)(l_2 + l_1)$
E $\frac{1}{2}(F_2 + F_1)(l_2 - l_1)$.

(**Cambridge**: *and all other Boards except SEB*)

20   The diagram shows a horizontal plane OXY with axes OX and OY at right angles.

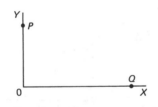

Which one of the following directions for a current in a straight conductor will produce a magnetic flux density at O in the direction $\overrightarrow{OX}$ (i.e. from O towards X)?
A vertically downwards at P.
B vertically upwards at P.
C vertically downwards at Q.
D vertically upwards at Q.
E horizontally above OX in the direction $\overrightarrow{OX}$.

(**London**: *and all other Boards except SEB*)

21   Using the potentiometer circuit shown, the balance (null) point is at X. A balance point to the left of X would be obtained by:
1   increasing the resistance of R.
2   increasing the resistance of S.
3   replacing cell P by a cell with a greater emf.

A 1, 2, 3 all correct.
B 1, 2 only correct.
C 2, 3 only correct.
D 1 only correct.
E 3 only correct.

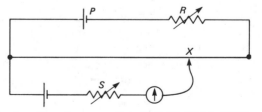

(**AEB** Nov 80: *and all other Boards except O and C Nuffield, SEB*)

22 Two capacitors of capacitance C and $2C$ are charged to pds V and $2V$, respectively. If the two positive plates are connected together and the two negative plates are connected together then this system of capacitors:
A gains charge but loses energy.
B gains energy but loses charge.
C loses both energy and charge.
D loses charge but not energy.
E loses energy but not charge. **(London:** *and all other Boards except SEB)*

23 At 0 °C the value of the density of a fixed mass of an ideal gas divided by its pressure is x. At 100 °C this quotient is:

A $\dfrac{100}{273}x$ B $\dfrac{273}{100}x$ C $\dfrac{273}{373}x$ D $\dfrac{373}{273}x$ E x

(NICCEA: *and all other Boards except O and C Nuffield)*

24 For the construction of a thermometer, one of the essential requirements is a thermometric substance which:
A remains liquid over the entire range of temperatures to be measured.
B has a property that varies linearly with temperature.
C has a property that varies with temperature.
D obeys Boyle's law.
E has a constant expansivity. **(Cambridge:** *all other Boards except SEB)*

25 Young's modulus of elasticity has dimensions of:
A $[M][L][T]^{-2}$.   B $[M][L]^{-1}[T]^{-1}$.
C $[M]^{-1}[L]^{-1}[T]^{-2}$.   D $[M]^{1}[L]^{-1}[T]^{-2}$.
E $[M][L]^{2}[T]^{-1}$.   *(–: all Boards except SEB)*

26 In the circuit diagram C is a 1 μF capacitor holding charge of 10^{-5} C and R is a 10 Ω resistor. If the switch is suddenly closed, the initial current flowing in the circuit will be: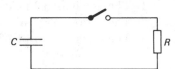
A zero   B 10^{-5} A   C 10^{-1} A
D 1 A   E 10 A   **(London:** *Oxford*, O and C* and all other Boards)*

27 When a 240 V rms ac supply is connected to the terminals PQ in the circuit shown, the fuse F breaks the circuit when the current just exceeds 13 A rms. If the supply were replaced with a 120 V dc source, an identical fuse would break the circuit when the current, in amperes, just exceeds:
A 13/2   B $13/\sqrt{2}$ C 13   D $13\sqrt{2}$ E 26

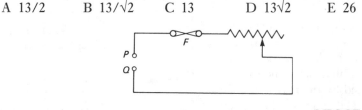

(NICCEA: *all other Boards)*

28 The temperature of a hot liquid in a container of negligible thermal capacity falls at a rate of 2 °C per minute just before it begins to solidify. The temperature then remains steady for 20 min by which time the liquid has all solidified.

The quantity $\dfrac{\text{specific heat capacity of liquid}}{\text{specific latent heat of fusion}}$ is equal to:

A 1/40 K^{-1} B 1/10 K^{-1} C 1 K^{-1} D 10 K^{-1} E 40 K^{-1}

(Cambridge: *and London, NEAB, AEB, Oxford and WJEC)*

29   The diagram shows the line spectrum from a hot gas. Which of the following reasons can account for line Y appearing much brighter than lines X and Z?

A Line Y has the biggest wavelength.
B Line Y has the biggest frequency.
C Line Y originates in the hottest part of the gas.
D Line Y is the result of electrons making a much larger energy jump than those responsible for lines X and Z.
E Line Y is the result of more electrons making that particular energy jump than in the other two lines.

(**SEB**: *and all other Boards*)

30   The uranium series of radioactive decays starts with an isotope of uranium, of mass number (atomic mass) 238 and of atomic number 92. What is the mass number and atomic number of the isotope, in the series, reached after a chain of decays involving a total of 3 α particles and a β particle starting from a U^{238} nucleus?

	A	B	C	D	E
Mass number (atomic mass)	226	226	230	230	231
Atomic number	85	87	85	87	86

(–: *all Boards*)

31   When a spacecraft enters the gravitational field around a planet, which of the following quantities changes with distance from the centre of the planet according to the inverse rule (i.e. $1/d$):
1 The gravitational field strength acting on the spacecraft.
2 The **total** energy of the spacecraft.
3 The gravitational pe of the spacecraft.
A 1, 2, 3 correct.   B 1, 2 correct only.
C 2, 3 correct only.   D 1 correct only.
E 3 correct only.   (–: *all Boards*)

32   Some astronomers think that there was once a planet in circular orbit about the Sun at a distance of 2.8 × the distance from Earth to Sun. What would have been its time period, T (i.e. time to complete 1 orbit around the Sun), in Earth-Years?
A 0.21   B 1.00   C 1.80   D 2.80   E 4.70

(–: *all Boards*)

33   In the Hall-effect experiment, a current-carrying sample is placed in a magnetic field, as shown in the diagram. The charge carriers, which cause the current flow, are deflected by the magnetic field and a pd (the Hall voltage) builds up across the sides of the sample. Which of the following statements is/are correct?

1 The magnetic force (Bqv) is balanced out by the electric field force (qV_{H}/d) once the Hall voltage (V_{H}) has built up.
2 The Hall voltage is proportional to the magnetic field strength B.
3 The Hall voltage is proportional to the current I.
A 1 only.   B 1 and 2.   C 1, 2 and 3.   D 2 and 3.   E 2 only.

(–: all *Boards except Oxford, SEB and NICCEA*)

34 Which of the graphs below best represents the current when a battery connected in series with a large self-inductor is switched on (at p) and then off (at q)?

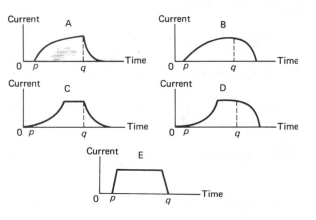

35 An ac source supplies an output of constant peak voltage and variable frequency f. Which of the graphs A–E shows the variation of current in the circuit as the frequency of the ac is increased?

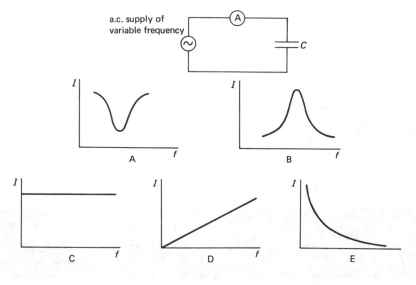

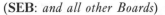

(**SEB**: *and all other Boards*)

36 The electrons in a cathode ray tube are accelerated from cathode to anode by a pd of 2000 V. If this pd is increased to 8000 V the electrons will arrive at the screen with:
A twice the kinetic energy and four times the velocity.
B four times the kinetic energy and twice the velocity.
C four times the kinetic energy and four times the velocity.
D twice the kinetic energy and twice the velocity.
E twice the kinetic energy and the same velocity.

(**SEB**: *and all other Boards*)

37 A radioactive isotope X initially contains 10^{20} atoms of X and none of its daughter product Y, which is stable. Each atom of X that decays releases 8×10^{-13} J of energy. Given that the half-life of X is 4 hours, the total energy released in the first 12 hours is:
A 8×10^7 J
B 4×10^7 J
C 6×10^7 J
D 7×10^7 J
E 14×10^7 J

(–: *all Boards except SEB*)

38 In the Rutherford scattering experiment a collimated beam of alpha particles was scattered by a thin gold foil, and the distribution of the scattered particles was studied. Which of the following statements is(are) a **direct deduction** from the results of this experiment?

1 The atoms in the foil are arranged in a regular lattice.
2 Alpha particles have an associated wavelength.
3 Atoms have nuclei that are much smaller than the atom itself.

A 1 only. B 3 only. C 1 and 3 only.
D 2 and 3 only. E 1, 2 and 3.

(**NICCEA**: *and all other Boards except Oxford and O and C*)

39 The two insulated conductors shown have the same cross-sectional area but Y is twice as long as X, and has twice the thermal conductivity. The junction temperature, in °C, will be:

A 20 B $33\frac{1}{3}$ C $66\frac{2}{3}$ D 25 E 50

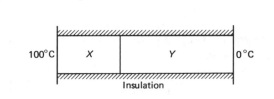

(−: *all Boards except SEB*)

40 A sphere of radius a moving with a velocity v under streamline conditions in a viscous fluid experiences a retarding force given by $F = kav$, where k is a constant. The dimensions of k are:

A $M\,L^{-2}\,T^{-1}$ B $M\,L^{-2}\,T^{-2}$ C $M\,L^{-1}\,T^{-1}$
D $M\,L\,T^{-1}$ E $M\,L\,T^{-2}$

(**Cambridge**: *and all other Boards*)

TEST YOUR KNOWLEDGE QUIZ ANSWERS

1 C Electrical conduction arises from the nonrandom drift of charge carriers under the influence of an electric field.

2 D You must consider the motion in two separate parts. Use E1.5 to calculate the speed at the end of the first part of the motion. Since the speed at the end of the first part is the same as the speed at the start of the second part, you can use E1.5 again (with new force and initial speed values) to determine the speed at the end of the second part.

3 B The slope of X is 9 times that of Y. The maximum stress for X is 3 times that of Y. The area under a stress–strain line gives the energy stored per unit volume.

4 C See E7.4 and its explanation for 1. For 2, how is 1 eV of energy defined? See 8.1 if necessary. Electrons lose energy to the lattice ions through collisions. For 3, see 7.2 if necessary.

5 C See E4.3.

6 E Use E4.3 and remember red light has a longer wavelength than green.

7   C   The natural frequency of fundamental vibrations is given by $f = \dfrac{1}{2L}\sqrt{\dfrac{T}{\mu}}$. Use this formula when you consider each statement.

8   B   Use $v^2 = u^2 + 2as$ to calculate impact speeds, and follow up with impact momentum (i.e. just before impact) if necessary. You can use your speed values to determine ke values if necessary, or use ke gain = pe lost. Remember that all freely falling objects accelerate at the same rate in a uniform gravitational field.

9   E   Remember that the resultant of any two of the forces is balanced out by the third force when the body is in equilibrium.

10   D   The force is vertical and the displacement is horizontal.

11   C   See Fig. 4.6 and E4.8. Note that the question refers to the 'second'-order beam.

12   C   **Without** the extra resistor calculate the total resistance of the circuit and then determine the current through the meter. Remember that this current gives a deflection of 20 units.
     With the extra resistor, calculate the total resistance of the new circuit and then determine the new current. Assuming that the current is proportional to the deflection, calculate the new deflection.

13   D   In circuit J, the ammeter records the current through R alone, whereas the voltmeter records the pd across $R + A$. In circuit M, the ammeter records the current through R and V, while the voltmeter records the pd across R only. See 7.3 if necessary.

14   D   See 4.2 and Fig. 4.5 if necessary.

15   C   For 1, see the comments before E8.5. For 2 and 3, consider E8.6.

16   B   See 8.4, Fig. 8.7 and Fig. 8.8. Remember that a tungsten filament lamp gives a continuous spectrum whereas a sodium discharge lamp gives a line spectrum.

17   A   See E1.8.

18   B   A sketch will help you visualise the situation. Both P and Q will have the same *angular* velocity, ω, but different *linear* velocities. See E1.11 if necessary.

19   E   Consider the following force–length graph for the wire. The work done in stretching the wire is given by the shaded area of the graph. Clearly, that area depends upon the difference of the lengths, not the sum. Also, the area depends upon the average force.

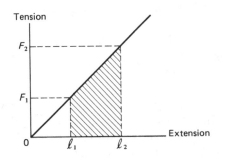

20   B   Try sketching the magnetic field due to a straight wire at P and Q. Use the corkscrew rule to find the direction of the field in each case.

21 **E**   For 1 and 3, consider if each change in turn increases or decreases the pd across the potentiometer wire. Since the test cell emf is unchanged, then is more or less balance length required to give the same pd across the balance length as before?

For 2, remember that at the balance point, there is no current through the meter. See 7.4 if necessary.

22 **E**   See 7.6. As the two capacitors initially have different potentials, there will be a pd when they are connected which will cause a current to flow. This will dissipate energy.

23 **C**   The quotient is density/pressure. Since density = mass/volume, x can be expressed as $\dfrac{\text{mass}}{\text{volume} \times \text{pressure}}$. Mass is fixed but pressure $\times$ volume is proportional to the absolute temperature; thus the quotient is inversely proportional to the absolute temperature.

24 **C**   You cannot assume that any particular type of thermometer is relevant; in other words, alternative A is essential only for liquid–in–glass thermometers but not for any other type. Choose the alternative that applies to all types.

25 **D**   EITHER use E5.1 to obtain an expression for Young's modulus in terms of load, extension, etc., and then work out the dimensions of each term and combine them; OR remember that Young's modulus has the same units as stress so work from there!

26 **D**   Calculate the pd across the charged capacitor C from E7.12. Initially, this will also be the pd across R. See 7.6 and Fig. 7.17 if necessary.

27 **C**   How is rms current defined? Disregard the voltage values, and think about the basic meaning of rms current. See 7.7 if necessary.

28 **A**   Assume mass = m, specific heat capacity of liquid = c and specific latent heat of fusion = l.

Write down an expression for the heat loss per minute of mass m of liquid when cooling at 2 °C per minute. See E6.3 if necessary. From your expression, write down the heat loss in 20 min of mass m of liquid cooling as above.

When the liquid solidifies, assume its rate of loss of heat is the same as just before solidifying started. Then the heat loss is the same as above, and you can equate your expression to ml, etc.

29 **E**   The brightness of an individual spectral line depends upon the number of photons of that particular wavelength emitted. Remember that each emitted photon is produced when an electron makes a particular energy jump.

30 **B**   See E8.11 and E8.12 for the effects upon the nucleus of α emission and β emission.

31 **E**   For 1 and 3, refer to the list of equations in 2.5. For 2, remember that ke gain is equal to pe loss as the craft approaches the planet; you must consider the total energy (i.e. ke + pe).

32 **E**   The centripetal force on a planet (mv^2/R) is provided by the gravitational attraction between the planet and the Sun (GMm/R^2). Since the speed of the planet v = circumference ($2\pi R$)/time period (T), then it follows that $T^2 = 4\pi^2 R^3/GM$. (M = mass of Sun, R = orbit radius of planet.)

Since GM is the same for the 'mystery' planet as for the Earth, use the equation in the form $T = \text{constant} \times R^{3/2}$.

33 **C**   See 2.11 and E2.22.

34 A Referring to E2.29, the 'back emf' opposing the build-up of current must gradually decrease as more current flows, so from the equation it follows that dI/dt must similarly decrease as the current increases, i.e. the slope of the graph must gradually get *less* with time. When the battery is switched off there will be an 'open circuit' so the current will decrease rapidly, i.e. a *large negative* dI/dt (or slope).

35 D See E7.17 if necessary. Express the peak current I in terms of the peak pd V, the frequency f and the capacitance C. Given that C and V are constants, choose the graph which best shows your expression for I in terms of f.

36 B See 8.1 if necessary. If the accelerating voltage is increased by $\times 4$, then consider first the factor for the increase of ke. Then, remembering that the ke is proportional to (velocity)^2, consider the factor for the increase of velocity.

37 D How many atoms of X remain after 12 hours? Hence calculate the number that have decayed after that time has elapsed, and then the total energy released during that time. Watch out for simple arithmetical errors – division of 10^{20} by 2 does *not* give 10^{10} (answer is 5×10^{19} of course).

38 B This refers to the experiment carried out by Geiger and Marsden for Rutherford (see 8.5). Note that the question does not ask whether the numbered statements are true or false, but whether they follow directly from results of the experiment.

39 E Since the heat flow/second through X = heat flow/second through Y, equate $\dfrac{kA\Delta\theta}{L}$ (see E6.11 for meaning of symbols if necessary) for X to the corresponding expression for Y using an assumed interface temperature θ. Hence calculate θ.

40 C Not all Boards write dimensions inside square brackets; indeed occasionally square brackets are used to analyse units as well. The method of balancing an equation to obtain dimensions is explained in 9.1.

PROGRESS ANALYSIS

Place a tick next to those questions you got right.

Question	Answer	Question	Answer	Question	Answer	Question	Answer
1		11		21		31	
2		12		22		32	
3		13		23		33	
4		14		24		34	
5		15		25		35	
6		16		26		36	
7		17		27		37	
8		18		28		38	
9		19		29		39	
10		20		30		40	

My total mark is: _____ out of 40

If you scored 1–10

You need to revise basic concepts, learn the key equations and definitions and memorise essential knowledge in all topics before you progress further.

If you scored 11–20

Identify which broad topic areas need more work. Concentrate on these broad areas as above before you go any further.

If you scored 21–30

Identify individual topics that need attention and bring them up to the same standard as the rest of your work. You may find that you went wrong on a particular style of question such as recognising graph shapes so sort out any such weaknesses before you progress to the Mock Exam Questions.

If you scored 31–40

Don't get too confident – there's still much to do! Identify any individual topic weaknesses and remedy them before you move on to the Mock Exam Questions.

MOCK EXAM QUESTIONS

SHORT-ANSWER QUESTIONS

There are 12 questions arranged in two sets. For each set, you should be able to complete the questions in about 1 hour.

SET 1

1   In driving a pile into the ground, a hammer of mass 500 kg falls freely from rest through a height of 5.0 m onto a pile of mass 1500 kg. The pile and hammer then move together as the pile is driven 0.12 m into the ground.

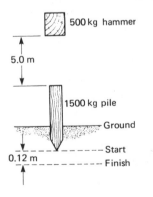

(a) Determine the speed of the hammer just before it hits the pile.
(b) (i) Using the principle of conservation of momentum, calculate the common speed of pile and hammer immediately after the collision.
    (ii) State one assumption which you must make to justify your application of momentum conservation in part (b)(i).

(c) From the moment just after the collision until the system comes to rest, what is the change in: (i) the total kinetic energy of pile and hammer; (ii) the total potential energy of pile and hammer?

(d) By considering these energy changes, or otherwise, calculate a value for the average resistive force which the ground offers to the motion of the pile during its movement into the ground.

(**SEB**: *all other Boards*)

2   A preliminary stage of spacecraft Apollo 11's journey to the moon was to place it in an Earth parking orbit. This orbit was circular, maintaining an almost constant distance of 189 km from the Earth's surface. Assuming the gravitational field strength in this orbit is 9.4 N kg^{-1}, calculate:
(a) the speed of the spacecraft in this orbit; and
(b) the time to complete one orbit.
(Radius of the Earth = 6370 km.)   (**London**: *all other Boards except SEB*)

3   (a) Optical interference can be observed by the *superposition* of light waves from *coherent* sources. Explain the meaning of the words in italics.
(b) Describe, with the aid of a diagram, one experimental arrangement for producing two coherent light sources, explaining why the sources are coherent.

(**NEAB**: *Oxford* and all other Boards
except O and C Nuffield and SEB*)

4   A bathroom is a cavity which produces resonance of sound with very little damping. This has the effect of improving the sound made by some singers.
(a) What is heard by a singer which suggests that a bathroom gives rise to *resonance*?
(b) (i) State what is meant by *damping*.
(ii) Give *one* reason why there is very little damping of sound waves in a typical bathroom.
(c) A small bathroom has walls which are 3.0 m apart. Standing waves are set up in the air in the bathroom with vibration nodes at these walls.
(i) Sketch diagrams to represent the standing waves produced at the two lowest resonant frequencies.
(ii) Calculate the lowest resonance frequency assuming that the speed of sound in the bathroom is 340 m s^{-1}.

(**AEB** June 93: *all Boards*)

5   Describe the path traversed by an electron when it is projected at right angles to (a) a uniform magnetic field, (b) a uniform electric field.
A uniform electric field is superimposed on a uniform magnetic field of 5.0×10^{-2} T so that the two fields allow an electron to travel through them in a straight line with velocity 4.0×10^{6} m s^{-1}. What is the strength of the electric field? State clearly the relation between the directions of the two fields and the direction of motion of the electron.   (**SUJB**: *all Boards except SEB*)

6   A closed square coil consisting of a single turn of area A rotates at a constant angular speed ω about a horizontal axis through the mid-points of two opposite sides. The coil rotates in a uniform horizontal magnetic flux density B which is directed perpendicularly to the axis of rotation.
(a) Give an expression for the flux linking the coil when the normal to the plane of the coil is at an angle α to the direction of B.
(b) If at time $t = 0$ the normal to the plane of the coil is in the same direction as that of B, show that the emf E induced in the coil is given by $E = BA\omega \sin\omega t$.
(c) With the aid of a diagram, describe the positions of the coil relative to B when E is (i) a maximum (ii) zero. Explain your answer.

(**NEAB**: *all other Boards except SEB*)

Set 2

7 In answering the questions below you will have to estimate various quantities and then use your estimates in order to obtain the required answers.
Show clearly your estimates and all the steps in your calculations. Always include units in estimates and in answers.
The men's world high jump record is about 2.3 m.
(a) How much gravitational potential energy would an athlete have to supply in order to jump this height?
(b) State two other factors that must be taken into account in order to obtain a more accurate value for the total energy required.
(c) Accepting the answer in (a)
(i) What would be his vertical take-off speed?
(ii) What would be the average vertical force he exerted on the ground during take-off?

(**O and C Nuffield**: *all other Boards*)

8 A uniform rod *AB* of length 50 cm and mass 10 kg is freely hinged at *A* to a fixed point, and is supported in a horizontal position by a string *BC* inclined at an angle of 60° to the horizontal. Calculate the tension in the string.
What is the work done when the string is pulled so as to raise the rod to the position *AB'*, where it is inclined at an angle of 60° to the horizontal?

(**SUJB**: *all Boards*)

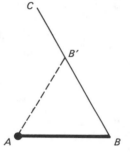

9 (a) Sketch a graph showing how the product *pV* varies with *θ*, where *V* is the volume occupied by one mole of an ideal gas at a pressure *p* and a Celsius temperature *θ*. Explain the significance of the gradient and of the intercept on the temperature axis.
How, if at all, would the graph change if (i) a second mole of the same gas were added to the first, (ii) the original gas were replaced by one mole of another ideal gas having half the relative molecular mass of the first?
(b) An ideal gas has a relative molecular mass of 4.00. The total translational kinetic energy of the molecules of a certain mass of this gas is 374 J at a temperature of 27 °C. Calculate (i) the total translational kinetic energy of the molecules at a temperature of 127 °C, (ii) the mass of gas present.
Molar gas constant = 8.32 J mol^{-1} K^{-1}.

(**NEAB**: *all other Boards except O and C Nuffield*)

10 A certain solenoid has a resistance of 6.0 Ω and an inductance 2.5 H.
(a) Draw sketch graphs to show how the resistance *R* and the reactance *X* of the solenoid depend on frequency *f*.
(b) At what frequency is the resistance of the solenoid 1% of its reactance?

(**Cambridge**: *all other Boards*)

11 The rear window of a car, shown in the diagram, has a heating element on the inner surface. The element consists of 12 identical strands of resistance wire.

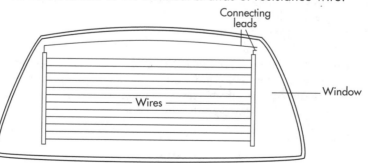

When the heater operates, the current from the 12 V battery is 25 A.

(a) Determine: (i) the total power dissipated by the heater; (ii) the resistance of each strand of resistance wire.

(b) A misted window has a layer of water which is 0.040 mm thick covering the inside surface of area 1.5 m².

The energy required to vaporise 1 kg of water in these conditions is 2.5×10^6 J. The density of water is 1000 kg m⁻³.

(i) Determine the mass of water on the window. (ii) Determine the minimum time for the heater to vaporise this mass of water. (iii) State *two* reasons why the time taken to demist the window will be considerably longer than this.

(**AEB** June 94: *all Boards*)

12 (a) (i) Describe briefly the Rutherford–Bohr model of a hydrogen atom. (ii) State one feature of this model that cannot be explained by the laws of classical physics.

(b) Give a simplified sketch of the energy levels of the electron in this atom. Use your diagram to explain why the spectrum of light from a hydrogen discharge tube contains lines.

(c) The ionisation potential of hydrogen is about 14 eV. (i) Estimate the longest wavelength of electromagnetic radiation which could cause the ionisation of hydrogen. (ii) A hydrogen atom may also be ionised by an inelastic collision with an electron of suitable energy. Estimate the speed of the slowest electron that could cause ionisation.

(Planck's constant $h = 6.6 \times 10^{-34}$ J s; speed of light $c = 3.0 \times 10^8$ m s⁻¹; electron charge $e = 1.6 \times 10^{-19}$ C; electron mass $m_e = 9.1 \times 10^{-31}$ kg.)

(**NICCEA**: *all other Boards except SEB*)

LONG-ANSWER QUESTIONS

The questions are in two sets. For each set, choose three questions to answer in one session. Allow 30 minutes per question except for Nuffield questions which are allocated 45 minutes.

Set 1

1 A rocket is caused to ascend vertically from the ground with constant acceleration a. At a time, t_s, after leaving the ground the rocket motor is shut off.

(a) Neglecting air resistance and assuming that acceleration due to gravity, g, is constant, sketch a graph showing how the velocity of the rocket varies with time from the moment it leaves the ground to the moment it returns to ground. In your sketch, represent the ascending velocity as positive and the descending velocity as negative. Indicate on your graph (i) t_s, (ii) the time to reach maximum height, t_h, (iii) the time of flight, t_f.

(b) Account for the form of each portion of the graph and explain the significance of the area between the graph and the time axis from zero time to (i) t_s, (ii) t_h, (iii) t_f.

(c) Either by using the graph or otherwise, derive expressions in terms of a, g and t_s for (i) t_h, (ii) the maximum height reached, (iii) t_f.

(**NEAB**: *all other Boards*)

2 This question is about explaining ideas in physics.

Choose one of the three subjects, (a) (b) or (c) given on the next page. For the one that you choose you should give a careful explanation of each of the three topics listed. Your explanation should be suitable for a friend studying A-level physics who missed the teaching of this particular subject.

Show also how your explanations could help your friend to understand one everyday application of the subject.

Subjects:
(a) Electronics   Topics: block diagrams, logic, feedback.
(b) Thermodynamics   Topics: number of ways, temperature, entropy.
(c) Radioactivity   Topics: decay, radiation, isotopes.
                    (**O and C Nuffield**: *(c) only for all other Boards*)

3   In the model of a crystalline solid the particles are assumed to exert both attractive and repulsive forces on each other. Sketch a graph of the potential energy between two particles as a function of the separation of the particles. Explain how the shape of the graph is related to the assumed properties of the particles.

The force F, in N, of attraction between two particles in a given solid varies with their separation, d, in m, according to the relation

$$F = \frac{7.8 \times 10^{-20}}{d^2} - \frac{3.0 \times 10^{-96}}{d^{10}}$$

State, giving a reason, the resultant force between the two particles at their equilibrium separation.
Calculate a value for this equilibrium separation.

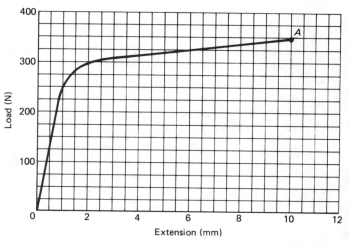

The graph displays a load against extension plot for a metal wire of diameter 1.5 mm and original length 1.0 m. When the load reached the value at A the wire broke. From the graph deduce values of
(a) the stress in the wire when it broke,
(b) the work done in breaking the wire,
(c) the Young's modulus for the metal of the wire.
Define *elastic* deformation. A wire of the same metal as the above is required to support a load of 1.0 kN without exceeding its elastic limit. Calculate the minimum diameter of such a wire.

                    (**O and C**: *all other Boards except SEB*)

4   State *Kirchhoff's laws*. Explain how each is based on a fundamental physical principle.
Use the laws to deduce values of the currents I_a, I_b, I_c and I_d as shown in the circuits below (Figs. 1 and 2).

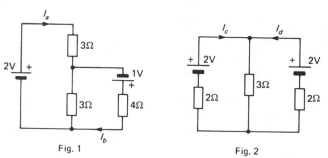

Fig. 1   Fig. 2

A battery, switch and uniform wire of variable length l are connected in series. A bar magnet is suspended vertically above the wire by a fibre fixed to a torsion head fitted with a scale to measure angular rotation. The magnet is initially aligned along the wire. When the switch is closed, the magnet twists but it may be restored to its original position by rotation of the torsion head through a known angle φ. Values of φ for total lengths l of the wire are given.

| φ (°)... | 33.2 | 20.1 | 14.3 | 11.1 |
| l (m)... | 1.00 | 2.00 | 3.00 | 4.00 |

(a) Explain why the magnet twists when current flows.
(b) Explain why φ increases with increasing current.
(c) Plot a graph of $1/\varphi$ against I and find the intercept on the I-axis.
(d) Discuss the significance of this intercept with reference to the battery.

(**Cambridge**: *all other Boards except SEB*)

5 (a) The electromagnetic spectrum may be classified into a number of different regions. Make a table to show clearly the name and an approximate wavelength of radiation in each main region.
(b) (i) What is meant by *polarisation*?
(ii) Why is it impossible to polarise sound waves?
(iii) Describe how the light from a filament lamp could be plane polarised. How could you then test that the light had, in fact, been polarised?
(c) Light is said to exhibit both wave and particle aspects.
(i) Show how the wave aspect of light helps to explain the interference pattern observed in the Young's slits experiment.
(ii) Explain what is meant by *photoelectric emission*.
Sketch a graph to show how the maximum kinetic energy of photoelectrons depends on the frequency of the incident radiation. How does the particle aspect of light account for the form of this graph?
(d) The photon energy of a certain electromagnetic radiation is 1.66×10^{-15} J.
(i) Calculate the wavelength of this radiation.
(ii) To which region of the electromagnetic spectrum does this radiation belong?
(iii) Outline the principle of a method to produce such radiation.

(**NICCEA**: *all Boards*)

Set 2

6 Two bodies are in thermal equilibrium. What does this statement mean? State the Zeroth law of thermodynamics and explain its importance in relation to the use of a thermometer to measure temperature.
(a) The temperature of a hot liquid, measured on the empirical centigrade scale of a certain resistance thermometer, is 68.4°.
(i) What is meant by describing the resistance thermometer scale as *empirical*?
(ii) Write down an equation which defines the centigrade scale of the thermometer in terms of resistance readings.
(iii) Draw a diagram of a simple circuit for use in resistance thermometry. How may the resistance be determined accurately?
(b) The element of the resistance thermometer in (a) is of mass 0.013 kg and has a specific heat capacity of 4.5×10^2 J K^{-1} kg^{-1}. Initially, it was at room temperature, for which a reading of 17.1° was obtained. It was then completely immersed in 0.30 kg of liquid of specific heat capacity 2.5×10^3 J K^{-1} kg^{-1}, giving an equilibrium reading of 68.4°.
(i) What was the temperature of the liquid just before the thermometer was immersed? (For the range of temperature of the experiment, assume that the specific heat capacities of the thermometer and the liquid are independent of thermodynamic temperature and that the empirical scale of the resistance thermometer is linear with respect to thermodynamic temperature. Neglect the heat capacity of the container.)
(ii) How could the cooling effect of the thermometer be made less significant?

(**Cambridge**: *all other Boards except O and C Nuffield – (b) only for SEB*)

7 (a) (i) Explain how stationary transverse waves form on a stretched string when it is plucked.
(ii) State the factors that determine the frequency of the fundamental vibration of such a string and give the formula for the frequency in terms of these factors.

(b) When two notes of equal amplitude but with slightly different frequencies f_1 and f_2 are sounded together, the combined sound rises and falls regularly.
(i) Explain this, and draw a diagram of the resulting waveform.
(ii) Show that the frequency of these variations of the combined sounds is $f_1 - f_2$.

(c) A car engine has four cylinders, each producing one firing stroke in two revolutions of the engine. The exhaust gases are led to the atmosphere by a pipe of length 3.0 m.
(i) Assuming that vibration antinodes occur near each end of the pipe, calculate the lowest engine speed (in revolutions per minute) at which resonance of the gas column will occur.
(ii) What may happen at higher speeds?
(iii) Where, in the gases in the pipe, will the greatest fluctuations of pressure take place at resonance?
(Take the speed of sound in the hot exhaust gases to be 400 m s⁻¹).

(**Oxford**: *all other Boards except SEB*)

8 (a) (i) State the law of electromagnetic induction.
(ii) What is an induced emf? Define self-inductance and mutual inductance.

(b) The figure shows the current I through an inductor when the pd V is applied across it.
(i) Has the inductor any significant resistance? Give a reason for your answer.
(ii) Determine the value of the inductance L of the inductor.

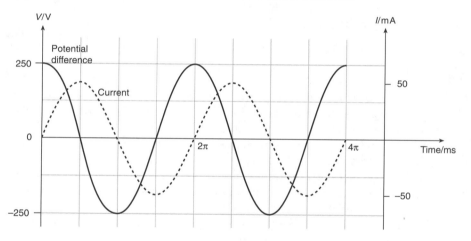

(c) A direct current through the inductor in (b) steadily increases from 0 to 120 mA in 6 s. The current stays constant at 120 mA after the 6 s.
(i) What is the induced emf at (A) 2 s, (B) 5 s, (C) 7 s?
(ii) What is the value of the flux ϕ at 5 s?
(iii) The area of cross section of the coil is 10^{-3} m² and it has a total of 40 000 turns.
What is the value of the magnetic field in the coil at 7 s?

(**WJEC**: *all other Boards except London*)

9 In this question you are asked to consider whether facts or theories come first in explaining pieces of physics.
Here are two opposing views about facts and theories.
❶ 'Facts come first. You can't make up any theories until you have a lot of facts to go on.'
❷ 'Theories come first. In practice, theories tell you which facts to collect, and without theories you get a heap of meaningless information.'

(a) Imagine you had to teach a new set of sixth formers about any **two** of the topics listed in (i) to (iv) below. Discuss the two examples you choose in detail, explaining the 'facts' and 'theories' involved at the various stages in each.

(b) For your two topics, explain briefly whether you think view 1 or view 2 is more appropriate, or whether you think it more sensible to adopt neither view, or a mixture of both views.

Topics

(i) Energy levels in atoms.

(ii) The wave/particle models of light.

(iii) Electromagnetic waves.

(iv) The nuclear model of the atom.

(**O and C Nuffield**: *all other Boards*)

10 Explain what is meant by capacitance.

The plates of a parallel-plate capacitor each have an area of 25 cm² and are separated by an air gap of 5 mm. The electric field intensity between the plates is 7×10^4 V m^{-1}. If one plate is at zero potential relative to earth, find the potential of the other plate and indicate on a sketch some equipotential surfaces in the gap. What is the potential half way between the plates? (Ignore end effects.)

What is the capacitance of the capacitor, and how much electrical energy is stored in it?

A slab of dielectric of relative permittivity 15 is introduced into the isolated capacitor so as to exactly fill the gap. What are the new values of (a) the potential difference, (b) the capacitance and (c) the energy?

Give a labelled sketch of a practical form of variable capacitor and indicate **one** use. ($\varepsilon_0 = 8.85 \times 10^{-12}$ F m^{-1})

(**WJEC**: *all other Boards except SEB*)

PRACTICAL PROBLEM QUESTIONS

Although these questions are all taken from Nuffield papers they will be useful to all students. In the examination 11 minutes per question is allowed, but this does include the time for taking the measurements. The measurements are shown in colour.

1 A piece of wood and a piece of metal are provided. Put a sheet of graph paper on the wood and then put a sheet of carbon paper sensitive side downwards, on the graph paper, as shown in the diagram.

(a) Release the glass marble from a height of 0.5 m so that it lands on the carbon paper and makes a mark on the graph paper. What is the area of the mark?

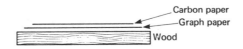

Area of mark = 20×10^{-6} m²

(b) Replace the wood by the metal and repeat (a) so that a mark is made. Measure the area again.

Area of mark = 3×10^{-6} m²

(c) Assume that the area is proportional to the depth of the dent produced and compare as quantitatively as you can the force exerted by the marble on the wood with the force exerted by the marble on the metal.

(d) State any assumptions you have made, in addition to that given, in doing part (c).

(**O and C Nuffield**: *all other Boards*)

2   The circuit shown in the diagram is set up on the bench. *X* and *Y* are two resistors, and *S1* and *S2* are two switches.

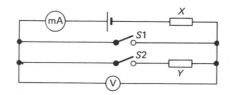

(a) Take readings to complete the following table.

		Voltmeter reading	Ammeter reading
(i)	*S1* open, *S2* open	2.0 V	0 mA
(ii)	*S1* closed, *S2* open	0 V	62 mA
(iii)	*S1* open, *S2* closed	0.5 V	46.5 mA

(b) Deduce a value for *X*, stating any assumptions you make.
(c) Deduce a value for *Y*, stating any assumptions you make.

(**O and C Nuffield**: *all other Boards*)

3   Masses totalling 200 g are secured to the end of a metre rule which is clamped to the bench. A stopwatch is provided.
(a) Make as accurate a measurement as possible of the period of oscillations of the masses on the end of the rule. Show your readings and working.
Three timings are made for 50 oscillations: 19.6 s, 19.5 s, 19.3 s.
(b) The period *T* of the oscillations is given by the formula

$$T = 2\pi\sqrt{\frac{m}{k}}$$

where *m* is the moving mass and *k* is the restoring force per unit displacement.
(i) Using a spring balance provided measure a value of *k* for this system as accurately as possible. Say what measurements you take and how you use them to obtain your value of *k*.
The amplitude of the oscillating masses is about 2 cm so the following results are taken:

Deflection/cm	+2.0	+1.0	−1.0	−2.0
Force needed/N	+1.8	+1.0	−0.9	−1.7

(ii) Substitute for *m* and *k* into the formula given to determine a value for *T*.
(iii) Comment on whether the measured value for *T* and the calculated value compare as you would expect.

(**O and C Nuffield**: *all other Boards*)

4   This question is concerned with the absorption of light by coloured filters. The circuit shown is set up on the bench, and a 'white light' lamp and two identical red filters are supplied. When light falls on the photoresistor, its resistance changes, causing the reading of the meter to change.
(a) With the lamp switched off, note the reading of the meter. 0 A.
(b) Switch on the lamp and, without

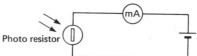

any filters in position, adjust its distanc from the photoresistor until full-scale deflection of the meter is obtained. Write down the value of this current. 10 mA.
(c) (i) Support one red filter in front of the lamp and note the reading. 4.5 mA.
(ii) Say why the reading of the meter falls when the filter is put in position.
(d) (i) Add a second red filter and note the reading. 3.3 mA.
(ii) Why is the change in the reading produced by adding the second filter less than the change produced by adding the first filter?

(e) Use the results obtained above to deduce what you think would be the effect of adding a third red filter.

(**O and C Nuffield**: *all other Boards*)

5 This experiment requires you to charge an electroscope and to do some calculations about capacitances. The apparatus shown is already set up on the bench. Plug the lead into the battery so that the bulb lights and casts a shadow of the gold leaf on to the screen. Make sure the electroscope is discharged by touching the cap with your finger.

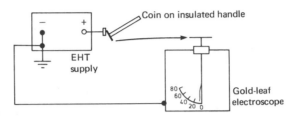

(a) Touch the 2p coin on to the positive terminal of the eht supply and transfer the charge to the cap of the electroscope. Repeat this three more times and note the total deflection of the leaf caused by the four transfers of charge. 65°.

(b) Discharge the electroscope again and transfer the charge in the same way using the $\frac{1}{2}$p coin this time until the same deflection, or as nearly as possible the same deflection, is obtained. Write down the number of transfers of charge needed and the deflection caused by the total charge. 6 transfers produce a deflection of 65°.

(c) Use the information you have obtained in (a) and (b) to estimate the ratio: (capacitance of a 2p coin)/(capacitance of a $\frac{1}{2}$p coin). Set out your reasoning clearly and state any assumptions you make.

(**O and C Nuffield**: *all other Boards*)

6 In this experiment you are asked to compare the masses of two trolleys by a dynamic method. Most of the marks are awarded for part (b).

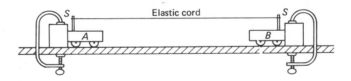

(a) Pull the trolleys apart until they reach the stops S, as shown in the diagram. The positions of S must not be changed throughout the experiment. Release the trolleys simultaneously and note how far the trolleys move before colliding. Make three sets of observations.

Distance moved by trolley A/cm	36.5	35.2	35.5
Distance moved by trolley B/cm	24.0	25.4	25.0

(b) Work out what you can about the relative masses A and B.

(**O and C Nuffield**: *all other Boards*)

COMPREHENSION PAPERS

There are two comprehension papers. The time allowed for each is 1 hour.

Paper 1 (1 hr) This article is about those aspects of noise implied in the well-known definition 'sound undesired by the recipient'. There is in principle no technical difficulty in expressing, in detail, the physical events that constitute a noise. This may be done to any desired degree of accuracy. For example, oscillograph records of the wave forms of the pressure variations at each ear of the hearer will contain all the information of an acoustic nature about the noise.

There are only three basic dimensions of acoustic sensation: pitch, loudness and quality. Thus two sounds may differ from each other with respect to one of these dimensions while being the same with respect to the other two. Loudness is a subjective quantity that corresponds rather closely with the physical intensity of sound. Loudness represents the size of the sensation and is not concerned with the pleasantness or unpleasantness of the sound.

Measurements of subjective values must be kept as simple as is consistent with the facts. One example is the simplified rule relating loudness, S, in sones to loudness level, P, in phons: $P - 40 = 33 \log_{10} S$. On the sone scale the number is proportional to the average person's estimate of the magnitude of the loudness sensation. On the phon scale the number depends on the physical intensity of a 1000 Hz reference tone which the average person judges to be of the same loudness as the noise.

The scale of noticeable differences is much coarser in the subjective world than in the physical. This has important consequences; for example, it takes a rather large physical reduction of a noise to achieve a modest degree of subjective improvement. At the same time, if difficulties in enforcing noise limits are to be avoided, these limits must be clearly defined in terms of meter readings. There must not be too great a tolerance on the maximum allowed reading.

The test of subjective measurability depends on checks of consistency. For example, if in the opinion of a listener A and B are equally noisy, and B and C are equally noisy, then one can verify experimentally that he also finds A and C equally noisy. In a similar way, the possibility of measurement on the sone scale depends on the experimental verification of propositions such as this: if A is three times as loud as B, and B is twice as loud as C, A must be six times as loud as C. In practice we exact a further condition of measurability, namely a reasonable consensus of opinion between different 'normal' listeners, because we are ordinarily concerned with decisions taken on behalf of communities of people rather than individuals. Loudness passes these tests of measurability.

Questions

1   Are the following quantities physical or subjective? Give a brief explanation of each of your answers.
   (a) The variation of sound pressure at the ear.
   (b) The energy transmitted by the sound wave.
   (c) The loudness level of a sound.
   (d) The 'annoyingness' of a sound. (10 marks)
2   Explain in your own words what you have learned from the article about the following:
   (a) loudness.
   (b) the sone scale. (6 marks)
3   Give an example of a physical and a subjective quantity from a branch of physics not included in the article. (4 marks)
4   Explain in your own words the meaning of the phrase 'the scale of noticeable difference is much coarser in the subjective world than in the physical'.
   (3 marks)
5   Sketch the kind of oscillograph record you would expect to obtain from:
   (a) a short burst of a pure tone.
   (b) starting the engine of a lorry. (4 marks)
6   The loudness level of a sound x is 73 phons.
   (a) What is the loudness of x?
   (b) If the loudness of x is four times the loudness of a sound y, what is the loudness of y? (3 marks)
   (**London**: all other Boards)

Paper 2 First read carefully the passage below and afterwards answer the questions that follow.

Thermoelectric cooling

A thermocouple develops an emf if the junctions of the two wires of dissimilar metals composing it are maintained at different temperatures. This is the Seebeck effect and an important term concerning it is thermoelectric power which may be defined as the emf developed per degree difference in temperature between the junctions. It is possible to relate graphically the thermoelectric power to the temperature difference between the junctions. This has been done for a copper–iron thermocouple in Fig. 1 and the expected emf for a given temperature difference can be calculated from the area under the graph.

The converse of the Seebeck effect is the Peltier effect in which heat is given to, or absorbed from, the surroundings if an electric current is passed through a junction between two dissimilar materials. The Peltier effect can be made the basis of a heat pump, i.e. a device that transfers heat from a cold junction to a warmer one. This is best done by using specially prepared substances known as n-type and p-type semiconductors as the effect is then so large as to make thermoelectric cooling of practical interest where small size and absence of mechanical movement are desirable. Fig. 2 shows a single cooling unit consisting of such semiconductor elements, n and p, joined with suitable contacts. T_0 is the ambient temperature, i.e. the temperature of the surroundings. When a current I flows as shown, heat is pumped from the common junction which reaches a temperature T which is less than the ambient temperature by an amount which we shall denote as ΔT.

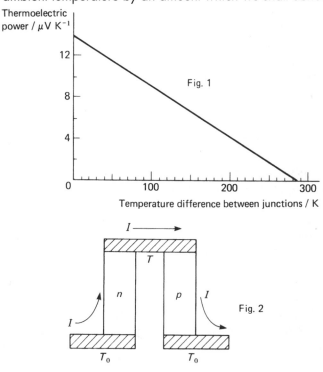

The magnitude of the Peltier cooling effect is reduced by heat conducted down the elements and by the normal heating effect which accompanies an electric current (Joule heating). In the absence of external heat sources, the cooling effect is balanced by these heat losses and it can be proved that the relevant equation is:

$$QTI = \lambda\Delta T + \tfrac{1}{2}I^2R \qquad \text{(i)}$$

where Q is the thermoelectric power, λ the thermal conductance and R the total electrical resistance of the semiconductor elements. From equation (i) it can also be proved that the maximum value of ΔT is given by:

$$\Delta T_{max} = \frac{Q^2 T_{max}^2}{2\lambda R} \qquad \text{(ii)}$$

where T_{max} is the temperature of the common junction when ΔT is a maximum. In an experiment with a thermoelectric cooling unit using n-type lead telluride and p-type lead telluride the following results were obtained:

I/A	1.0	2.0	2.5	3.0	3.5	4.0	5.0
T/K	4.9	9.3	11.0	11.3	11.0	10.1	7.9

In such a cooling arrangement, a figure of merit, F, based on the expression for the maximum temperature difference, ΔT_{max}, may be defined as:

$$F = \frac{Q^2 \sigma}{k}$$ (iii)

where σ is the electrical and k is the thermal conductivity. For the semiconductor material lead telluride typical values are $\sigma = 10^5\ \Omega^{-1}\ m^{-1}$ and $k = 4\ W\ K^{-1}\ m^{-1}$ and the predicted maximum temperature difference, ΔT_{max}, is 11.5 K. An even more promising thermoelectric material is bismuth telluride for which a figure of merit of $3 \times 10^{-3}\ K^{-1}$ has been obtained, leading to a maximum temperature difference of 80 K. High values of ΔT_{max} are necessary since additional heat loads, such as would be experienced in actual devices, act to reduce the temperature difference.

Semiconducting materials having high thermoelectric powers can also be made into effective generators of electricity by applying the Seebeck effect. These are particularly attractive with high-F materials since the same figure of merit applies. In fact, however, it is usual for different materials to be used for power generation since they are required to withstand higher temperatures.

1 (a) Explain in your own words what is meant by the statement that the Peltier effect is the converse of the Seebeck effect. (3 marks)

(b) What does the passage describe as the practical advantages of thermoelectric cooling? (2 marks)

2 (a) What would be the effect on the thermoelectric cooling unit shown in Fig. 2 if the n-type semiconductor were replaced by a p-type element equivalent to that already present? (2 marks)

(b) What would be the effect of reversing the current through the unit in Fig. 2? (2 marks)

(c) What is meant by Joule heating? (2 marks)

3 By consideration of dimensions or otherwise, derive the units of λ, thermal conductance (equation (i)), and explain the relationship between thermal conductance and thermal conductivity, k. (4 marks)

4 Use Fig. 1 to estimate the emf of a copper–iron thermocouple when the cold junction is at 0 °C and the hot junction is at (a) 285 °C, (b) 200 °C.

(4 marks)

5 (a) For lead telluride, plot a graph of the quoted experimental values of ΔT against the corresponding values of current. Include on your graph the predicted value of ΔT_{max} and comment briefly on whether your graph fits in with the accepted theory. (4 marks)

(b) Rewrite ΔT in terms of the other quantities given in equation (i) and hence show that the maximum value of ΔT occurs when $I = QT/R$. (2 marks)

6 If the figure of merit for a unit constructed of lead telluride is only $\frac{1}{12}$ of that for bismuth telluride, use the numerical information given in the passage to calculate the thermoelectric power of lead telluride. (3 marks)

(**NEAB**: *all other Boards except SEB*)

DATA ANALYSIS PAPERS

Paper 1 $(1\frac{1}{2}$ hours)
Read the following account of an experimental investigation and then answer the questions at the end.

Corresponding measured values of potential difference, V, and current, I, for a semiconductor diode are given in the Table.

Potential difference, V/V	Current, $I/\mu A$
0.255	0.40
0.315	1.60
0.345	3.6
0.385	8.9
0.410	18.2
0.455	52.2
0.475	90.3
0.495	140
0.505	182
0.515	223
0.530	310

Questions

1 Using the values in the Table, plot a graph with I as ordinate against V as abscissa. Determine, for the point on the graph corresponding to $V = 0.500$ V,
   (a) the rate of change of I with V, and hence
   (b) the percentage change in I corresponding to a 1% change in V. (12 marks)

2 The following theoretical equation (the 'rectifier equation') applies for certain types of semiconductor diode:
$$I = I_0(e^{aV} - 1) \qquad (1)$$
where I_0 and a are constants. If V is sufficiently large,
$$I \approx I_0 e^{aV} \qquad (2)$$
so that $\ln I \approx \ln I_0 + aV$
or $\log_{10} I \approx \log_{10} I_0 + 0.434 aV$.
From an appropriate table of values plot either a graph of $\ln(I/\mu A)$ as ordinate against V as abscissa or a graph of $\log_{10}(I/\mu A)$ as ordinate against V as abscissa. From your graph derive values for I_0 and a. (11 marks)

3 Explain for which value of potential difference in the Table the approximation made in equation (2) will be most serious. Using the values of I_0 and a derived in Question 2, calculate the current at this pd using the exact equation (1). ($e = 2.72$.) Plot the corresponding point on your second graph.

   State, giving your reasons, whether you consider that use of the approximate equation (2) was justified in analysing the results in the Table. (7 marks)

   (**London**: *all other Boards*)

Paper 2 (45 min)

ISOTOPE	R in cm	E in MeV	T in s	lg R	lg T
$^{228}_{90}$Th	4.02	5.38	5.98×10^7	0.604	7.78
$^{224}_{88}$Ra	4.35	5.68	3.15×10^5	0.638	5.50
$^{220}_{86}$Em	5.06	6.28	5.46×10^1	0.704	1.74
$^{216}_{84}$Po	5.68	6.77	1.60×10^{-1}	0.754	−0.80
$^{212}_{83}$Bi	4.79	6.05	3.61×10^3	0.680	3.56

The table contains data referring to a number of α-emitting isotopes in the thorium series. R is the range in air of the alpha particles emitted by each isotope, E is their energy and T is the half-life of the isotope.

Geiger and Nuttall suggested the following relationship:

$$\lg T = m \lg R + B \qquad (i)$$

where m and B are constants.

(a) Plot a graph of lg T (y-axis) against lg R (x-axis) and determine the gradient of the line.
(b) Determine values for m and B in equation (i). Show how you arrived at your answers.
(c) Geiger suggested another possible relationship:
$$R = aE^{3/2} \qquad (ii)$$
State, with reasons, the quantities which you would plot on a graph to test this relationship.
(d) Draw up an appropriate table of values for the graph you have chosen.
(e) Plot this graph.
(f) Explain whether this graph has confirmed equation (ii).

(**AEB** June 83: *all other Boards*)

MOCK EXAM SUGGESTED ANSWERS

SHORT-ANSWER QUESTIONS

1 (a) 10 m s^{-1}. Just before impact, all the initial pe is converted into ke. Hence v can be found.

   (b) (i) 2.5 m s^{-1}. See 1.3 if necessary.

   (c) (i) 6250 J. Use your value for the common speed to calculate the ke of the system just after impact. Since the final ke is zero, then the change of ke is numerically equal to the ke of the system just after impact.

      (ii) 2400 J. For the change of pe from just after impact to the rest position, remember that the system falls 0.12 m after impact.

   (d) 7.2×10^4 N. Using the relationship: average force $\times$ distance moved $=$ change of energy, the average resistive force can be found.

2 (a) 7.85 km s^{-1}. The centripetal force (mv^2/r) to keep the satellite in orbit is supplied by the gravitational attraction (mg), giving $g = v^2/r$.

   (b) 5250 s. Simply use: time = distance round one orbit/speed.

3 (a) You should firstly explain the meaning of 'coherent sources'; See 4.1. Superposition is discussed in 3.5.

   (b) See 4.1 for a description of Young's double-slit experiment. The arrangement of apparatus is shown in Fig. 4.2. In addition to describing the experimental arrangement, you must explain why the two slits of the double-slit arrangement act as coherent emitters of light waves.

4 (a) Increased loudness.

   (b) (i) See p.99.

      (ii) Tiled walls reflect sound waves very effectively, thus reducing the absorption of sound waves.

   (c) (i) See p.107.

      (ii) For walls 3.0 m apart, the fundamental wavelength would be 6.0 m. Hence calculate the frequency corresponding to this wavelength. See E3.5 if necessary.

5 (a), (b) See Fig. 8.1 and Fig. 8.2 for the paths, if necessary.

   $2 \times 10^5 \text{ V m}^{-1}$. An electron will travel straight through only if its velocity is such that the electric force balances the magnetic force. See 8.1 and E8.3 if necessary.

6 (a) Start by making a quick sketch of the coil and field lines, as seen from along the coil axis. Mark the normal to the coil and label angle α. The flux per turn is given by the coil area $\times$ B-component along the normal; hence, derive the required expression.

   (b) Substitute $\alpha = \omega t$ into your expression for total flux from (a). Then use E2.25 to obtain the expression for the induced emf.

   (c) Maximum emf is when the sides of the coil cut directly across the field lines.

7 (a) Estimate the athlete's mass. Then calculate the gain of pe.

   (b) Does the athlete's centre of gravity also rise by about 2.3 m? Why does the athlete take a run up to the high jump bar?

   (c) (i) Assume ke is converted to pe. (ii) You need to estimate the distance moved by the force which gives the initial 'lift-off'. Assume work done by the force goes to ke of 'lift-off', and so calculate the force.

8   57 N. Start by making a sketch diagram showing the three forces acting on the rod, which are its weight (acting at the C of G), the tension in the string and the reaction at the hinge. To calculate the tension, take moments about A so eliminating the reaction from the equation. Remember that the moment of a force about a point is given by force × *perpendicular* distance from the point to the line of action.

21.7 J. Note that the three forces acting on the rod form a triangle, each side representing a force vector. It is a general rule, sometimes called the **closed polygon** rule, that the force vectors for an object in equilibrium form a closed polygon – in this example, the closed polygon is a triangle because there are only three forces acting on the rod.

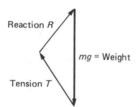

Reaction R

mg = Weight

Tension T

To calculate the work done, calculate the height gain of the centre of gravity of the rod, then calculate the gain of pe of the rod for the work done.

9   (a) Use the ideal gas equation modified by substituting $\theta + 273$ for T, the absolute temperature. Then consider the equation in the form $y = mx + c$ with $y = pV$ and $x = \theta$, and so give the required sketch graph, etc.
(i) With two moles, use E6.13 to consider how the equation would change, and how the new gradient relates to the gradient for one mole. However, does the intercept on the temperature axis alter?
(ii) E6.13 does not include the relative molecular mass, so if the number of moles is the same, the equation is the same – therefore, is the graph the same?
(b) (i) 499 J, (ii) 4×10^{-4} kg. See 6.6. Remember to convert to kelvins!

10 (a) See 7.8 if necessary. The reactance graph is a straight line through the origin (y = reactance, x = frequency) with a slope which you can calculate. Alternatively, calculate the coordinates of one point other than the origin, and so mark each axis with the origin and each coordinate of the chosen point.
(b) 38.2 Hz. Use E7.19 if necessary.

11 (a) (i) 300 W; see E7.4 if necessary.
(ii) 5.8 Ω; the wires are in parallel hence the current through each wire is 25/12 A and the pd across it is 12 V. See E7.7 if necessary.
(b) (i) 0.060 kg. The volume of the water layer is the thickness × the surface area. Use mass = volume × density to calculate the mass.
(ii) 500 s. Calculate the energy needed to vaporise 0.06 kg of water, given 2.5×10^6 J is needed to vaporise 1 kg. The heater supplies 300 J per second. Hence calculate the time taken to supply the necessary amount of energy.
(iii) Heat loss to the outside will reduce the amount of energy available to vaporise the water layer. Also, further condensation may occur.

12 (a) (i) Rutherford's main contribution was showing that almost all the mass of an atom is concentrated in a small positively charged nucleus with orbiting electrons accounting for most of the atom's volume. Bohr's main contribution was showing that electrons could move only in certain fixed orbits, and energy was absorbed or emitted when an electron jumped from one allowed orbit to another. See your textbook if necessary. (ii) This is concerned with the existence of discrete orbits. According to classical theory, electromagnetic radiation should be emitted continuously by an electron on a circular path.

(b) See 8.3.

(c) (i) 88 nm. Since the value given is in eV (electronvolts), it is strictly speaking, the ionisation **energy**. See 8.1 for the definition of an electronvolt. This indicates that the ionisation energy can be expressed in joules by multiplying by the value of electron charge e. The energy given to an electron by electromagnetic radiation can be estimated from hf_0 where h is Planck's constant and f_0 is the minimum frequency required. Since $v = f\lambda$, then v/λ (or in this example c/λ_0) can be substituted for f_0 making the energy hc/λ_0. Equating this to the ionisation energy (in joules) gives a value for λ_0. (ii) 2.2×10^6 m s^{-1}. The ionisation energy must be supplied by the electron that collides with the atom. Equate the required ionisation energy in joules to the kinetic energy of the electron, and hence calculate speed.

LONG-ANSWER QUESTIONS

1 (a) Before sketching the graph, you should try to establish the situation clearly by breaking the motion into three parts:

① From $t = 0$ to t_s: the rocket speed increases steadily from rest.

② From $t = t_s$ to t_h: the rocket still moves upwards, but its speed falls steadily until it comes to momentary rest at its highest point.

③ From $t = t_h$ to t_f: the rocket falls to earth with steadily increasing speed.

Your sketch graph should show y = velocity, x = time. Remember that the slope of each part is equal to the acceleration in that part ($+a$ in 1, $-g$ in 2 and 3).

(b) In your account, two key points should be applied to each part of the motion:

① Slope = acceleration.

② Area under line = displacement. Since the rocket falls back to earth at $t = t_f$, then the area above the x-axis should equal the area beneath the x-axis (since area above represents +ve displacement, etc.) up to $t = t_f$. For the significance of the area up to $t = t_s$ and up to $t = t_h$, remember what t_s and t_h represent.

(c) GRAPHICALLY: Write down an expression for the speed when the motors are shut off in terms of a and t_s: since this speed drops to zero in time $(t_h - t_s)$ at a rate $-g$, then you should be able to write an expression for $(t_h - t_s)$ in terms of a, g and t_s. For the maximum height, remember that the triangle area of parts 1 and 2 gives the displacement to maximum height; triangle area = $\frac{1}{2}$ base $\times$ height. For t_f use the fact that the area under part 3 = the area under the other two parts. Remember to give your answers in terms of a, g and t_s.

OTHERWISE: Apply the dynamics equations E1.1–1.4.

2 There is a temptation for candidates answering this style of question to write long, rambling, unplanned answers that cover their entire knowledge of the required subject area. This question is intended to test knowledge, but perhaps of greater importance is the ability to explain information logically and also to identify terminology and explain its meaning. This is the idea behind the stipulation that the answer is 'for an A-level student who missed the teaching of the subject'.

3 See 2.4 and Fig. 2.5(b) in particular.

0.281 nm. One term of the equation represents the repulsive interaction, the other the attractive interaction so at equilibrium the two terms balance one another out. Hence calculate d at equilibrium.

(a) 217 MN m^{-2}. See 5.1 if necessary. (b) 3.0 J. The area under the curve gives the work done, so count squares – also, you need to work out how much work each square corresponds to. (c) 1.40×10^{11} N m^{-2}. Use E5.1 if necessary but remember to use the section of the graph from 0 to the limit of proportionality. 3.0 mm. Elasticity is discussed in 5.2. From the graph, read off the load corresponding to the elastic limit, then calculate the stress at that point. Hence calculate the area needed to support 1 kN, and so determine the diameter of the second wire.

4 Kirchhoff's 1st law is that the sum of the currents into a junction is equal to the sum of the currents out of the junction. Use 'conservation of charge' to explain the law. The 2nd law is that for a closed loop of a circuit, the sum of the emfs round the loop is equal to the sum of the pds across the resistances of the loop. Because emf and pd are both defined in terms of energy per unit charge, use 'conservation of energy' together with the meaning of emf and pd to explain the second law.

$I_b = 4/11$ A, $I_a = 17/33$ A, $I_c = I_d = 1/4$ A. Fig. 1: Assume the current through the lower 3 ohm resistor is downwards, and write this current in terms of I_a and I_b. There are three possible loops but because there are just two unknown quantities (i.e. I_a and I_b), it is only necessary to consider two of the three loops. Take care to sum pds in accordance with the current directions. With two equations, solve for I_a and I_b.

Fig. 2: Start by writing an expression for the current through the 3 ohm resistor in terms of I_c and I_d. Then consider two loops, and for each loop write an equation using Kirchhoff's 2nd law. Hence solve for I_c and I_d.

(a) In an external magnetic field, a bar magnet will always try to align itself along the field line direction – with its N-pole trying to move in the same direction as the field, and its S-pole in the opposite direction. What is the direction of the magnetic field around a straight wire? Hence give your explanation.

(b) Increased current will cause the magnetic field of the wire to alter. How? A bigger angle is necessary if the torque on the bar magnet due to the wire is bigger.

(c) −1.51 m. 'Plot' means use graph paper – as opposed to 'sketch' when you can use axes drawn on ordinary paper.

(d) φ is proportional to the magnetic field strength which is proportional to the current I: l is proportional to the *wire* resistance R. For the given circuit, $E = I(R + r)$ where E is the battery emf and r is its internal resistance. The intercept on the l-axis therefore gives a length of wire equal to which resistance?

5 (a) See p.108 if necessary.

(b) (i) See p.104 if necessary. (ii) Sound waves are longitudinal waves and therefore cannot be polarised. (iii) See p.104 if necessary.

(c) (i) See p.119 if necessary. (ii) See p.222 and Fig. 8.3 if necessary.

(d) (i) 0.119 nm; see E8.8 if necessary. (ii) Use the table you made for part (a). (iii) See p.226 for the principle of the X-ray tube.

6 Thermal equilibrium is discussed in 6.1.

(a) (i) The resistance thermometer scale is given by E6.1, so explain why it is empirical (i.e. based on experiment). (ii) See E6.1. (iii) See 7.5 if necessary.

(b) (i) 68.8°. When the thermometer is put into the liquid, the thermometer gains heat energy from the liquid until their temperatures become equal. Let the initial temperature of the liquid be θ, and write down expressions for the heat loss of the liquid in terms of θ, and the heat gain by the thermometer. Assume heat loss from the liquid equals heat gain by the thermometer, and hence determine θ. (ii) Consider how your calculation in (i) would differ if the initial temperature of the thermometer was greater.

7 (a) (i) Explain how the arrangement produces two travelling waves passing through one another, give the conditions for the formation of stationary waves, and explain how these conditions are met here. See 3.4 if necessary. (ii) See Chapter 3, question 19 if necessary.

(b) (i) See 3.6 for an explanation of 'beats'. (ii) Suppose the two notes are in phase at $t = 0$, so giving a 'beat'. For each complete cycle of the higher note, the lower note falls behind in phase by a set amount per cycle of the higher note. Eventually, the lower note is one complete cycle behind the higher note so bringing the two notes back into 'zero phase difference' and the next beat. Let T = time between successive beats, so write down the number of cycles completed by each note in that time, and use the fact that one completes one cycle less than the other in that time. Hence derive an expression for the beat frequency $1/T$.

(c) (i) 33.3 Hz. The situation is like Fig. 3.13(a). (ii) See 3.6. (iii) Pressure is proportional to density so where does the density vary most?

8 (a) See 2.12 if necessary.

(b) (i) Since the current and the pd are 90° out of phase, what does this tell you about the resistance? (ii) 5.9 H. Work out the rate of growth of current when the pd is at its peak of 250 V. Then use E2.29 to work out L.

(c) (i) (A) 0.12 V, (B) 0.12 V, (C) zero. In the first 6 s, the induced emf is constant and can be worked out using E2.29. After 6 s, the induced emf is zero because the current is constant. (ii) 0.59 Wb. The total flux $\phi = LI$. Hence work out ϕ (iii) 0.18 mT. The magnetic flux density $B = \phi/NA$, where A is the area of cross-section and N is the number of turns.

9 Some comments on the four topics may help you in planning your answer.

(i) The first ideas about energy levels arose from experiments (by Franck and Hertz) involving excitation potentials in gases. The theory derived from this work was confirmed by further experiments involving emission and absorption atomic spectra (see 8.3 and 8.4). Further evidence arises from studying behaviour of electron streams in thyratrons (anomalous drops in tube current can be due to excitation); see Nuffield Unit 2.

(ii) The original theory for light was a particle model which had to be modified to a wave theory as a result of experiments (such as Young's 'slits') that proved wave properties (see 4.1). However, the photoelectric effect (see 8.2) showed that light had also to be explained using a particle model; hence the development of a joint theory (De Broglie's equation E1.18 and Planck's equation E1.16) that treats particles and waves as equivalent.

(iii) It was some time before it was realised that light, ultraviolet, infrared, X-rays, etc. were all members of the same family of radiation. It was experiments that showed them all to have the same speed (and also to be transverse waves) and thus provided the link. The method of production of electromagnetic waves was understood theoretically long after the practical discovery of the 'family' of waves. There were some early theories of particle streams that had to be discounted by showing that the radiation had no 'mass'. See 3.7 for a brief discussion of the nature of the waves.

(iv) The original model of the atom was the 'currant bun model' proved wrong by Rutherford's alpha-particle scattering (see your textbook). The theory had to be modified to a 'nuclear' theory as a result of this experiment. Later studies of radioactivity suggest that the nucleus must consist of discrete particles experimentally supported by the discovery of the neutron and the proton. Electrons had been identified earlier but initially were not understood to be 'bits' of atoms.

10 Capacitance is explained and defined in 7.6.

350 V; 175 V. The electric field between charged parallel plates is uniform. See 2.8 for a full discussion of uniform electric fields. Use E2.9 to calculate V (the plate pd). Fig. 2.1 shows the field pattern for parallel plates. Note that electric field intensity is the same as electric field strength.

4.43 pF; 2.71×10^{-7} J. To calculate the capacitance of the parallel plates, use E2.12. The equation for energy can be calculated from E2.14, using the values for capacitance (C) and pd (V) between the plates which you have already calculated.

(a) 23.3 V, (b) 66.5 pF, (c) 1.82×10^{-8} J. With the dielectric and the isolated capacitor, you should first realise the significance of the word 'isolated' here; it means that the charge on the plates is fixed at the value it has in the first part (calculated from your values for C and V with equation E2.11). The problem is most easily tackled by calculating the new capacitance first; remember that filling the spacing with dielectric of relative permittivity k increases the capacitance to $C_0 k$ (C_0 = original capacitance). Then, calculate the new pd across the plates, using the value of charge as above and the new value of capacitance; see E2.11 if necessary.

Finally, use the new values of pd and capacitance to calculate the new energy stored; see E2.14 if necessary.

For a practical form of a variable capacitor and its use, consult your textbook.

PRACTICAL PROBLEM QUESTIONS

1 (c) Force on metal is about 7× that on wood. Because the glass marble is dropped from the same height in each experiment it will have the same ke to lose each time it hits the carbon paper. This ke loss can be measured as work done (force × distance) where the distance involved is the depth of the dent. It should be easy now to compare the forces involved in the two cases; note that the deeper dent does not represent the bigger force!

(d) The following points may help you. The 'work done' formula applied to constant forces only; will the force be constant in these experiments? Does the bouncing of the ball matter? (E1.5 may be applied to help solve this.) Is all the ke of the ball converted into helping create the dent?

2 (b) 32.2 ohms. The key is to understand the significance of the readings (i) (ii) and (iii) in part (a). It may help to redraw the circuit diagrams (in rough) simplifying them to allow for the effects of opening and closing the switches; an example for (iii) is given below. One of the sets of readings of (a) gives a value for the emf of the battery (assuming the voltmeter draws a negligible current). Another gives a value for the

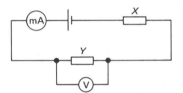

current through X when the full emf of the battery (assuming that the internal resistance of the battery and the resistance of the ammeter are negligible) is dropped across X. Hence the value of X can be determined. Note that closing S1 connects the voltmeter across a wire which, having effectively no resistance, will have zero pd across it so giving a zero reading on the voltmeter. As long as the ammeter's internal resistance and the internal resistance of the battery are negligible, then the only resistance in the circuit is X which must therefore have the full battery emf across it.

(c) 10.8 ohms. The third experiment of part (a) gives the current through X and Y when they are connected in series (same current) and the pd across Y. Here the internal resistance of the cell and the resistance of the ammeter will not affect the result. However, if the voltmeter's resistance is not negligible then the ammeter will read incorrectly.

3 (a) 0.39 s. When calculating T avoid the temptation to quote your answer to too many significant figures. Use the variations in timings of 50 oscillations to guide you (see also 9.3).

(b) (i) 90 N m⁻¹. Note k is the force needed to displace the masses by unit distance (i.e. 1 metre), but the displacements are measured in centimetres! (ii) 0.30 s. Convert the masses into kg! Quote the answer to a sensible number of significant figures. (iii) Do the results agree? If not, why not? Do they agree after allowing for the likely experimental errors? Can the values of m and k be wrong? What about the mass of the ruler – would this affect the results in the way you have found in your calculations?

4 Obviously the resistance of the photoresistor drops when light shines on it, hence the current in the circuit will increase. In part (b) white light (ROYGBIV) is allowed to hit the photocell. However, the red filter absorbs most of the colours apart from red (OYGBIV). So with only red light hitting the photocell a current of 4.5 mA is recorded.

A second filter reduces the brightness of this red light so the photocell reads 3.3 mA. (e) 2.4 mA. If the current reading were proportional to the brightness of the light then introducing a filter cuts the red light brightness by about a quarter (from 4.5 to 3.3 mA). So a third filter should reduce the brightness by a quarter again. Beware the trap of wrongly suggesting that as the current dropped by 1.2 mA (4.5 – 3.3) it will do so again.

5   The scale on the electroscope could be calibrated in volts; i.e. the electroscope is charged up to the same voltage in parts (a) and (b). (c) 6:4. Using $Q = CV$ we can see that the same charge must be transferred to the electroscope leaf on each occasion. So four transfers of charge from the 2p coin are equivalent to six from the $\frac{1}{2}$p coin. But coins behave as capacitors and they are also charged up to the same voltage each time (the eht supply voltage) before their charge is transferred to the gold leaf. This makes it easy to compare the capacitances of the two coins. However, the comparison can only be made if all the charge from the coins is transferred to the electroscope; under what conditions will this happen? Also what about the effect of charge leaking from the coins before the transfer is made – could this affect the results?

6   The trolleys both travel for the same length of time (t) before they collide. Furthermore the trolleys experience the same strength forces but acting in opposite directions (action and reaction are equal and opposite); this is due to the elastic cord that connects the trolleys. Remember that the starting velocity (u) of each trolley is zero and use $F = ma$ and $s = ut + ...$ (E1.3) for each of the trolleys to get a value for the ratio of their masses. Although the force and acceleration are NOT constant as the second formula requires, it is still reasonable to assume that the distance travelled is proportional to the acceleration.

COMPREHENSION PAPERS

Paper 1

1   Measurement of a physical quantity is independent of whoever makes the measurement; however, the measure of a subjective quantity varies from one person to another. With this point in mind, answer (a) to (d) in turn. Your background knowledge of pressure and energy should enable you to explain your answers to (a) and (b). For (c), the last sentence of the third paragraph is the key to your explanation – judgement is involved! For (d), everyday experience ought to enable you to give a brief explanation.
2   (a) Look closely at the second and final paragraphs before you answer. (b) See the third paragraph. Remember you must explain in your own words.
3   Medical physics is one branch which has many examples of a physical stimulus giving a subjective response. Think about one of the human senses other than 'sound'.
4   Give a simple example to illustrate your answer; if the intensity of a sound is doubled, will its subjective measurement also double? . . . or will it be more or less than double? A much coarser scale means greater difficulty in its measurement.
5   (a) Remember that a pure tone is a single frequency note. (b) The sound would contain a range of frequencies, and as the lorry 'revs up' (engine speed increases), the higher frequencies would predominate.
6   (a) 10 sones. Use the equation given in the passage.
    (b) 2.5 sones. Read the final paragraph before answering the question.

Paper 2

1   (a) The Seebeck effect is described at the beginning of the passage, the Peltier effect in the second paragraph. Do not forget to put the explanation into your *own* words.
    (b) See the second paragraph.
2   When current passes through n-type material, the end at which the (conventional)

current leaves the material becomes cooler; for p-type material, the end at which the current enters becomes cooler.
   (a) Would the new piece of p-type material try to make the junction cooler or warmer?
   (b) See initial comments.
   (c) See the text before equation (i); you should explain what 'normal' electrical heating is.

3   W K^{-1}. Use equation (i); each term of the equation is a power term (e.g. $\frac{1}{2}I^2R$ is in watts).

4   (a) 1.995 mV, (b) 1.800 mV. The y-axis is in μ V K^{-1}; the x-axis is in K. What does the area under the line represent?

5   (a) Use equation (ii) to calculate a value for ΔT_{max}.
   (b) Differentiate the expression for ΔT with respect to I. What is the value of $d(\Delta T)/dI$ when ΔT has its maximum value? Hence prove $I = QT/R$.

6 1.0×10^{-4} W A^{-1} K^{-1}. Use equation (iii).

DATA ANALYSIS PAPERS

Paper 1

NB Before starting, read through *all* the questions so that you know exactly what you have got to do.

1   Make sure you draw a *smooth curve* through the points.
   (a) 3.7×10^{-3} Ω^{-1}. Draw a tangent at 0.500 V and extend it to give you a *large* triangle from which to measure the gradient (which has units!).
   (b) 12%. Find 1% of V (i.e. 1% of 0.500 V), then from your value of the gradient at this point, find the corresponding change in I. Express this *change* in I as a % of the value of I at 0.500 V.

2   24 V^{-1}, 1 nA. The question says *from an appropriate table of values*, so you must show a table of V, I and $\ln(I/\mu A)$ or $\log_{10}(I/\mu A)$. Before drawing the graph, read question 3 very carefully!
   $\ln I = \ln I_0 + aV$ can be rearranged as
   $\ln I = aV + \ln I_0$, i.e. of the form $y = mx + c$.
   The equation will be more accurate for the larger values of V, so the slope in this region should be found to give a. The scale of your graph would be too small if you were to include the origin and the intercept c, so c must be calculated by substitution of suitable values (i.e. large V) of y and x into $y = mx + c$. Remember that c will then be $\ln I_0$, not I_0.

3   The approximation is *best* for *large* values of V. Some indication of the % error incurred by using the approximation should be included in your discussion.

Paper 2

   (a) −55.0 and (b) −55.0, 40.8. Since equation (i) is of the form $y = mx + c$, your knowledge of graphs and their equations should enable you to predict the shape of the graph of lg R against lg T if the suggested relationship is correct. The y-value is from 0.604 to 0.754 so do you need to include $y = 0$? Use the slope (i.e. gradient) and y-intercept to obtain values for m and B – see 9.2 if necessary.
   (c) See E9.10 if necessary. By plotting a graph of $y = $ lg R against $x = $ lg E, explain how you could then test the '3/2' relationship.
   (d) and (e) Your y-scale must cover the values 0.604 to 0.754, and the x-scale is only over a limited range as well. Do you need to include $x = 0$, $y = 0$, and if you did, how would it affect the accuracy of your test?
   (f) What value ought the slope have if the '3/2' relationship is correct? Hence measure the slope as accurately as possible to test the relationship.

Multiplying factor	To the nearest order of magnitude	1	2	3	4
$\geqslant 10^{24}$	n charge carriers (m^{-3}) in metals ($\sim 10^{28}$)		M_s mass of the Sun (2×10^{30} kg)		W_s Sun's radiated power (3.9×10^{26} W)
$\times 10^{23}$					
$\times 10^{22}$					
$\times 10^{21}$		M mass of Earth's oceans (kg) (1.42)			
$\times 10^{20}$	f_γ typical gamma-ray frequency (Hz)				
$\times 10^{19}$			V volume of the Moon (m^3) (2.2)		
$\times 10^{18}$					
$\times 10^{17}$		age of the Earth (s) (1.4)			
$\times 10^{16}$					
$\times 10^{15}$	f frequency of light (Hz)				
$\times 10^{14}$					
$\times 10^{13}$	f atomic vibration frequency (Hz)				
T $\times 10^{12}$	R typical input impedance of a dc electrometer (Ω)				
$\times 10^{11}$	E Young's modulus of metals (Nm^{-2})	distance from Earth to Sun (m) (1.496)	$T_{1/2}$ half life of carbon -14 (s) (1.8)		
$\times 10^{10}$	age of the universe in years (y)				one Curie (s^{-1}) (3.7)
G $\times 10^{9}$					V pd of lightning flash (V)
$\times 10^{8}$	v_β typical speed of β-particles (ms^{-1})			c speed of light in a vacuum (ms^{-1}) (3.00)	
$\times 10^{7}$		T Temperature at Sun's centre (K) (1.36)	P_c critical pressure of water (Pa) (2.2)	seconds in a year (s) (3.15)	h height of a parking orbit (m) (3.6)
M $\times 10^{6}$	v_α typical speed of α-particles (ms^{-1})		l_v sp. lat. ht. vap. of water (J kg^{-1}) (2.3)	E breakdown field of air (V m^{-1})	
$\times 10^{5}$	ρ resistivity of pure silicon (Ωm)	P_0 atmospheric pressure (Pa) (1.013)		l_f sp. lat. ht. fus. of water (J kg^{-1}) (3.3)	
$\times 10^{4}$		ρ density of lead (kg m^{-3}) (1.13)	f highest audible sound frequency (Hz)		
k $\times 10^{3}$	k typical ionic bond constant (N m^{-1})	ρ_w density of water (kg m^{-3}) (1.0)	T_m melting point of steel (K) (1.63)	P svp of water at at 25°C (Pa) (3.17)	c sp. ht. cap. of water (J kg^{-1} K^{-1}) (4.2)
$\times 10^{2}$	c sp heat capacity of metals (J kg^{-1} K^{-1})	field in Earth's atmosphere (V m^{-1}) (1.3)		v velocity of sound in air (ms^{-1}) (3.3)	
$\times 10^{1}$	k thermal conductivity of alloys (W m^{-1} K^{-1})	g acceleration of free fall (ms^{-2}) (0.981)	f lowest audible sound frequency (Hz)	monatomic molar ht. cap. (J K^{-1} mol^{-1}) (2.5)	
(×1) $\times 10^{0}$	λ typical VHF radio wavelength (m)	ρ_a density of air (kg m^{-3}) (1.3)	γ c_p/c_v for helium (1.66)	π (3.142)	

5	6	7	8	9	Multiplying factor
	M_E mass of the Earth (6×10^{24} kg)				$\geqslant 10^{24}$
	L Avogadro constant N_A (mol^{-1}) (6.02)				$\times 10^{23}$
		M_m mass of the Moon (kg) (7.3)			$\times 10^{22}$
					$\times 10^{21}$
					$\times 10^{20}$
					$\times 10^{19}$
M mass of Earth's atmosphere (kg)	number of eV in one Joule (eV) (6.25)				$\times 10^{18}$
					$\times 10^{17}$
					$\times 10^{16}$
				metres in 1 light year (m) (9.46)	$\times 10^{15}$
					$\times 10^{14}$
		E energy released in fission (J kg^{-1})			$\times 10^{13}$
					tera $\times 10^{12}$
					$\times 10^{11}$
					$\times 10^{10}$
				f atomic clock frequency (Hz) (9.192 631 770)	**giga** $\times 10^{9}$
		R_S radius of the Sun (m) (6.96)			$\times 10^{8}$
E energy released by burning coal (J kg^{-1})	f UHF TV channel frequency (Hz)				$\times 10^{7}$
	R_E radius of the Earth (m) (6.5)				**mega** $\times 10^{6}$
					$\times 10^{5}$
		P svp of ether at 25°C (Pa)			$\times 10^{4}$
v velocity of sound in steel (m s^{-1})	ρ_E mean density of Earth (kg m^{-3}) (5.51)				**kilo** $\times 10^{3}$
	T_c critical temperature of water (K) (6.47)	μ_r relative permeability or iron (7.0)	c_p for carbon dioxide (J kg^{-1} K^{-1}) (8.2)	number of MeV in 1 amu (MeV) (9.31)	$\times 10^{2}$
$T_{1/2}$ half life of radon -220 (s) (5.4)	k bond constant for steel atoms (N m^{-1})				$\geqslant 10^{1}$
	ε_r relative permittivity of porcelain		R molar gas constant (J K^{-1} mol^{-1}) (8.31)	breaking strain of rubber	($\times 1$) $\times 10^{0}$

Multiplying factor	To the nearest order of magnitude	1	2	3	4
$\times 10^{-1}$	λ typical microwave or radar wavelength (m)	B field at end of bar magnet (T)			1 mph in metres per second (ms^{-1}) (4.5)
$\times 10^{-2}$	ρ typical semiconductor resistivity $(\Omega\,m)$		volume of a mole at stp (m^3) (2.24)		
m $\times 10^{-3}$		η viscosity of water (Pa s) (1.0)		k_W Wien displacement constant (mK) (2.9)	temp. coeff. of resistance of Cu (K^{-1}) (+ 4.3)
$\times 10^{-4}$	λ typical infrared wavelength (m)				
$\times 10^{-5}$	B strength of Earth's magnetic field (T)		η viscosity of air (Pa s) (1.7)		
μ $\times 10^{-6}$	C max. capacitance of non-electrolytics (F)	μ_0 magnetic constant $(H\,m^{-1})$ (1.26)			
$\times 10^{-7}$	ρ typical resistivity of metals $(\Omega\,m)$		F force used to define the ampere (N) (2.0)	λ_B wavelength of blue light (m)	
$\times 10^{-8}$	λ typical ultraviolet wavelength (m)				
n $\times 10^{-9}$	F typical force exerted by electron beam (N)		D diameter of an oil molecule (m)		
$\times 10^{-10}$	λ typical X-ray wavelength (m)	D diameter of a carbon atom (m) (1.3)			
$\times 10^{-11}$	C typical parallel plate cap. in air (F)				
p $\times 10^{-12}$	λ typical gamma-ray wavelength (m)				
$\times 10^{-13}$					
$\times 10^{-14}$	D diameter of medium-size nucleus (m)		E min. ke of 'fast' neutron (J)		
f $\times 10^{-15}$	E typical X-ray (1 keV) photon energy (J)				
$\times 10^{-16}$					
$\times 10^{-17}$					
a $\times 10^{-18}$					
$\times 10^{-19}$			e charge of an electron (C) (− 1.6)		
$\leqslant 10^{-20}$	ρ density of outer space ($\sim 10^{-21}$ kg m^{-3})	k Boltzmann constant (1.38×10^{-23} J K^{-1})	m_n neutron or proton m_p mass (1.67×10^{-27} kg)		E typical ke of slow neutron ($\times 10^{-21}$ J)

5	6	7	8	9	Multiplying factor
γ surface tension of mercury (N m^{-1}) (5.0)	k thermal conductivity of water (Wm^{-1} K^{1}) (6.0)	tyre/dry road kinetic friction coefficient		therm. cond. of glass (Wm^{-1} K^{-1}) (0.93)	$\times 10^{-1}$
		γ surface tension of water (N m^{-1}) (7.3)		ρ density of hydrogen (kg m^{-3}) (9.0)	$\times 10^{-2}$
					milli $\times 10^{-3}$
					$\times 10^{-4}$
					$\times 10^{-5}$
					micro $\times 10^{-6}$
	σ Stefan's constant (W m^{-2} K^{-4}) (5.67)		λ_R wavelength of red light (m)		$\times 10^{-7}$
					$\times 10^{-8}$
					nano $\times 10^{-9}$
		G gravitational constant (N m^2 kg^{-2}) (6.67)			$\times 10^{-10}$
				ε_0 electric constant (F m^{-1}) (8.85)	$\times 10^{-11}$
					pico $\times 10^{-12}$
					$\times 10^{-13}$
					$\times 10^{-14}$
					femto $\times 10^{-15}$
					$\times 10^{-16}$
					$\times 10^{-17}$
					atto $\times 10^{-18}$
		h Planck constant (6.63 $\times 10^{-34}$ J s)		m_e electron mass (9.11 $\times 10^{-31}$ kg)	$\times 10^{-19}$
					$\leqslant 10^{-20}$

INDEX